T0298702

Fractional Calculus View of Complexity

Tomorrow's Science

Fractional Calculus View of Complexity

Tomorrow's Science

Bruce J. West

Information Sciences Directorate
Army Research Office
Research Triangle Park, NC

CRC Press is an imprint of the
Taylor & Francis Group, an **Informa** business

A SCIENCE PUBLISHERS BOOK

CRC Press
Taylor & Francis Group
6000 Broken Sound Parkway NW, Suite 300
Boca Raton, FL 33487-2742

© 2016 by Taylor & Francis Group, LLC
CRC Press is an imprint of Taylor & Francis Group, an Informa business

No claim to original U.S. Government works

Printed on acid-free paper
Version Date: 20150728

International Standard Book Number-13: 978-1-4987-3800-2 (Hardback)

Visit the Taylor & Francis Web site at
http://www.taylorandfrancis.com

and the CRC Press Web site at
http://www.crcpress.com

Foreword

Change, evolutionary or revolutionary, helpful or harmful, is how we track events through time to our universe. In this volume, Bruce West confronts change within science and shows us ways to look at the future of science that are different than the pathways of its past. Science in itself may be an evolutionary process, but the ways humans perform science, utilize it, and understand it can be and, according to West, are currently revolutionary. West shows the revolutionary fervor of modern science by describing some of the dramatic changes in scientific needs, utilizations and methodologies. Even though Bruce West does not exactly say this, I will—we are in the midst of a paradigm shift in science. The revolution in science has started. Elements of this shift include moving from little science (single investigators) to big science (multi-disciplinary teams of scientists), the increased role of complexity, and connections in understanding the informational and networked nature of science, and the human utilization of science within the context of the rest of society.

Bruce West has an exceptional way of weaving science history, theory, application, philosophy, and management into one cohesive story. One of the main characters in West's story is the mathematics of fractional calculus. West makes many compelling cases to convince readers of the power and upcoming role of fractional calculus methodology. Much like Isaac Asimov who saw his role as a futurist "to reconnoiter the territory up ahead so that humanity, in its travel through time, may have a better notion of what to aim for and what to avoid" [1], West provides both an aim point and a path for future science with fractional calculus and complexity playing major roles. In the last chapter of the book, the future nature and philosophy of science are clearly shown as only a futurist like West can describe. Most exciting for me is West's analysis of the underlying philosophy and principles of tomorrow's science. His four hypothesized changes in scientific principles in the last chapter are profound and impactful and my favorite element of the book.

I recently heard a colleague explain the interdisciplinary role of mathematics in science: "Mathematics is part of all science, but it is also only a part of any science". I agree with West and my colleague—modern science is inherently interdisciplinary. It relies on combining concepts, models, methods, knowledge, and perspectives from all angles, dimensions, and far out regions of scientific inquiry. The collective science enterprise, as it begins to encompass large cooperative organizations and teams to form the concept of big science, is where future science will make its mark. In the past, these kinds of large-scale interdisciplinarity were limited to outsized projects like the Manhattan Project, or building the Internet, or reserved for government-led studies conducted by the National Academy of Science, or the National Research Council [4], but today big science is more and more the norm of industrial research centers and university laboratories. While science education continues to use a more narrow disciplinary focus to produce organized sets of departments, courses, and topics that are accessible, structural, sequential, and assessable for education, real science is very different. In the research world, science is completely unaware of these artificial disciplinary boundaries. Through use of vivid examples and an insightful description of complexity, West describes this holistic and interconnected view of the modern science enterprise. As West clearly demonstrates in his historical anecdotes and future projections that science contains many complex, layered, multi-scaled, dynamic, multi-dimensional, and contextual components, it is the collective endeavors that will make further progress possible. Edward Teller (1980) wrote back in the industrial age that even though science should pursue simplicity in its models and theories in order to provide clarity and understanding for society, there were many elements in science (like the understanding of life itself) that are inherently and purposefully complex [5]. The difference between Teller's industrial age view of science and West's more modern and futurist view is in the expanse of the complexity and connectivity of science. Teller saw complexity filling a limited role and connectivity only needed for few large-scale efforts, whereas West sees complexity and connectivity as central elements of science.

As the book unfolds, it becomes more and more obvious that there is much more to performing science than structured processes or strictly empirical and quantitative foundations and hierarchies. According to Bruce West, research in science must shift its principles to allow for incorporation of modern perspectives that lead to inventive processes where established methodologies and knowledge are combined with novel procedures and entirely new contexts. While this assembly can lead to more chaotic, qualitative and emergent methods, West explains how society and science will benefit.

Modern science is handling the paradigm shift by building connections among disparate concepts, fields, and contexts to construct explanations of the world in ways that are not possible through strictly disciplinary means.

Scientists synthesize these connections to deepen understanding of underlying scientific principles. An example is the entry of information and computer science into the fold of science over the past half-century. These new elements will play significant roles in the future, and as Denning and Martell explain, "the laws of information reveal new possibilities and constraints that are not apparent from the laws of physics." [2] West lays out a future path showing how science will construct a system of shared knowledge, appropriate levels of abstraction, and an enlightened human context that enables science to engage society. As to whether science is unified into one science with many modes of thought or there are many different, yet connected science disciplines may be mostly semantics, but what is clear and foundational are the strengths and necessities of the connections and integrations.

It is no accident that this move towards integration of science coincides with globalization and the proliferation of information networks. Integration has become sciences' answer to the frustrations of science in the fragmented industrial-age where each discipline was hamstrung as a self-contained entity. The previous disciplinary, linear, empirical, assembly-line science has been challenged and overwhelmed by the dynamic societal changes at work in the information age. The much more collective form of modern science embraces the complexity and gathers information and ideas from all of science so that powerful scientific processes are created, stored, accessed and shared. Some of the largest, most successful mass collaborations in science are made through entirely open, virtual and social networks [3]. In place of disciplinary limitations, modern science has built a collective power that is more creative, more original, and more effective than any and all single disciplinary perspectives of the past. Bruce West's excellent book embraces the complexity, connectedness, and fractional calculus mathematics to assemble future aim points and pathways for science to follow and scientists to contemplate.

6 April 2015

Chris Arney
Professor of Mathematics
Chair of Network Science
United States Military Academy
West Point, New York

References

[1] Asimov, I., Change! Seventy-one Glimpses of the Future, Houghton Mifflin (1981).

[2] Denning, P.J., C.H. Martell, Great Principles of Computing, MIT Press (2015).

[3] Nielsen, M., Reinventing Discovery: The New Era of Networked Science, Princeton University Press (2012).

[4] Oleson, A., S. Brown, The Pursuit of Knowledge in the Early American Republic, Johns Hopkins Press, (1976).

[5] Teller, E., The Pursuit of Simplicity, Pepperdine University Press (1980).

Preface

This book is based, in large part, on a lecture I gave at the *Network Frontier Workshop* at Northwestern University in December of 2013. The lecture was a tutorial intended to explain the quantitative reasoning entailed by the fractional calculus applied to complex physical, social and biological phenomena. The reason to study the fractional calculus is its inextricable link to complexity and that it provides a way to think systematically about complex phenomena in general and complex networks in particular. The ordinary calculus provided ways to organize our thinking about such physical phenomena as acoustic and electromagnetic waves, diffusion and even quantum mechanics. However, it does not do well with the complexity of earthquakes, avalanches, turbulence, cognition and stock market crashes. For these and other complex phenomena the notions of scaling and fractals were introduced and through them new ways of thinking were developed.

As is sometimes the case, the talk was so well received that I decided to write it up and submit it for publication. The *Colloquium* section of the *Reviews of Modern Physics* was interested in the manuscript and after a round of responding to reviewer comments and suggestions it was accepted for publication [22]. What became clear in responding to the reviewers' concerns was the existence of a wide interest in the material. After completing the paper it was obvious that an even more extended version of the presentation might have some value for the broader scientific community.

In fleshing out the paper to make the book reasonably self-contained an unintended perspective of science emerged. The view point that kept asserting itself is a resurgence of the *Natural Philosophy* of the seventeenth and eighteenth centuries. A perspective in which the pursuit of scientific knowledge and understanding has wisdom as its ultimate goal, and not merely the ordered accumulation of empirical facts necessary for a rational model of the world. The scientist that most personifies this ideal, in my view, is Leonardo da Vinci. As a scientist he was able to incorporate his artistic skills in his pursuit of understanding anatomy, botany and physics. As an engineer his application of the arts and sciences into the planning of entertainments at the dinner parties of his benefactors and his equally gifted development of fortifications and armaments for their city states were unparalleled. Above it all was his singular ability to record his observation on how the world worked.

Herein I present the case for science regaining its position as *Natural Philosophy* and for reasserting its integrated nature, thereby pulling away from the disciplinary constructs devised to make the understanding of scientific growth and the complexity of phenomena intellectually manageable. The approach is unlike a text devoted to a pedagogical presentation of a specialized topic or a monograph focused on an author's area of research, the method of the present book is an effort to accomplish both these things while providing a rationale for why the reader may be interested in learning more about the fractional calculus. This book is for the researcher who has heard about many of these scientifically exotic activities but could not see how they fit into their own scientific interests or how they could be made compatible with the way they understand science. It is also for the novitiate who has not yet decided where their talents could be most productively applied.

I sent a draft of the manuscript for this book to a number of colleagues for comments. I am very grateful to those that responded, who include: Professors Chris Arney, Mario Bologna, Richard Magin and Michael Shlesinger. They provided multiple suggestions that vastly improved the book, but as always, any residual errors of fact or flaws in presentation are entirely my own.

Acknowledgement

I wish to acknowledge the love and encouragement of my wife Sharon and to thank her for her good humor regarding what must seem to be a never ending stream of scientific thinking.

Nomenclature

AMI	:	atrial myocardial infarction
AR	:	allometry relation
ARMA	:	auto-regressive integrated moving average
ARIFMA	:	auto-regressive integrated fractional moving average
CBF	:	cerebral blood flow
CLT	:	central limit theorem
CTRW	:	continuous time random walk
DMM	:	decision making model
ERH	:	echo response hypothesis
IPL	:	inverse power law
FDR	:	fluctuation-dissipation relation
FDWE	:	fractional diffusion-wave equation
FPE	:	Fokker-Planck equation
FFPE	:	fractional Fokker-Planck equation
FGn	:	fractional Gaussian noise
FK	:	fractional kinetics
FLE	:	fractional Langevin equation
FLogE	:	fractional logistic equation
FME	:	fractional master equation
FP	:	Fokker-Planck
FPSE	:	fractional phase space equation
FPU	:	Fermi, Pasta, Ulam
FRW	:	fractional random walk
FSH	:	fractional search hypothesis
GLE	:	generalized Langevin equation
GWF	:	generalized Weierstrass function
HRV	:	heart rate variability
LE	:	Langevin equation
LFD	:	local fractional derivative
ML	:	Mittag-Leffler

MLF	:	Mittag-Leffler function
MLM	:	Mittag-Leffler matrix
MLMF	:	Mittag-Leffler matrix function
MRI	:	magnetic resonance imaging
MRL	:	modified Riemann-Liouville
MW	:	Montroll-Weiss
NMR	:	nuclear magnetic resonance
OP	:	operations research
PCM	:	principle of complexity management
PDF	:	probability density function
PSE	:	phase space equation
RG	:	renormalization group
RGK	:	renormalization group kinetics
RL	:	Riemann-Liouville
RRW	:	Rayleigh random walk
SCLT	:	stochastic central limit theorem
TBI	:	traumatic brain injury
TBM	:	total body mass

Contents

CHAPTER 1

The Challenge of Complexity

Imagine a vast sheet of paper on which straight Lines, Triangles, Squares, Pentagons, Hexagons, and all other figures, instead of remaining fixed in their places, more freely about on or in the surface, but without the power of rising above or sinking below it, very much like shadows—only hard and with luminous edges—and you will then have a pretty correct notion of my country and countrymen. Alas, a few years ago, I should have said "my universe": but now my mind has been opened to higher views of things.

E.A. Abbott [1]

This book proposes a new way of thinking. A way of thinking made necessary by the demands of contemporary science to understand complexity in multiple disciplines. Complexity entails the qualitative, as well as, the quantitative richness of phenomena, so that its scientific understanding must encompass both. In order to garner such understanding in a systematic way requires going beyond the traditional scientific method as the only way of knowing. Consequently the mathematical models of reality have been pushed to their limits and beyond, those which were developed and applied so successfully to the explanation and understanding of physical phenomena in the nineteenth and twentieth centuries are no longer adequate to describe the emergent phenomena of the twenty-first century. Herein we propose to adopt a fresh perspective entailed by the use of the fractional calculus.

The scientific method, with its dependence on the experimental answering of questions, is often attributed to Francis Bacon [2] and the formulation of the theoretical interpretation in terms of mathematics stems from Galileo, who was a contemporary of Bacon. These gentlemen along with Newton, who was born the year Galileo died, ushered in the modern era of scientific investigation. However, a full century before they made their contributions one man recorded nature's workings in his notebooks on the following subjects: anatomy, acoustics, architecture, astronomy, atmosphere, botany, casting, comparative anatomy, colour, geology, flight, flying machines, human proportions, hydraulics, inventions, landscape, light and shade, mathematics, medicine, movement and weight, music, natural history, naval warfare, optics, perspective, philosophy, physical geography, sculpture, the nature of water, and warfare. The person was, of course, Leonardo da Vinci, perhaps the greatest painter who ever lived.

One of the remarkable things to be noted of the thirty chapter headings listed for da Vinci's notebooks [7] is that the contributions are nearly three to one on the side of science and engineering. Of course his contributions to each heading area are not of equal significance either scientifically or artistically. However his art has penetrated popular culture, for example, the Vitruvian Man depicted in Figure 1.1 has become such an icon for humanity that a rendition of it has been enclosed in a capsule and launched into space as a silent ambassador to extraterrestrial beings. Of most interest here is the fact that all these various topics were treated by him in essential the same way. Since he never attended a university, he taught himself Latin, and had no formal schooling in mathematics all his science was accomplished through observation, natural intellect and the application of his artistic skills. He investigated the world in its natural setting and attempted to understand the phenomena being observed, whether it was the turbulence of rushing water, the flight of birds or the despair on an aging face, by capturing their essential aspects in his sketches. These observations were recorded in scrupulous detail, without approximation or simplification. He was arguably the last of the natural philosophers to seek understanding of the complexity of empirical phenomena without the use of simplifying models.

Even in a casual reading of his notebooks one is struck by da Vinci's interweaving of science and art. He gives the same attention to detail in his discussion of how to sketch the folds in drapery as he does describing the operation of a pulley or lever. Today's gap between art and science identified and decried by C.P. Snow [16] is completely absent from da Vinci's work. Snow first articulated what he perceived to be the problem in a talk he gave at Cambridge in 1959, which formed the basis of his classic work [16]. The heart of his argument is captured in an all too familiar quote from his essay:

> A good many times I have been present at gatherings of people who, by the standards of the traditional culture, are thought highly educated and who have with considerable gusto been expressing their incredulity at the illiteracy of scientists. Once or twice I have been provoked and have asked the company how many of them could describe the Second Law of Thermodynamics. The response was cold: it was also negative.

Yet I was asking something which is the scientific equivalent of: Have you read a work of Shakespeare's?

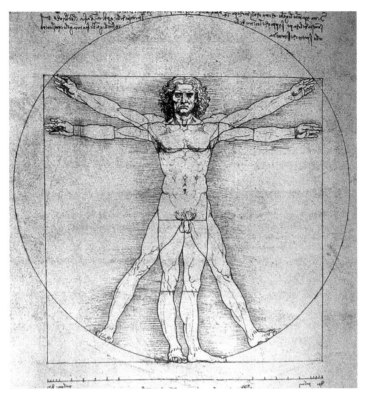

Figure 1.1 The proportions of the human figure. Probably the most well known icon of the human body: http://en.wikipedia.org/wiki/Vitruvian_Man.

Related questions concern how the gap between the two arose and why the two are even further apart today then when Snow first articulated the problem in the last century? The final question is how do we close the gap?

We start with da Vinci because although he was a scientist, engineer and inventor, he thought in a fundamentally different way than a modern day scientist. I argue that is a good thing. In his own words [12]:

> I am fully conscious that, not being a literary man, certain presumptuous persons will think that they may reasonable blame me; alleging that I am not a man of letters. Foolish folks! do they not know that I might retort as Marius did to the Roman Patricians by saying: That they, who deck themselves in the labors of others will not allow me my own. They will say that I, having no literary skill, cannot properly express that which I desire to treat of; but they do not know that my

subjects are to be dealt with by experience rather than by words; and [experience] has been the mistress of those who wrote well. And so, as mistress, I will cite her in all cases.

In a present day context da Vinci's word 'experience' could be replaced with 'observation', but the meaning would remain the same. It is unfortunate that his notebooks were mislaid, lost, suppressed and otherwise unavailable to the scientists and engineers of the next few centuries. Consequently his approach to the quantitative understanding of the world had no impact on the subsequent development of science. Consider his general introduction to the book on painting [12]:

> Among all the studies of natural causes and reasons Light chiefly delights the beholder; and among the great features of Mathematics the certainty of its demonstrations is what preeminently (tends to) elevate the mind of the investigator. Perspective, therefore, must be preferred to all the discourses and systems of human learning. In this branch [of science] the beam of light is explained on those methods of demonstration which from the glory not so much of Mathematics as of Physics and are graced with the flowers of both. But its axioms being laid down at great length arranging them on the method both of their natural order and of Mathematical demonstration; sometimes by deduction of the effects from the causes, and sometimes arguing the causes from the effects; adding also to my own conclusions some which, though not included in them, may nevertheless be inferred from them....

A very different view of light was given by Newton who analyzed sunlight by directing it through a prism in his laboratory to discover that white light is not a single color, but consists of a mixture of all the colors of the rainbow. Newton's theory was rejected by the poet and scientist Goethe [20]. Goethe could not make the aesthetic principles associated with the effects of color compatible with the experimental results and critiqued Newton's theory with phrases such as: "incredibly impudent"; "mere twaddle"; "ludicrous explanation"; "admirable of school-children in a go-cart"; "but I see nothing will do but lying, and plenty of it". These observations and more may be found in a popular lecture on Goethe by von Helmholtz in 1853, who best summarizes Goethe's view as:

> Just as a genuine work of art cannot bear retouching by a strange hand, so he would have us believe Nature resists the interference of the experimenter who tortures her and disturbs her: and, in revenge, misleads the impertinent kill-joy by a distorted image of herself.

As a poet and a scientist Goethe not only shared da Vinci's world view but also his method for exploring the frontiers of what is known. He found the separation between art and science imposed by the isolating experiments of the forerunners of the modern scientist to be antithetical to a real understanding of the world. Although ignorant of da Vinci's work, Goethe fully appreciated the schism that was being imposed on what constitutes knowledge and rallied against what was being lost in this widening gap.

The methods of Newton and subsequent generations of scientists have been successful for three hundred years and are the basis for the industrial society of the nineteenth and twentieth centuries, as well as the information society of the twenty-first century. It is only with the transition from the industrial to the information society that the lack of understanding of the complexity of the networked world in which we live has become increasingly evident. It is as if the very phenomena modern science is seeking to understand can only be realized in their natural settings and removing them to the laboratory destroys them.

1.1 Little Science, Big Science

In the middle of the last century Derek de Solla Price published his seminal book *Little Science Big Science* [11] in which he introduced the metrics of science to a large audience. In this book he traced the exponential growth of science through the eighteenth, nineteenth and well into the twentieth century. A prominent feature of his book was its emphasis on the immediacy of science, which is due in large part to that exponential growth. Price pointed out that a peculiar aspect of the exponential growth of science is that most of science activity had occurred within the author's lifetime and this will continue to be true as long as the process remains exponential. Phrased differently he emphasizes that 80 to 90 per cent of all scientists that have ever lived were alive at the time of publication and that is probably not too far off from today's estimate of the number of scientists.

Price [11] explained that Little Science is the science done by the individual; often being done as a gentlemanly diversion because doing science was not considered a vocation until well into the twentieth century. Even the person who was arguably the greatest of the modern scientists, Sir Isaac Newton, was initially a gentleman farmer, then an academic and later in life Master of the Mint, but he never received any direct income from his legendary scientific contributions.

Why did science develop in the way outlined by de Solla Price? The reason has to do at least in part with the successes of the in-depth investigations of the physical and life sciences by Newton, his contemporaries and scientific progeny. These successes could be listed from the chapters of the encyclopedic work on mathematical physics by Morse and Feshbach [8]. This work could be used to tell the story of the theoretical understanding of the physical sciences from its past until well into the present. Add to that tale the successes in the studies of the life and social sciences and it is not surprising that no one person could master it all, much less make meaningful research contributions in more than one or two fields.

However a significant number of crucial contributions to science would be missed in such a telling of history. The pathological functions that concerned the nineteenth century mathematicians such as Weierstrass and his student Cantor are not mentioned in [8], nor are the probability densities that violate the central limit theorem and studied by Lévy [6] in the early twentieth century. Also absent is the central paradox of mathematics uncovered by Gödel [4] in the middle twentieth century, and the chaos observed in nonlinear dynamics by Poincaré [10] in the late nineteenth century is no where to be found. This is not a criticism, but is intended to make explicit the type of phenomena with which these authors of traditional scientific text books and the scientific community in general were concerned, and what was thought to be pathological and therefore outside the domain of study.

The strategy devised to understand the continuing explosion in science over the past few centuries was specialization, exclusion and the introduction of an ever increasing number of scientific disciplines. Excluding phenomena from the mainstream narrative served to strengthen the story line of the phenomena that science could explain. In addition, increasing the number of disciplines and subdisciplines assisted in restricting the domain of responsibility of the individual scientist to a particular speciality. Specialization did not make solving scientific problems any easier, in fact, as the complexity of problems being considered increased, the skills of the specialist became less useful for their solution. This is primarily the result of complexity violating the assumptions on which the traditional methods are based; one such assumption would be differentiability.

The success of government funded laboratories, where scientists worked in research groups on large-scale problems, was the harbinger of how science could and would be done in the second half of the twentieth century. This was no where more evident than in the building of the atom bomb during World War Two; from the equivalence between mass and energy recognized in 1905 by Einstein working in isolation at a patent office in Switzerland to the over 130,000 people employed by the Manhattan Project from 1942 to 1946 leading to the experimental verification of that equivalence in the Trinity Test at Alamogordo Bombing and Gunnery Range in New Mexico on July 16, 1945.

Price also established that the exponential growth of science could not be sustained and at the time of the publication of his book science was entering a period in which that growth was slowing and would eventually saturate as depicted schematically by the logistic (sigmoidal) curve in Figure 1.2. The question therefore arises as to what form science will take as we emerge beyond this region of saturation; beyond Big Science. I maintain that all the indicators point to science reasserting its integrated nature and regaining its position as *Natural Philosophy*. In the process of accomplishing this rebirth it must

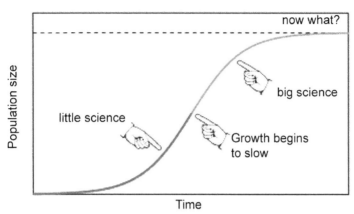

Figure 1.2 The logistic increase in the number of scientists in the Western World versus time as invisioned by de Solla Price [11] is depicted. From the *Little Science* of the individual to the *Big Science* of the collective into the unknown sociology of science in the future.

necessarily pull away from the disciplinary constructs devised to make the understanding of rapid scientific growth intellectually manageable.

1.2 Complexity

Investigations into the importance of complexity in scientific investigations have been swirling about in science and the philosophy of science for over a century without reaching consensus on a definition. Each discipline has its own way of probing phenomena to determine its level of complexity, but in each case complexity is related to the difficulty in understanding the phenomenon being investigated. In the physical sciences complexity is often involved in discussing such things as phase changes from gases to liquids to solids, as well as the statistical fluctuations in everything from the flow of water in a turbulent stream to the diffusion of aromatic molecules in a confined space. Social phenomena, on the other hand, are rarely simple so the challenge for the social scientist has always been to constrain the investigation sufficiently to study the matter quantitatively without suffocating the property of interest.

Complexity is invoked at the boundary between what we can model and predict mathematically and what we cannot. The social and life sciences often lack the mathematical rigor found in the physical sciences, but the intent is to have a comparable level of predictability. Herein we intend to pull together the common features of what has been identified as complexity in a variety of disciplines and address these common aspects from a single mathematical/scientific perspective.

1.2.1   Networks

In the second half of the twentieth century it became clear to the scientific community that the deterministic clockwork universe of Newtonian physics was not only unsuitable for the complex networks in the social and life sciences but it was not even appropriate for complex physical networks. Without belaboring the mathematics here the traditional analysis of mechanical networks was replaced with nonlinear dynamics, which in the 1970s and 1980s became commonly known as chaos theory. If these results had been confined to physical processes it may not have had the ubiquitous impact found throughout science. For better or worse it was found that most if not all phenomena are nonlinear and what had been so mysterious about biological and sociological processes in the past could often be traced back to the misapplication of linear concepts to their dynamics and consequently to incorrect predictions and interpretations.

The application of chaos theory to social and biological networks in the 1980s and 1990s lead to the development of the discipline of complex adaptive networks. For some this latter discipline provided the scientific proof that social phenomena are not predictable in the same way physical processes are. For others the restrictions found in the predictions of complex physical phenomena have freed the imagination from the rigid structures of strict deterministic physics to a less certain, but richer tapestry of possible futures more compatible with what is observed in the social and life sciences. Consequently, complexity became a balance between the regular and stochastic and measures of this balance are on the forefront of research.

Complex adaptive networks as a discipline has morphed into the fledging network science in the last decade, which incorporates the giant strides made by its predecessors into a new scientific perspective. The criticality that has become evident in complex networks is traced to the formation of consensus in social networks, to the discharge avalanches in neuronal networks and finally to the phase transitions in physical networks. We are now poised to address the emergent properties of real world situations where these three functionally distinct networks interact with one another, as well as with others.

Networked science exists in the physical, informational, social and cognitive domains of the society of which it is a part. Consequently the overarching goal of research within this intellectual construct is to develop the fundamental principles and procedures that can enable modeling, design, analysis, prediction and control of the behavior of communication, sensing, and decision making networks and to optimize the interactions within this complex network of networks. Many research barriers exist to realizing this goal, including the lack of an adequate mathematical formalism, which hinders the development of efficient simulation and modeling tools, and thus inhibits exploration of the engineering design space. The basic research advances over the past decade have been building the basis of a science by developing a mathematical/scientific framework with proper metrics, language, structures, and processes to provide a constitutive understanding of networks that have a broad spectrum of functionalities.

The existing research strategy has been to develop an unifying scientific framework that merges multiple theory and application disciplines that had not previously been connected and whose merger has produced some remarkable results. However this approach has also identified gaps in our understanding as well as in our disciplinary development. In this regard a grand challenge for the future is to construct a formal description of the flow of data/information through dynamic complex networks, whether secure messages over a communication channel; neuron impulses within the brain; or orders within a military organization. In the experiential world information occurs in intermittent bursts in both space and time; propagates across links that vary in quality and that connect elements with multiple properties; each qualifier in turn invalidates a foundational assumptions made in the past, as we subsequently explain. Put simply, these networks are not topologically homogeneous, nor temporally isotropic, and they are typically non-ergodic, which means that time averages need not have the same numerical value as the averages over statistical distribution functions.

Tomorrow's science will be unlike the traditional scientific disciplines in that it identifies a particular structure, that of a network, as the basis for a science. On second thought, this concentration on the formal structure of networks is perhaps analogous to the identification of fields in nineteen century physics. Since structural webbing exists within and among all scientific disciplines, the over arching goal of this nascent science is to develop a body of principles that enables modeling, design, analysis, prediction, and control of the behavior of the underlying phenomena in physical, information, social, and cognitive networks, to name only a few. The emphasis is not restricted to the network structure within a given discipline, but is focused on networked phenomena containing many disciplines and how such complex and functionally different networks interact with one another. As a rudimentary science, its underlying principles are still being formulated, so that a fundamental challenge for this effort has been to create and organize the fundamental knowledge upon which to base these principles. Moreover, tomorrow's science is intended to bridge the knowledge gap between disciplines and to break down the artificial barriers that have been created between disciplines. In order to accomplish this ambitious undertaking a new kind of thinking is needed and correspondingly a new kind of mathematics necessary do this thinking is required.

1.2.2 Information

The nature of science is to understand both the natural and the synthetic, that is man-made, phenomena; the nature of engineering is to control these phenomena through intervention. This distinction between science and engineering is, in part, the reason politicians often view science as a luxury and engineering as a necessity. The imposed separation of science and engineering is even more artificial than that between art and science since their true nature is symbiotic. The interdependence of these two ways of gaining knowledge becomes apparent when it is observed that the rewards of control cannot be maintained in the absence of understanding. The benefits are transient because things change and are uncertain so that, without understanding that enables adaptability, control is only temporary. In order to exercise proper control the engineer does not need to know the details of the process by which the information is delivered to her, but she must be confident in the output of that process.

The process begins with identifying the sources of error in the raw data, along with their size and how often they occur. The sources of error can be traced to the noise in sensor outputs, or to the inconsistency and ambiguity of human reporting, for these and other reasons data itself lacks fidelity and is never 100% accurate. The question remains how to determine and turn to advantage the level of uncertainty in the data. Put more pointedly: Is the observed variability in the data the 'signal' or is it 'noise' covering up the signal? Quantifying how the uncertainty in the data corrupts and/or masks the desired information is the next step in the process. When are the observed patterns indicative of real information and when are they artifacts of the method of analysis? When are the random fluctuations part of the 'signal' and the slow smooth modulation part of the 'noise'? A traditional approach to this control involves filtering and feedback.

Engineers (and some scientists) know how to control and therefore filter linear dynamic systems, and as Jervis [5] points out, linearity involves two propositions: (i) systems whose changes in output are proportional to changes in input and (ii) systems whose total change in output results from the sum of the changes in individual inputs. Jervis notes that we often, if not always, expect linear response. For example, if a little foreign aid slightly increases economic growth, then additional aid should stimulate even more growth, not accounting for such nonlinear properties as saturation or threshold effects. In fact the linear properties of proportionality and additivity of observables are violated in all complex systems. This wisdom is encapsulated in such homilies as "the straw that broke the camel's back" to indicate the last of a large number of linear additive actions results in a qualitative change in a phenomenon. Mathematically such a qualitative change is described as a bifurcation or catastrophe.

1.2.3   Unintended Consequences

Changes made in a complex network introduce planned-for effects that are desirable and unplanned surprises that are not so desirable. The latter have been called "side effects" by physicians in order to have a label for the unwanted and sometimes fatal response to changes in medication. In a complex network it is impossible to illicit a single response to a change and consequently there is an ever present need to anticipate the unpredictable, since we inevitably modify networks in our pursuit of control.

Jervis [5] provides a long list of unintended consequences entailed by the complexity of various phenomena, including: in order to kill insects we inadvertently put an end to the singing birds; in cleaning the water in our harbors we promote the growth of mollusks and crustaceans that destroy wooden piers and bulkheads; pesticides often destroy the crops that they are designed to save by killing the pests' predators. In short, without a clear quantitative modeling of a complex phenomenon introducing a change to realize a particular outcome *always* results in unintended consequences. There is perhaps a corollary to this theorem and that is: *the more significant the intended outcome of an intervention the more overwhelming the potential side effects.*

Rosenau [13] in his pessimistic discussion of the deterioration of nation states points out that national landscapes are giving way to ethnoscapes, mediascapes, ideoscapes, technoscapes and finanscapes. The observations he makes are compelling even if the conclusions he draws are open to some skepticism. For our purposes we recognize how these new "-scapes" or perhaps "-spheres" entail a change in the way we think about how scientific research is realized; the linear predictable trajectory of the past story of scientific advance involving the iterative process of theory→prediction→experimental test→modification of theory→prediction and so on, must be replaced with an erratic nonlinear behavior to capture the true dynamics of the process. He goes on to say regarding nations, but the observation applies equally well to the broader area of scientific research:

> ...the long-standing inclination to think in either/or terms has begun to give way to framing challenges as both/and problems. People now understand, emotionally as well as intellectually, that unexpected occurrences, that minor incidents can mushroom into major outcomes, that fundamental processes trigger opposing forces even as they expand their scope, that what was once transitional may now be enduring, and that the complexities of modern life are so deeply rooted as to infuse ordinariness into the surprising development and the anxieties that attach to it.

He [13] goes on to observe that complexity theory is unlikely to ever provide a method for predicting particular events. This in itself suggests a reexamination of the way in which science and the quantitative models it provides are used by those that implement the theory in the experiential world. The modern theory serves the purpose of cautioning the scientist/engineer against looking for the panacea, the simple linear solution to the problem, and to seek the more circumspect and less certain alternatives and to "embrace the complexity"; the halting step ought to replace the confident stride.

1.2.4   Mathematical Framework

The reader should know at the start that this is not a monograph on mathematics, although there will be ample use of a certain kind of mathematics in the following chapters. It is neither a text in any one of the specific disciplines we have mentioned. It is more of an essay intended to suggest what science might be like in the future as a heterogeneous mixture of the various disciplines, much as it was for da Vinci. In the nineteenth century the experiments into electricity, magnetism, and acoustics lead to the notion of fields and the propagation of disturbances as waves. The development of partial differential equations and the extraordinarily successful description of physical phenomena using analytic functions provided a mathematical infrastructure with which to understand these aspects of the world and quantify phenomena ranging from the specular reflection of the sun from the ocean surface to the changing pitch in the whistle of a passing train. The marriage of mathematics and physical experiments matured into a strategy for doing science, both in the development of theory and in the design of new experiments with which to test those theories. That same strategy is applied herein, but in a contemporary setting it leads in a number of unexpected directions. We follow one of these directions in these pages and embark on a path to which relatively few have been attracted. There have been so few travelers on this road, in part, because science is a social activity and most scientists need companionship for long trips.

Multiple-scales and cross-scale couplings are recurrent themes in the study of physical, informational, and social/cognitive complex phenomena. Existing formalisms address one, two or a continuum of scales, but the underlying problems consisting of a finite number (>2) of interacting scales remains open for investigation. Existing mathematical approaches offer insights into the quantitative aspects of complex phenomena with multiple scales (no characteristic scale) in stationary and near equilibrium networks, but what is now being developed is a way to describe the dynamics of complex phenomena/networks when the underlying processes are neither stationary, nor in equilibrium. Network characteristics cannot be deduced from the properties of individual components; they emerge during the formation and evolution of the network. Consequently, a mathematical framework is being developed to characterize the interactions between the dynamic network components, the temporal evolution of the network, and its response to external stimuli, including attacks. The framework is taking into account heterogeneity, non-stationarity, even non-ergodicity, as well as conflicting constraints and objectives. The notion of ergodicity is found throughout the physical sciences and rests on the assumption that an average has two equivalent definitions; one in terms of a probability density and the other in terms of time. Today there is Ergodic Theory, which is a branch of mathematics framed around the conditions necessary for the time and ensemble averages of a process to be equal. This is the last time we mention this theory.

One way to characterize the dynamics of complex phenomena is by understanding how information is propagated within and across complex adaptive networks: how information flows across networks so as to maximize utility, for example, as perceived by a design engineer, under constraints, such as timeliness and resource usage. Tools from statistical physics such as continuum percolation theory, stochastic geometry, and modern statistics are being used to deal with non-Markovian behaviors, heavy-tailed phenomena, and the mutual interactions between functionally different networks. To cope with multiple, potentially conflicting, constraints and objectives, a generalized control theory of locally reacting multiple agents is being developed through the mimicking of self-repair and replication identified in biological networks. Each of these mathematical tools addresses different pieces of the puzzle.

A set of common concepts for complexity science does not exist that is defined across the various scientific disciplines. There are discipline-specific nomenclatures that have been developed for specialized needs, but the equivalence of these terminologies for a discipline-independent characterization of networks and complex phenomena has not yet been established. The mathematical formulation of these experimentally based concepts may well provide a nascent language for such a science.

Mathematical rigor across the component disciplines is promoted herein by focusing on ubiquitous aspects of the underlying complexity, such as the appearance of non-stationary, non-ergodic, and renewal statistical processes. These properties are manifest through the empirical inverse power-law (IPL) probability density functions (PDFs) that appear in physical, informational, and social/cognitive networks, as will be subsequently explained. An approach that has provided preliminary success in understanding these complex networks and indicates potential for further advances in the future is the adaptation and extension of the methods of non-equilibrium statistical physics. These techniques have been used to characterize the dynamics of complex physical networks and have been extended to the study of informational and social phenomena, particularly decision-making with incomplete information in an uncertain environment [21]. The successes to date have been a combination of analytic and advanced computational results with the expectation that dramatic advances in the form of "principles for information exchange between complex networks" will soon be established in general settings [17, 19].

1.3 Chapter Overviews

The general notion of complexity is arguably at odds with many of the traditional methods used to solve dynamical physical problems, particularly the linear methods we review in Chapter 2. The waves that propagate on and in water, in air and through solid materials are lost in the non-responsiveness of such materials as rubber shock absorbers. The solid support of packed gravel roads slips away in the clinging mud after a heavy rain. The single voice of the operatic singer is lost in the din of the murmuring audience. The forces between particles in the three phases of matter become unreliable in those in-between phases where objects are neither solid, liquid, nor gas. Jello looks solid and may be lifted without changing shape (a small cube), but it cannot support a load. Its shearing strength is more like that of a liquid, which is to say that it does not have any. But even these properties may become lost in the presence of fluctuations.

Chapter 2 provides a sketch of how complexity has been modeled in the physical sciences adopting a linear perspective. This is done in part because it is useful to know one's location before embarking on a journey. It is also useful to know the limitations of what it is that we think we know. Mark Twain once wrote:

> The trouble with the world is not that people know too little, but that they know many things that ain't so.

In a scientific context this could be interpreted to mean that what prevents us from understanding data from a new phenomenon are the forgotten assumptions made in the discipline used to interpret the data. One example might be assuming that the observed fluctuations in a given data set have Normal statistics. How I think about the process is very different if I believe the average value of a variable is finite and is representative of a process when in fact it diverges and is not characteristic of the process. All the models of complexity discussed in this second chapter have such limiting fundamental assumptions that we make explicit and subsequently abandon.

The two primary strategies for modeling complex dynamical systems based on linear dynamics are introduced. The first uses a large network of coupled harmonic oscillators; one for the system of interest, and a large number of others to represent the environment. The equations of motion are explicitly solved and yield a generalized Langevin equation, which contains the dynamics for the oscillator of interest coupled to the environment. The coupling consists of two parts; a linear dissipation and a random force, the ratio of the two determines the temperature of the environment; the Einstein relation. The second strategy discussed is based on the evolution of the probability density for the momentum and displacement of the oscillator of interest. The resulting Fokker-Planck equation is shown to be consistent with the dynamic description given by the Langevin equation.

However it would be counter productive to restrict our thinking about linear models as approximations to complex phenomena. We demonstrate that finite-order nonlinear dynamical systems have an equivalent infinite-order linear representation. There is, in fact, a mathematical theorem due to Carleman [3] that proves the existence of such a relationship. We demonstrate the utility of this linear representation by using it to solve a nonlinear equation exactly, that being the logistic equation. Of course there are simpler ways to solve the logistic equation, but constructing the solution from an infinite-order representation does demonstrate how to use the method, which we employ latter to explicitly solve a previously unsolved nonlinear equation. Chapter 2 closes with a brief review of the use of dimensional analysis to prepare the reader for the scaling used in subsequent chapters.

A new way of thinking in science is introduced in Chapter 3; a way that builds on the notion of fractals introduced by the late Benoit Mandelbrot in the middle of the last century. I had the good fortune to attend the University of Rochester for my graduate work in physics in the late 1960s. Elliott Montroll held the Einstein Chair in the Department and he would invite his friend Mandelbrot, who at the time was an IBM Fellow, to be a department colloquium speaker. Mandelbrot would typically lecture on the large fluctuations in commodity prices and pose remarkable questions such as why the night sky is not uniformly bright. In my memory his answers always focused on the limitations of modeling in science, whether physical or economic, stressing their reliance on the underlying assumptions such as continuity, homogeneity and isotropy. Later these concerns matured into his concepts of geometric and statistical fractals, which we discuss in due course.

Mandelbrot's analyses have subsequently been absorbed by two generations of scientists to the point where it is sometimes difficult to explain what was truly innovative about his perspective. Scaling, self-similarity and fractals all seem to merge together in many discussions, but what is often lacking is an appreciation for what fractals imply about the dynamics of the underlying process. The dynamics of fractal phenomena remains largely unexplored and in this chapter we offer at least a partial explanation as to why.

Chapter 3 introduces the idea that one way a system or structure can be complex is through its variation in size and duration of dynamic events. The notion of a fractal function is introduced through Mandelbrot's generalization of the function Weierstrass introduced into mathematics a century and a half ago. The generalized Weierstrass function is shown to be self-similar, that is, it has structure at all scales with no one scale dominating. The derivative of this function diverges and consequently it cannot be the solution to an equation of motion, but it does solve a renormalization group relation. Richardson suggested it as a model for the non-differentiable velocity field of the turbulent wind dispersing smoke belching from a London chimney. This idea is developed in a subsequent chapter.

The notions that variations are coupled across scales in both space and time is captured through the concept of a fractal dimension making the structure self-similar. A complex fractal dimension ties the geometric coupling across scales to a log-periodic modulation of the variability. This scaling is described by means of renormalization group theory, which was devised to explain the scaling observed in multiple physical systems that undergo phase transitions. It is not only physical systems that have observables that display modulated scaling [22]. Complex physiological systems such as the size of the branchings in bronchial airways; the variability in cerebral blood flow and the statistical fluctuations in the profit is stock market prices are all known to manifest this log-periodic behavior. These exemplars are intended to indicate the wide range of phenomena whose structure and/or dynamics are captured by fractal dimensions with real and imaginary parts.

The ground work is laid in these early chapters for introducing the formal concepts of fractional integrals and derivatives. In Chapter 4 we begin this introduction by reiterating how the notion of randomness enters into our discussion of complexity. In physics the idea of randomness is often first encountered through the image of a random walker taking discrete steps. Subsequently, the limit of vanishingly small steps is used to obtain the continuum representation of this process. So it seemed that the new concept of a fractional derivative could be most readily introduced using the device of a fractional random walker and again consider the limit of vanishingly small steps. This is the precursor to such phenomena as anomalous diffusion in which the variance may increase as a power law in time, different from the integer exponent found in classical diffusion.

The simplest fractional derivatives often yield results that are counter intuitive such as the fractional derivative of a constant being non-zero. Such results are part of the reason that a new way of thinking is necessary to understand complex phenomena, where such fractional derivatives have been shown to appear in the equations of motion. Chapter 4 begins the journey through this technical labyrinth. One immediate benefit of the fractional calculus is the result that it is shown to be the natural way to describe the dynamics of fractal processes. Physical examples of when the fractional calculus and fractals dovetail are viscoelastic materials in which stress relaxation is no longer exponential and the familiar description of classical diffusion by Brownian motion of a heavy particle in a fluid of lighter particles is reexamined. Brownian motion is shown to have three distinct time domains: the microscopic, the mesoscopic and the macroscopic. The theory, resulting in equations, for the dynamics in the mesoscopic domain has existed for over a century, but it is only recently that experiments have been able to probe the short and intermediate time domains and interpret their data in terms of fractional differential equations.

The solution to linear rate equations is the exponential function. The solution to linear fractional rate equations is the Mittag-Leffler function (MLF), which is a natural generalization of the exponential, reducing to it in the appropriate limit. In Chapter 5 we show that just as the solution to systems of linearly coupled rate equations can be expressed as an exponential matrix, the solution to systems of linearly coupled fractional rate equations can be expressed as a Mittag-Leffler matrix. Here again the appropriate limit of the Mittag-Leffler matrix solution coincides with the exponential matrix solution. A few examples of low dimension are worked out in detail, such as the fractional linear harmonic oscillator and its time-dependent properties are analyzed.

In order to avoid getting lost in the formalism of the fractional calculus we periodically reestablish contact with "real world" phenomena such as the cooperative behavior in social, ecological and physical phenomena. A physical example of cooperative, even if random, behavior discussed in Chapter 6 is that of homogeneous turbulence with an example of data taken from wind gusts over the deep ocean. Unlike the case of fluctuations with Gaussian statistics these fluctuations in the wind velocity are observed to have the more general Lévy statistics as first observed in the smoke plumes from chimneys by Richardson. Turbulence is one of the great mysteries of physics and consequently every theory attempting to explain its properties begins with a list of assumptions and/or approximation and the present discussion is no different. Here a relatively simple model of homogeneous isotropic turbulence shows that the observed Lévy statistics can be explained by means of a fractional diffusion equation in space.

Stepping from physics to physiology we find that the linear, dissipative and diffusive phenomenological Bloch equation description of magnetic precession and relaxation of nuclear spins used to describe magnetic resonance imaging (MRI) is generalized to a fractional diffusion equation in space and time based on recent experimental data. The dynamics of the nuclear spins has been called "spin turbulence". The fractional calculus is here seen to provide deep insight into a range of phenomena that would not be accessible using more traditional thinking.

An ecological exemplar discussed in Chapter 6 is foraging, which is not unlike the search strategies used to hunt submarines in the Second World War. The discipline of Operations Research [9] (OR), which was designed to use mathematical methods to optimize complex real-world structures and processes [15], grew out of the need to solve such problems by the military. More benign applications of OR involve searching for people lost at sea, or lost in the wilderness, and to even codify birds searching for prey. We discuss the case of albatross searching for food in some detail and find conditions for when the simple random walk is the appropriate strategy, and when it is not, using arguments based on actual data. In general it has been determined that foraging using Lévy statistics is optimal in ecological applications. The connection of the Lévy statistical distribution with the fractional calculus is established in the discussion regarding homogeneous turbulence and is used to advantage in searches, whether it is predators searching for prey, or me trying to remember where I left my lecture notes.

This chapter, in closing, explores a connection between complex networks whose dynamics are members of the Ising universality class and the dynamics of an individual member of the network. The response of the individual to the network dynamics is shown, using a subordination argument, to be expressible as a fractional Langevin equation. The solution to the average fractional

Langevin equation is shown to yield the probability of an individual changing states to be determined by a MLF.

The most familiar way to understand complex phenomena has historically been through the replacement of vast arrays of ordinary differential equations describing the dynamics of individual particles with the phase space equation for a probability density function (PDF). In this way the fluctuating single particle trajectories of the Langevin equation have traditionally given way to Fokker-Planck equations. In Chapter 7 this replacement is expanded to replace the chaotic trajectories of the fractional Langevin equations (FLEs) with fractional kinetic equations (FKEs), that is, phase space equations for the PDF with fractional derivatives in both space and time. The necessity for such a description was first encountered in the attempts to understand the phenomenon of anomalous diffusion using the fractional calculus [14].

Chapter 7 may be the most mathematically challenging; it is certainly the richest in physical/mathematical concepts. It provides the foundation for much of the applications that have preceded it. It was not presented first because such ordering may very well have discouraged all but the most dedicated. Hopefully the applications presented throughout the earlier chapters will have whetted the reader's appetite sufficiently to pursue, if not to master, the subsequent material.

A summary of the arguments we have presented in support of the fractional calculus view of complexity is in presented in Chapter 8. This is followed by an encapsulation of what we have learned in the form of four non-traditional scientific truths.

1.4   After Thoughts

A number of attempts have been made to develop a way of doing science that is respectful of the complexity of the phenomena being studied. Examples of such efforts that come to mind include Cybernetics, Systems Theory, Complexity Theory and their subsequent generalizations. The common element of these and other such efforts is the recognition that complex phenomena, whether natural or artificially constructed, ought to be treated as a whole and not selectively dissected and once understood stitched back together. This book does not seek to accomplish this Herculean task, but has the more modest goal of juxtaposing the disparate contributions made by a number of gifted scientists into a single strategy for gaining understanding and acquiring new knowledge. This strategy, if successful, will be an application of da Vinci's approach to understanding for those without his obvious gifts and it forms the basis of what I have called tomorrow's science, which in reality is five centuries old.

The complex phenomena that concern today's scientists systematically violate the traditional simplifying assumptions made in all the standard disciplines. The dynamics of physical, social, and biological processes are not linear and their statistical fluctuations are not Normal. While this statement is perhaps too strong in general, it is intended to point out that the assumptions of linearity and Normality are often lurking in the background of most theory used to interpret empirical data. The experiments used to support these assumptions in the past were often constrained in such a way as to yield the desired results, rather than to test for them. However the time scales that can be realized in today's physical experiments are so short that the coarse graining arguments typically used in modeling microscopic processes are no longer tenable. Social media provide a tsunami of data on the time scales of communications showing that the assumption of Poisson statistics for messages, so dear to information theory, is clearly violated. Linear models, whether deterministic or random, can no longer explain the high resolution, short time, data rich phenomena of interest today and in the foreseeable future.

The elegant analytic mathematics of the nineteenth and twentieth centuries, so successful in explaining the field theories of physics, are not appropriate for describing the dynamics of the scale-free complex phenomena so apparent in the social and medical sciences. This is not to say that these methods are no longer of value, but rather they must be implemented in new ways. The development of nonlinear dynamics has made this all too obvious. The systematic inclusion of chaos into the evolution of complex systems, through the scaling behavior of fractional diffusion equations is a relatively new idea that requires continuing development. The discontinuous nature of fractal processes, both geometrical and statistical, require a reexamination of equations of motion to incorporate fractional operators. The fractional calculus provides a new way to model the influence of an erratic background on a system of interest and this description dovetails with that of nonlinear dynamics.

Tomorrow's science will merge the various strategies developed to address intriguingly difficult problems long thought to be isolated, but ultimately determined to be fundamental. The nonlinear dynamics and chaos of Poincaré, the fractal time series and statistics of Mandelbrot, the information theory of Shannon, and the infinite-order exact linear representation of Koopman and Carleman all converge to provide a particular kind of understanding of complexity. An example may help clarify these general remarks.

A generic problem is how to characterize the response of complex networks to external perturbations and to understand the manner in which information is shuttled back and forth between such networks; ultimately whether or not there exist general principles that guide the flow of information. It has been determined that information transfer in an information-dominated process is guided by the *principle of complexity management* (PCM) [17] that stipulates the conditions for maximum information exchange between complex networks. Under certain well-defined mathematical conditions it was proven that the information transfer between two complex networks is maximized when the nature of the complexity of the two are matched [19, 21]. A temporally complex network is assumed to be scale-free with an IPL in the time intervals between events. The degree of complexity is specified in terms of the IPL index, which when the index is greater than two the first moment of the PDF is finite and the statistics are ergodic. When the index is less than two the first moment diverges and the statistics are non-ergodic. The response of an ergodic network to stimulation by a non-ergodic network vanishes in time, resulting in no net information transfer. On the other hand, the response of a non-ergodic network to stimulation by a non-ergodic network produces a complete domination of the perturbed network and consequently there is a maximum transfer of information between the two.

The PCM explains why a repetitive stimulus of unvarying amplitude and frequency content (ergodic stimulus) induces a response that fades quickly since no new information is being presented to the brain. The lack of new information allows the brain to shift its focus from the more to the less familiar, the latter providing new information that may have survival value. This is the process of habituation [18] and its understanding clarifies why we are attentive when a speaker moves around the stage, waves his arms and modulates his speaking voice in both volume and accent, but we nod off when the speaker stands in one spot and speaks in a monotone with his arms hanging lifelessly from his sides.

For a somewhat less controversial example consider the sound of splashing water from a leaky faucet. The sequence of water drops can set your teeth on edge and lead to tossing and turning throughout the night. Experiments have determined that the leaky faucet stimulus has a distribution in time intervals between splashes with a power-law index establishing the acoustic events to be non-ergodic. Consequently, the response of the brain to the intermittent splashes is a maximum. Of course it is not just annoying stimuli that refuse to fade away. Classical music has been shown to manifest IPL (non-ergodic) behavior as well and to resonate with human cognition, leaving strains of melody running through your head long after the music stops.

These and other lingering influences are not dependent on the mechanism coupling the two complex networks together. The sustained response is again a consequence of the PCM. The network dynamics produces the complexity associated with network topology and the complexity associated with the intermittency of events, both of which in turn determine the efficiency of the information exchange between the networks. The collaboration within a network has been shown to be maximal at the critical value of the control parameter that determines the strength of the interaction between the elements within a network. But these latter considerations anticipate what is discussed subsequently in this book.

References

[1] Abbott, E.A. 1952. *Flatland: A Romance of Many Dimensions*, Dover Pubs, NY.

[2] Bacon, F. 1620. *Novum Organum.* English translation, Thomas Fowler (Ed., notes, etc.) 1878, McMillan and Co., Clarendon Press, Oxford.

[3] Carleman, T. 1932. *Acta Mathematica* **59**, 63.

[4] Gödel, K. 1931. *Monatshefte Math. Physik.* **38**, 173.

[5] Jervis, R. 1997. In Alberts D.S. and T.J. Czerwinski, 1997, Editors, *Complexity, Global Politics and National Security*, National Defense University, Washington, D.C.

[6] Lévy P., 1925. *Calcul des probabilities*, Gauther-Villars, Paris.

[7] MacCurdy, E., Ed. 1960. *The Notebooks of Leonardo da Vinci*, Konecky & Konecky, Old Saybrook, CT.

[8] Morse, P.M. and H. Feshbach. 1953. *Methods of Theoretical Physics*, McGraw-Hill Book Comp., New York.

[9] Morse, P.M. and G.E. Kimball. 1956. In *The World of Mathematics, Vol. Four*, Ed. J.R. Newman, Simon and Shuster, New York.

[10] Poincaré, H. 1888. *Mémoire sur les courves définies parles equations différentielles, I–IV, Oevre 1*, Gauthier-Villars, Paris.

[11] Price, D.J. de Solla. 1963. *Little Science, Big Science .... and Beyond*, Columbia University Press, New York.

[12] Richter, J.P. 1970. *The Notebooks of Leonardo da Vinci, Vol. 1*, Dover, New York; unabridged edition of the work first published in London in 1883.

[13] Rosenau, J.N. 1997. In Alberts D.S. and T.J. Czerwinski, Editors, *Complexity, Global Politics and National Security*, National Defense University, Washington, DC.

[14] Seshadri, V. and B.J. West. 1982. *Proc. Nat. Acad. Sci. USA* **79**, 4501–4505.

[15] Shlesinger, M.F. 2006. *Nature* **443**, 281–282.

[16] Snow, C.P. 1959. *The Two Cultures*, London: Cambridge University Press.

[17] West, B.J., E.L. Geneston and P. Grigolini. 2008. *Phys. Rept.* **468**, 1–99.

[18] West, B.J. and P. Grigolini. 2010. *Physica A* **389**, 5706.

[19] West, B.J. and P. Grigolini. 2011. *Complex Webs; Anticipating the Improbable*, Cambridge University Press, UK.

[20] West, D. and B.J. West. 2012. *Int. J. Mod. Phys. B* **26**, 1230013-1.

[21] West, B.J., M. Turalska and P. Grigolini. 2014. *Network of Echoes; Immitation, Innovation, and Invisible Leaders*, Springer, New York.

[22] West, B.J. 2014. *Rev. Mod. Phys.* **86**, 1169–1178.

CHAPTER 2

Yesterday's Science

Techniques developed for measuring complexity have jumped from one area of study to another, like a smooth flat stone skipping over calm water, often leaving native investigators dizzy and ill equipped to apply the new methodologies at any depth or to connect the points of contact. Perhaps the growth of science has always appeared this way to those involved in its development. This was the conclusion reached by de Solla Price [27]; after studying centuries of data on the number of papers published, the number of new journals formed and the number of citations to those publications:

> Science has always been modern: it has always been exploding into the population, always on the brink of its expansive revolution. Scientists have always felt themselves to be awash in a sea of scientific literature that augments in each decade as much as in all times before.

The exponential increase in the number of papers published, the number of new journals created, both in paper and more recently in cyberspace, overwhelm the typical working scientist. In the attempt to manage and give structure to this torrent of information, new disciplines were systematically generated, as well as new journals to capture those aspects of the deluge the individual could master. He argued that the number of scientists at the time of writing his book was becoming saturated and in response to passing through the transition point, science was being redefined in terms of phenomena rather than disciplines. He reasoned that unconstrained exponential growth is unphysical and consequently the growth of science must eventually saturate following a logistic curve. However there need not be a single saturation level.

An exemplar of this type of growth with multiple levels of saturation using the rate of discovery of chemical elements was given by Price [27] and those data were interpreted to be a sequence of logistic curves stitched together. The argument was that as an old technology became exhausted and reached its saturation level, a new technology emerged to take its place. In the case of the discovery of new chemical elements each burst of discovery was predicated on the invention of a new technology necessary for its measurement. Or said the other way around when the number of elements discovered using an existing technology was exhausted, a new technology was applied to the problem resulting in further discoveries.

A different data set is depicted in Figure 2.1 in which the cumulative number of mammalian species discovered over time from 1760 to 2003 is plotted. A single logistic curve is used to fit the overall data, but it is clear that the data can also be interpreted in terms of an inter-connected sequence of logistic sub-processes [20]. This arrangement in terms of logistic sub-processes has more to do with the environment and the development of methods to discover new species than they do with the ecology of the species.

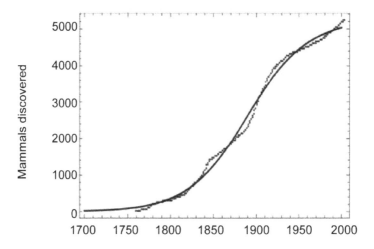

Figure 2.1 Cumulative number of mammalian species discovered over time from 1760 to 2003.(from [1] with permission).

This argument is more than a metaphor for the development of science, but is a paradigmatic representation of how the understanding of complex phenomena grows over time in response to changes in the supporting theoretical/technological infrastructure. Many believe, that in the last half of the twentieth century, we passed through one of the plateaus predicted by de Solla Price, and are now in the process of developing a new way of thinking about science. We have passed through what can be mastered by the formation of separate and distinct scientific disciplines and now enter a phase where synthesis and integration in the manner of da Vinci are necessary for further progress. I hope to convince the reader that this new way of thinking is facilitated by an old, but not widely known, form of quantitative reasoning, that being the fractional calculus.

2.1 Simple Linearity

Let us briefly review how traditional physical reasoning is tied to linearity [30] and begin with arguably the greatest of the classical physicists, Sir Isaac Newton. Ignoring the forms of his three laws of motion for the moment, which are the foundation of mechanics, let us examine how he answered one of the outstanding physics questions of his day, that being the reason the speed of sound in air has the value it does. Our focus will be on how Newton the natural philosopher thought about difficult problems for which he had no exact solution, or indeed no equation of motion. We do this because it is instructive to see how the success that emerged from his style of thought or strategy of model construction motivated future generations of scientists to think in the same way.

Sound propagates as a wave. This was known to Newton because of the observation of such effects as reflection, refraction and diffraction. What was not known to him was that there existed an equation to describe the propagation of waves, or perhaps he did know that such equations were possible, but the proper mathematical description of the phenomenon did not exist in 1686. Newton argued that a standing column of air consisted of air molecules of equal masses arranged along a line and locally interacted with neighboring air molecules by means of a linear elastic force that kept the molecules localized. He determined that a sound wave consists of periodic rarefactions and compressions of the air molecules and, as it turned out, is an ideal physical system for being modeled by a chain of linearly coupled oscillators.

The system of coupled oscillators was a natural model for the inventor of the calculus (fluxions) to develop because it required individual air molecules (oscillators) to undergo periodic motion during the passage of a disturbance without any net motion of the molecules themselves. Thus, the entity of a wave propagates, but without the transport of matter. The linear character of the model enabled Newton to reason to the solution of the equation of motion without explicitly writing it out. He was able to deduce that the speed of sound in air is $v = (\mathcal{P}/\rho)^{1/2}$ where $\mathcal{P}$ is the pressure of the wave and ρ is the mass density of the column of air. Using the isothermal volume elasticity of air, that is, the pressure itself, in his equation he obtained a speed of sound of 945 feet/second, a value 17% smaller than the observed value of 1142 feet/second.

This deviation in the prediction of the speed of sound from the experimental value stimulated the imagination of numerous scientists who attempted to construct more complex theories. This included Lagrange who in 1759 constructed and solved an explicit chain of linearly coupled oscillators:

$$m\frac{d^2\xi_n(t)}{dt^2} = k\left[\xi_{n+1}(t) - \xi_n(t)\right] - k\left[\xi_n(t) - \xi_{n-1}(t)\right] \tag{2.1}$$

where $\xi_n(t)$ is the displacement of the nth spring (oscillator) from its equilibrium position, m is the mass of the spring, k is the elastic spring constant and the frequency of oscillation is $\omega_0 = \sqrt{\frac{k}{m}}$. The restoring force is given by Hooke's Law, the difference in the force exerted by each of the nearest neighbors of a given spring. The local interaction occurs all along the chain of springs yielding an equation of motion Eq.(2.1) that is continuous in time but is discretely indexed in space along the chain by n.

This set of n linearly coupled differential-difference equations was solved analytically by Lagrange. The speed of sound in air is given by the speed at which the phase of the wave propagates and this is the ratio of frequency over the wavenumber of the wave, which reduces to Newton's result in the continuum limit. A result that Lagrange no doubt found upsetting, since he obtained the same value as Newton. In 1816 Laplace observed that by replacing the isothermal with the adiabatic elastic constant, arguing that the compressions and rarefactions of a longitudinal sound wave takes place adiabatically, obtained essentially exact agreement with experiment. Thus, one hundred and thirty years after his erroneous prediction Newton's reasoning was vindicated [30]:

> The success of the application of a harmonic chain to the description of a complicated physical process established a precedent that has evolved into the backbone of modeling in physics.

Thus, when the behavior of a system can be determined by the reasoning implied by linear models the system is said to be simple. But do not mistake the use of the term simple to mean that a phenomenon has only a few components or that it is uncomplicated and easily solved. A few-body system consisting of weakly interacting particles is simple, but so too is a many-body system consisting of an Avogadro's number of weakly interacting particles. Examples of linear but difficult problems that come to mind include: tracking the path of acoustic waves in the deep ocean, the scattering of light in fog, and the propagation of probability amplitudes in quantum mechanics. The simplicity of the individual interaction can be overcome by the sheer number of elements interacting in the process.

2.1.1   Linear Superposition

One of the major steps taken in the developing a linear view of natural phenomena was taken by the father (John Bernoulli) and son (Daniel Bernoulli) team through letter correspondence in 1727. They established that a system consisting of N one-dimensional point particles has N independent modes of vibration each of which is described by a unique frequency and unique harmonic function, these are the N eigenfrequencies and eigenfunctions for the system. The number of degrees of freedom describing the system's motion is equal to the number of entities that are coupled together. West explained Bernoullis' reasoning to be that if one has a system consisting of N one-dimensional point masses there are then N modes of vibration without making particular assumptions about how to construct the equations of motion as long as the coupling is weak. This was the first statement of the use of eigenvalues and eigenfunctions in physics.

Let us review the simplest of eigenvalue problems. Consider the N dimensional vector $\mathbf{X}(t) = (X_1, ..., X_N)$ whose dynamics are determined by the system of rate equations for its components

$$\frac{d\mathbf{X}(t)}{dt} = \mathbf{A}\mathbf{X}(t) \tag{2.2}$$

and $\mathbf{A}$ is an $N \times N$ matrix of time-independent constants. If one can define a similarity transform matrix $\mathbf{S}$ such that

$$\mathbf{Y}(t) = \mathbf{S}\mathbf{X}(t) \tag{2.3}$$

then the set of rate equations becomes

$$\frac{d\mathbf{Y}(t)}{dt} = \mathbf{\Lambda}\mathbf{Y}(t) \tag{2.4}$$

and the matrix $\mathbf{\Lambda}$ is diagonal

$$\mathbf{\Lambda} = \mathbf{S}\mathbf{A}\mathbf{S}^{-1} = \begin{pmatrix} \lambda_1 & 0 & \cdot & \cdot & \cdot \\ 0 & \lambda_2 & 0 & \cdot & \cdot \\ \cdot & 0 & \cdot & \cdot & 0 \\ \cdot & \cdot & \cdot & \cdot & 0 \\ \cdot & \cdot & 0 & 0 & \lambda_N \end{pmatrix}. \tag{2.5}$$

The spectrum of eigen or characteristic values is given by $(\lambda_1, ..., \lambda_N)$ and the N eigen or characteristic functions are $(Y_1, ..., Y_N)$. This results in the set of decoupled equations

$$\frac{dY_n(t)}{dt} = \lambda_n Y_n(t) \ ; \quad n = 1, ..., N \tag{2.6}$$

in terms of the characteristic behaviors of the system.

So what does this mean? Put simply it means that one can take a complicated system consisting of N coupled degrees of freedom and completely describe it by specifying one function and one constant per degree of freedom, for a periodic system the constant is a frequency. If the eigenfunctions and eigenvalues are known, that is all that is necessary to describe any general behavior of the system. The Bernoullis took a complicated system and partitioned it into readily identifiable and tractable pieces, each one being specified by an eigenfunction. Consequently, if one knows the eigenfunctions and eigenvalues then the contention was that one knows everything about the possible behavior of the system. Somewhat later Daniel Bernoulli [4] used the results of his correspondence with his father John to formulate the *Principle of Superposition*:

> The most general motion of a vibrating system is given by a linear superposition of its characteristic (eigen, proper) modes.

The importance of this principle cannot be overemphasized as a lens through which to view and understand complex phenomena. Up until the articulation of the principle of superposition all general statements in natural philosophy were concerned with mechanics, that is, they pertained to the movement of individual particles. The principle of superposition was the first formulation of a general law pertaining to a system of particles. Note that in 1755 it was asserted that the most general motion of a complicated system of particles is nothing more than the linear superposition of the motions of the constituent elements. This powerful view regarding how one can understand the evolution of a complex physical system strongly influenced how we understand the greater complexity in the social and life sciences.

It is worthwhile to note that not only did the superposition principle influence developments in the physical sciences, but it also influenced the description of the behavior of society and the nature of life itself. Consider another statement of the principle:

> A complex process can be decomposed into constituent elements, each element can be studied and understood individually and the elements reassembled to understand the whole.

The general form of the superposition principle was expressed in the mathematical formulation of Sturm-Liouville theory (1836–1838). The theory demonstrated that the space-time evolution of most physical phenomena known at the time could be represented by a class of differential equations and cast in the form of an eigenvalue problem. The solution to the equations of motion could then be represented as a summation over (superposition of) the eigen-motions of the physical system. Thus, the emerging disciplines of acoustics, heat transport, electromagnetic theory in the eighteenth and nineteenth centuries could all be treated in a unified way. In the twentieth century this would allow the problems of the microscopic world to be treated as well; all of quantum mechanics would yield to the linear theory in an infinite dimensional space.

In this view the whole can be no more than the sum of its parts, just like the angles in a Euclidean triangle. There can be no emergent properties;

if the property is not contained within the elements it cannot emerge from their superposition. This formed the basis of reductionistic science, where a physical observable could always be traced back to an equation of motion. As a consequence of this perspective all phenomena were thought to be understood by treating nonlinearities as perturbations, that is, as if they were small deflections from the course set by the linear system. The idea of perturbation theory is that the dominant behavior of a system is described by linear dynamics, and the response to a nonlinearity is to quantitatively change the evolution of the system without modifying its qualitative behavior.

2.2 Complicated Linearity

A disproportionate response of a system to a nonlinearity is known to occur from the irregular solutions of Newton's equations of motion for strongly interacting systems. The behavior of even weakly nonlinear dynamic systems is more complicated than the superposition principle would suggest. Consequently understanding the process as a whole cannot be accommodated by knowledge of the N eigenvalues and eigenfunctions alone. One such phenomenon for which this is true is that of friction (dissipation) and is a consequence of the physical world not being simply linear. Dissipation is the irreversible transfer of useful energy into waste heat. In this section we follow the arguments presented elsewhere to explain some of the difficulties with the linear world view.

2.2.1 FPU Problem

The physicist E. Fermi, mathematician S. Ulam and computer scientist J. Pasta [8] initiated an investigation into the irreversible process of heat transport in solids adopting Newton's strategy for calculating the velocity of sound in air. They considered a mathematical model of a solid based on the harmonic approximation, with all the particles represented as harmonic oscillators just as Newton had done. They intended to show how macroscopic irreversibility was a consequence of the deviation of the microscopic interactions from linearity. When I was a graduate student in the late 1960's the failure of these three to prove their conjecture was called the Fermi-Pasta-Ulam (FPU) problem.

The three-dimensional harmonic lattice is adequate for the description of many physical properties, but is at variance with one of the fundamental principles of classical physics: for a given temperature of the environment the system ought to equilibrate in such a way that each degree of freedom (normal mode) has the same amount of energy, that being, $1/2k_BT$. This is the equi-partition theorem and it is violated by a strictly harmonic lattice. A decade after the FPU study it was proven [9] that the equi-partition theorem does apply if one investigates the excitation of a single oscillator coupled to a harmonic environment in the limit of an *infinite* number of degrees of freedom in the environment.

FPU reasoned that for equi-partitioning to occur within a calculation using the harmonic lattice, say for a heat pulse to equilibrate in a metal by raising the overall temperature slightly, a nonlinear interaction must be included. Consequently, they included a weak anharmonicity in each of the oscillators and anticipated that a calculation of the dynamics would support the principle of equi-partition of energy. The average strength of the nonlinearity included was an order of magnitude weaker than the strength of the linear coupling. This was the first attempt to verify the equi-partition theorem dynamically, all previous arguments began from equilibrium statistical mechanics. It was assumed that the two arguments were equivalent. However, they were amazed when equilibrium was not achieved. In fact they did not publish their results since they had failed to establish energy equi-partition and in spite of this their report became the most famous unpublished piece of research ever done at Los Alamos National Laboratory or anywhere else for that matter.

Newton in 1686 constructed a discrete model of linear waves and Euler in 1748 invented the continuous wave equation; the two representation were physically equivalent and the properties of the solutions were the same. FPU in 1955 constructed a discrete nonlinear model for equi-partition and in 1965 Zabusky and Kruskal [35] developed a nonlinear wave equation that was the continuous form of the discrete set of equations used by FPU. The solutions to the nonlinear wave equation are called *solitons* and are localized propagating modes in which the effects of dispersion and nonlinearity balance one another resulting in a freely propagating coherent structure.

For the initial state of an anharmonic chain to be nearly that of a soliton the initial distribution of values of the linear mode amplitudes need to have a specific form. As the near-soliton propagates these values change in time. When the near-soliton is reflected from the end of the chain, it returns to its initial configuration. The energy flowing from one normal mode into another reverses direction upon reflection so that the initial distribution of energy among the normal mode amplitudes repeats itself periodically. This recurrence was the behavior FPU observed in their calculations in violation of the principle of energy equi-partition.

The change in the world view entailed by the influence of even weak nonlinearities is a loss of the principle of superposition. Consider the solution to a linear dispersive field equation as

$$u\left(x,t\right) = \int_{-\infty}^{\infty} \exp\left[i\left\{kx - \omega\left(k\right)t\right\}\right]\widetilde{u}\left(k,\omega\left(k\right)\right)\frac{dk}{2\pi}. \tag{2.7}$$

Here $\widetilde{u}\left(k,\omega\left(k\right)\right)$ is the Fourier amplitude of a wave of mode number k and frequency $\omega\left(k\right)$. A dispersion relation gives the frequency as a function of wavenumber thereby relating the spatial to the temporal behavior of the field.

For a narrow-band process centered on $k = k_0$ it can be shown that the superposition of waves given by the Eq.(2.7) integrates to

$$u(x,t) \backsim \frac{\exp\left[i\left\{k_0 x - \omega(k_0) t\right\}\right]}{\sqrt{t}} \qquad (2.8)$$

using the method of stationary phase. It is clear that the wave decays to zero as $1/\sqrt{t}$. If one has a linear dispersive wave field and energy is injected as a pulse at the central wavenumber then no matter how the energy is transferred within the field that wave amplitude decays as $t^{-1/2}$. If one starts with a bump eventually it will become flat. This IPL decay is subsequently observed in the fractional harmonic oscillator as well, but for ostensibly the same reason.

If however there is a certain kind of anharmonicity in the dispersive wave field one can start with a bump and under certain conditions that bump will persist forever. In order for this persistence to occur the strength of the nonlinear terms in the field equations must be such as to exactly balance the effect of linear dispersion. This delicate balance of linear dispersion driving the component waves apart and the nonlinear interactions pulling them back together provides for coherent structures such as solitons and has been observed in such systems as the gravity wave field on the ocean surface [6]. The world is filled with such subtle effects that violate the linearity assumption, even leaving aside the devastation produced by strong nonlinearities.

2.2.2 Langevin Equation

Scientists have wrestled with modeling complexity for centuries. One of the more successful models, in the sense of being able to quantify a number of consistently observed experimental relations, is that due to Langevin [15]. He reasoned that any system of interest can be significantly influenced by its environment. However, the detailed dynamics of the environment may be incorporated into the dynamics of the system of interest through the related notions of fluctuations and dissipation. This is not a new idea. It was anticipated by Titus Lucretius Carus in his ancient poem *On the Nature of Things;* who wrote a century before Christ. In a recent article Greenblatt [10] summarizes in modern language what is interest to us here:

> The stuff of the universe, Lucretius proposed, is an infinite number of atoms moving randomly through space, like dust motes in a sunbeam, colliding, hooking together, forming complex structures, breaking apart again, in a ceaseless process of creation and destruction. There is no escape from this process.

In 1908 Paul Langevin argued that the forces acting on a heavy particle moving through an ambient fluid of lighter particles consists of two parts: a dissipative force and a fluctuating force. The dissipative forces arises from the viscosity of the background fluid in which the heavy particle is embedded, which converts the kinetic energy of the heavy particle's motion into the thermal energy of the ambient fluid. The fluctuating forces arises from an instantaneous imbalance in the number of fluid particles impacting the surface of the large particle. He assumed the imbalance to be random and did not include any hydrodynamic response of the fluid to the motion of the heavy particle. The erratic imbalance on the surface of the heavy particle causes it to move this way and that, without rhyme or reason, changing directions in a random manner as depicted in Figure 2.2.

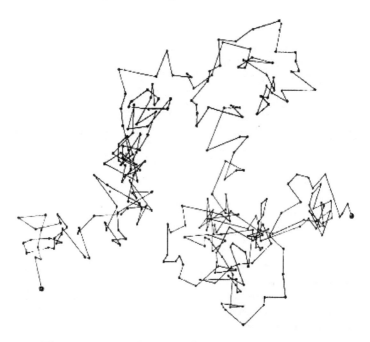

Figure 2.2 The coarse grained motion of a heavy particle in an ambient fluid of lighter particles is buffeted around in a random manner by thermal activity. Not unlike the motion envisioned by Lucretius, but these are the coarse grained paths observed by Perrin through a microscope [25].

The force law constructed by Langevin for the vector momentum of the heavy particle $\mathbf{P}(t)$ is

$$\frac{d\mathbf{P}(t)}{dt} = -\lambda \mathbf{P}(t) + \mathbf{f}(t) \qquad (2.9)$$

where λ is the rate of dissipation and is proportional to the fluid viscosity. The phenomenological fluctuating force $\mathbf{f}(t)$ is assumed to be completely specified by its statistical properties. Langevin assumed the components of the random force to have zero-centered Gaussian fluctuations and to be delta correlated in time. Denoting the average over an ensemble of realizations of the fluctuations by a bracket we have for the average force

$$\langle \mathbf{f}(t) \rangle = \mathbf{0} \tag{2.10}$$

and the components of the force are statistically independent as observed in the cross-correlation functions

$$\langle f_i(t) f_j(t) \rangle = 2D\delta_{ij}. \tag{2.11}$$

In general the fluctuations can be anisotropic with a different strength in each direction. That level of detail does not concern us here.

As Lindenberg and West [18] point out the strength of the collisions embodied in the parameter D in Eq.(2.11) is a manifestation of the kinetic energy of the fluid particles, which in turn is characterized by the thermodynamic temperature of the fluid. One would thus expect the level D of the fluctuations to increase with the ambient fluid temperature T. Langevin recognized that Einstein's fluctuation-dissipation relation (FDR) manifests itself in the present context as

$$D = \lambda k_B T \tag{2.12}$$

thereby making the strength of the fluctuations proportional to the temperature.

Of course Langevin's argument is phenomenological. The question is whether a physical model can be constructed that captures all the empirical features contained in his equation. Moreover, whether this model could be consistent with other physical principles, say with that of energy equi-partition. We could do worse than to follow the example of Newton.

We mentioned earlier in this chapter that the linear harmonic lattice could be used to dynamically verify the equi-partition theorem if it were infinite dimensional [9]. This approach was reviewed in a thermodynamic context by Lindenberg and West [18] for an idealized, closed Hamiltonian system consisting of a system of interest, the environment in which the system is embedded and a coupling between the two. An equivalent presentation was given by Tarasov [29], who reviewed a number of promising physical models that lent themselves to fractional generalizations. The model of the environment is a Hamiltonian for a set of linear harmonic oscillators with {displacements, momenta} denoted by $\{q_\nu, p_\nu\}$ $\nu = 1, 2, ..., N$ each with its mass m_ν and frequency ω_ν and in statistical physics the environment is called a heat bath for reasons that will become clear

$$H_b(\mathbf{q}, \mathbf{p}) = \sum_\nu \left[\frac{p_\nu^2}{2m_\nu} + \frac{m_\nu \omega_\nu^2}{2} q_\nu^2 \right]. \tag{2.13}$$

The isolated system of interest is modeled by a Hamiltonian with displacement Q, momentum P, unit mass and an arbitrary potential $U(Q)$:

$$H_s(Q, P) = \frac{1}{2} P^2 + U(Q). \tag{2.14}$$

The simplest form of the interaction between the system and heat bath (environment) is the Hamiltonian

$$H_{sb}(\mathbf{q}, Q) = -Q \sum_{\nu} \Gamma_{\nu} q_{\nu}, \tag{2.15}$$

with coupling constants Γ_{ν}. The total Hamiltonian for the combined system plus environment plus interaction is

$$H = H_b(\mathbf{q}, \mathbf{p}) + H_s(Q, P) + H_{sb}(\mathbf{q}, Q). \tag{2.16}$$

The solution to the equations of motion generated by the total Hamiltonian is presented in Appendix 2.5 resulting in the generalized Langevin equation

$$\frac{dP}{dt} + U'_m(Q) - \int_0^t P(\tau) K(t - \tau) d\tau = f(t). \tag{2.17}$$

Here we see the strategy that has been adopted to handle the complexity of a system described by a potential being coupled to the environment. Like Newton, the strategy is to replace the real world with a set of linear harmonic oscillators and determine its influence on the dynamics of the system of interest. The first influence is that the forcing function $f(t)$ is determined to be random. This is achieved by noting that the initial state of the environment is uncertain and is, in general, only determined by means of a distribution of initial conditions for the solutions to the equations of motion for the bath degrees of freedom. The function $f(t)$ is a sum of a large (infinite) number of independent random variables from the initial state of the bath. Consequently, applying the Central Limit Theorem results in the statistics of the random force being Gaussian.

The properties of such linear stochastic equations have been thoroughly explored by a number of investigators for physical, biological and social phenomena and we will not reproduce those investigations here, but we do note a number of general conclusions drawn from those studies. We already noted the Gaussian nature of the random force. The second general property concerns an observation first made by Einstein [7], that the strength of the fluctuations, the dissipation and the temperature are related by the FDR noted above Eq.(2.12). The FDR was subsequently generalized to include the memory kernel [14]. The average of the random force has been removed, to make the residual fluctuations zero centered, resulting in

$$\langle f(t)f(\tau) \rangle = k_B T K(t - \tau) \tag{2.18}$$

indicating that the fluctuations induced by the environment control the memory within the observable. The brackets denote an average over an ensemble of realizations of fluctuations, that is, realizations of the initial state of the environmental oscillators in this model. Note that Eq.(2.18) was rigorously established using the linear oscillators by a number of investigators [9, 18]. The average that was removed from the random force was added to the potential function in the GLE Eq.(2.17) and is the reason for the subscript on the potential.

The third general property of the harmonic model is that the spectrum of bath oscillations determines the nature of the memory. For example a very broad frequency spectrum localizes the random force in time, resulting in the memory kernel

$$K(t) = 2\lambda\delta(t),\qquad(2.19)$$

yielding the one-dimensional version of Eq.(2.11). The coefficient in Eq.(2.19) is the linear dissipation parameter introduced earlier.

The fourth general property is a consequence of the Langevin equation being linear in the fluctuations so that the statistics of system momentum is the same as that of $f(t)$. Thus, we can determine that the asymptotic statistics of $P(t)$ are described by a Gaussian distribution centered on the time-dependent average value determined by the potential and a variance determined by the FDR. Asymptotically the steady-state PDF of the system is the Boltzmann distribution [18]

$$P_{ss} = Z^{-1}\exp\left[-\beta H_m\right];\quad H_m = \frac{1}{2}P^2 + U_m(Q),\qquad(2.20)$$

where Z is the normalization, also called the partition function and the shifted potential is defined in Appendix 2.5. The time-dependent distribution is also available with a time varying average value and variance, but that level of detail does not concern us here.

The fifth and final general property, implicit in two of the others, is that the complexity of phenomena can be captured by replacing the detailed complex deterministic dynamic description by a simpler stochastic one. Moreover that the uncertainty in the stochastic description is not absolute. There are still laws that guide the evolution of complex phenomena, one being the Law of Frequency of Error in which the average value of the observable is assumed to be the correct value and deviations from that value are determined by a Gauss distribution, whose width determines the accuracy of the average in specifying the correct value of the observable.

This approach to modeling has been generalized to the treatment of nonlinear dynamical systems using large-scale computer codes. The computational approach typically goes by the name of Monte Carlo simulation, but a discussion of such techniques would take us too far from our path and so we do not mention them again.

2.2.3 Fokker-Planck Equation

The generalized Langevin equation constitute one way to take into account the complexity of many variables within a system. The description of the system dynamics uses such notions as particle trajectories and stochastic processes (time-dependent random variables). Another way to take complexity into account is by means of probability density functions (PDFs) that require a function to represent the behavior of an ensemble of trajectories in a phase space. In the latter representation the dynamic variable $Q(t)$ is replaced by the phase space variable q and the location along the trajectory is indexed with the time t. The location of a trajectory in phase space is given by the phase space density

$$\rho(q,t) = \delta(q - Q(t)) \tag{2.21}$$

which when averaged over an ensemble of trajectories, denoted by a bracket $\langle \cdot \rangle$ yields the PDF

$$P(q,t) = \langle \rho(q,t) \rangle. \tag{2.22}$$

This ties the PDF description to that of the dynamic variables. In a more complete notation the dependence of the PDF on the initial state of the variable $Q(0) = q_0$ would be indicated, as in the conditional PDF where $P(q,t \,|q_0)\, dq$ is the probability that the dynamic variable $Q(Q(0),t)$ lies in the interval $(q, q + dq)$ at time t, given that it had the initial value q_0. The evolution of the PDF in phase space is consequently determined by how the function varies with respect to both q and t [18, 19]. Here we present a straight forward approach to determining the phase space equation of evolution of the PDF that side steps a number of mathematical subtleties and which begins with the physical notion of diffusion.

We follow in part the presentation of Montroll and West [22] and note that the change in time of a stationary stochastic process using the conditional transition PDF for the dynamical variable $Q(t)$ to lie in the range $(q, q + dq)$ conditional on $Q(t') = q'$ is given by the chain condition

$$P(q,t \,|q_0,t_0) = \int_{\Omega} P(q,t \,|q',t')P(q',t' \,|q_0,t_0)dq' \tag{2.23}$$

where Ω is the domain of the dynamic variable (variate). This equation is often used as the mathematical starting point for the analysis of Brownian motion. The quantity $P(q,t \,|q_0,t_0)dq$ is the probability that the process undergoes a transition from the initial value q_0 at time t_0 to a final value in the interval $(q, q + dq)$ at time t in phase space through a continuous sequence of intermediate values indicated in the integral. These intermediate values locate the disruptions of the underlying trajectories tracing out the dynamics produced by the buffeting of the heavy particle by the lighter particles of the background fluid.

Equation (2.23) was introduced by Bachelier in 1900 in his Ph.D. thesis on speculation in the French stock market. The non-physical application of this equation to the diffusion of profit among stocks was in all likelihood the reason why his work went unnoticed for nearly 50 years. On the other hand, classical diffusion in a physical system was first successfully described in terms of a

PDF by Einstein [7] five years after Bachelier, and using the same equations [3]. In a subsequent paper Einstein conjectured that the observations through a microscopic of the erratic path of a pollen mote in water made in 1827 by the botanist Robert Brown might be a diffusive phenomenon. This off-hand comment was sufficient to insure Brown's scientific immortality in physics.

The chain condition Eq.(2.23) generally describes the evolution of the PDF for an infinitely divisible stable process, that is, a process that has the self-same statistical behavior regardless of the scale on which it is observed. The solution to this integral equation has the most general form of a Markov PDF. When the range of the variate is unbounded $\Omega = (-\infty, \infty)$ and the process under consideration has translational invariance, so the PDF is independent of the origin of the coordinate system $P(q, t | q_0, t_0) = P(q - q_0, t - t_0)$, the chain condition becomes

$$P(q - q_0, t - t_0) = \int_{-\infty}^{\infty} P(q - q', t - t')P(q' - q_0, t' - t_0)dq'. \qquad (2.24)$$

The stationary chain condition Eq.(2.24) is more simply expressed in terms of characteristic functions, the Fourier transform of the PDF. The Fourier transform is defined as

$$\widetilde{P}(k, t) = \mathcal{FT}\{P(q, t); k\} = \int_{-\infty}^{\infty} P(q, t)e^{-ikq}dq, \qquad (2.25)$$

which allows us to express the convolution in Eq.(2.23) as the product

$$\widetilde{P}(k, t - t_0) = \widetilde{P}(k, t - t')\widetilde{P}(k, t' - t_0). \qquad (2.26)$$

Montroll and West [22] explained that, since the PDF resulting from the characteristic function satisfies the product form its solution yields an infinitely divisible distribution. The most general form of the characteristic function for infinitely divisible distributions was first obtained by Paul Lévy in 1937 [17].

The traditional Gauss distribution for classical diffusion is

$$P_G(q, t) = \frac{1}{\sqrt{4\pi Dt}} \exp\left[-\frac{q^2}{4Dt}\right] \qquad (2.27)$$

and has a variance (dispersion) that increases linearly in time

$$\langle q^2; t \rangle = \int_{\Omega} q^2 P_G(q, t)dq = 2Dt. \qquad (2.28)$$

The Fourier transform of the Gauss distribution for $t \geq 0$

$$\widetilde{P}_G(k, t) = \mathcal{FT}\{P_G(q, t); k\} = \exp\left[-Dtk^2\right] \qquad (2.29)$$

Hence the Gauss distribution obeys the chain condition Eq.(2.26). The inverse Fourier transform of the time-derivative of the characteristic function yields:

$$\frac{\partial P_G(q,t)}{\partial t} = \mathcal{F}\mathcal{T}^{-1}\left\{\frac{\partial \tilde{P}_G(k,t)}{\partial t}; q\right\} = D\frac{\partial^2 P_G(q,t)}{\partial q^2} \tag{2.30}$$

the classical diffusion equation when q is interpreted as the velocity. This is the equation originally solved by Bachelier and subsequently by Einstein.

Lévy [16] found a number of PDFs that: 1) satisfy the chain condition Eq.(2.26); 2) have non-negative Fourier transforms and 3) are normalizable so that the PDF $P(q,t)$ conserves probability at all $t \geq 0$. The Gauss case is singular among the cases considered by Lévy in that it is the only one for which all the moments

$$\mu_n(t) = \int q^n P(q,t)dq \tag{2.31}$$

exist, being finite for all positive integer n. When these moments exist and have the properties enumerated below in the translationally invariant case the chain condition can be shown to give rise to the Fokker-Planck equation (FPE) for $P(q,t\,|q_0,t_0)$. We do not present the proof of the FPE here since it can be found in almost any text on statistical mechanics, but we record that,

$$\frac{\partial P(q,t\,|q_0,t_0)}{\partial t} = \frac{\partial}{\partial q}\left[-A(q) + B(q)\frac{\partial}{\partial q}\right]P(q,t\,|q_0,t_0) \tag{2.32}$$

which is solved subject to the initial condition

$$\lim_{t\to t_0} P(q,t\,|q_0,t_0) = \delta(q - q_0). \tag{2.33}$$

The two functions in the FPE are determined by the first moment

$$A(q) = \lim_{\Delta t\to 0}\frac{1}{\Delta t}\int (z - y)\,P(q,\Delta t\,|y\,)dy, \tag{2.34}$$

and by the second moment

$$B(q) = \lim_{\Delta t\to 0}\frac{1}{\Delta t}\int (z - y)^2\,P(q,\Delta t\,|y\,)dy, \tag{2.35}$$

respectively.

These last two equations tie the PDF to the dynamic description of the process. For example, in the case of simple Brownian motion without viscosity, for a unit mass particle in one spatial dimension, the Langevin equation is

$$\frac{dV(t)}{dt} = f(t) \tag{2.36}$$

where the statistical properties of the random force were discussed in the last section. Consequently the moment equations yield

$$A(v) = 0 \text{ and } B(v) = 2D \qquad (2.37)$$

and the FPE reduces to the classical diffusion equation Eq.(2.30).

The next more complicated process is one for which the first moment is a constant say V_0 and the second remains the same yielding the FPE

$$\frac{\partial P}{\partial t} = \frac{\partial}{\partial v} \left[-V_0 + 2D \frac{\partial}{\partial v} \right] P$$

whose solution under the boundary conditions

$$\lim_{v \to \pm\infty} P(v, t \,|\, v_0, t_0) = 0$$

is the traveling Gaussian packet

$$P(v, t \,|\, v_0) = \frac{1}{\sqrt{4\pi Dt}} \exp \left[-\frac{(v - v_0 - V_0 t)}{4Dt} \right] \qquad (2.38)$$

which propagates to the right with constant velocity V_0.

It is of course possible to generalize these arguments to multiple dimensions, to situations where the force is described by a potential function and the first moment is described by the gradient of the potential, to fluctuations having memory and spatial heterogeneity where the second moment is a function of space and time. But the complications very rapidly exceed the utility of the mathematical description and although there are a myriad of phenomena that are faithfully described by the FPE there is an ever increasing number of phenomena that cannot be so described. Even Brownian motion, where it all started, is subsequently shown to require a more subtle description when we are no longer focused on the microscopic or the macroscopic but observe the behavior of the Brownian particle somewhere in between these two familiar limits.

2.3   Nonlinear Dynamics

Nonlinear dynamics has to a large extent changed the way scientists view the world, which is strange given that it is a discipline defined by what it is not, that is, nonlinear dynamics is not linear dynamics. Setting this concern aside for the moment, it is reasonable to conclude that since linearity is only an approximation to reality, everything stated in this chapter must be considered to be conditional on linear dynamics and some results are to be taken with a grain of salt. The change demanded by nonlinear dynamics is not restricted to the physical world of material bodies, however, but also required new strategies for exploring the abstract world of social interactions, the mysterious domain of cognition, and the Pandora's box of medicine.

Over the past forty years there have been an astonishing number of books written about nonlinear dynamics and its implications. Some have painted with the fine grained detail demanded by researchers in this field [2, 33], others have been sensitive to the needs of the less mathematically skilled and extended the palette to include narratives on applications of the formalism [23, 32], still others have taken a broad brush approach, omitting details, to present the story of chaos as an attractive landscape [26, 28]. We adopt this last strategy in the few remarks included here on chaos and nonlinear dynamics.

The distribution of eigenvalues in the solution to the linear dynamical system has come to be known as the *spectral theory* of dynamical systems. Kowalski [13] pointed out that in 1932 Carleman [5], following the ideas of Poincaré and Fredholm, demonstrated that nonlinear systems of ordinary differential equations with polynomial nonlinearities can be reduced to infinite systems of linear differential equations. The Carleman embedding approach has successfully determined the solution to many nonlinear differential equations as presented and discussed by Kowalski and Steeb [12].

We cannot emphasize too strongly that the full complexity of nonlinear phenomena can be captured using an infinite dimensional linear representation. This is what we refer to as complicated linearity and has only been suggested by the all too brief analysis given here. The complete embodiment of nonlinear dynamics such as chaos resulting from the instability of dynamic orbits, has not yet been fully realized, but the general theory is in place [12] to develop a complete theory of stability using spectral methods.

2.3.1   A Little about Chaos

Phase space is the geometrical space in which a dynamical system lives. It consists of coordinate axes defined by the independent variables for the system and is useful for specifying a system's dynamics. Each point in such a space corresponds to a particular set of values of the dynamical variables that uniquely defines the state of the system. The point denoting the system moves about in this space as the phenomenon evolves and leaves a trail that is indexed by the time. This trail is referred to as the orbit or trajectory of the system and each initial state produces a different trajectory. It is often the case that no matter where the orbits are initiated they end up on the same geometrical structure in phase space. This structure to which all the trajectories are drawn as time increases is called an attractor. An attractor is the geometrical limiting set of points in phase space to which all the trajectories in the attractor's basin are drawn, and upon which they eventually find themselves. After an initial transient period, that depends on the initial state, an orbit blends with the attractor and eventually loses its identity. Attractors come in many shapes and sizes, but they all have the property of occupying a finite volume of phase space.

Whether or not a dynamical system is chaotic, is determined by how two initially nearby trajectories cover the system's attractor over time. As Poincaré stated, a small change in the initial separation of any two trajectories may produce an enormous change in their final separation (sensitive dependence on initial conditions)—when this occurs, the attractor is said to be "strange". One indicator of chaos is the distance between two nearby orbits increasing exponentially with time, which is to say, the distance between trajectories grows faster than any power of the time. But how can such a rapidly growing separation occur on an object, the attractor, which has a finite volume? Don't the diverging orbits have to stop after a while? The answer to this question has to do with the structure necessary for an attractor to be chaotic. The transverse cross section of the layered structure of a strange attractor is fractal and this property of strange attractors relates the geometry of fractals to the dynamics of chaos.

Chaos can be understood by watching a baker rolling out dough to make bread. The baker sprinkles flour on his breadboard, slams a tin of dough on the surface and sprinkles more flour on top of the pale ball. He works the rolling pin over the top of the dough, spreading it out until the dough is sufficiently thin; then he reaches out and folds the dough back over onto itself and begins the process of rolling out the dough again. The Russian mathematician Arnold gave a memorable image of this process using the head of cat inscribed in a square of dough, as shown in Figure 2.3. After the first rolling operation the head is flattened and stretched, that is, it becomes half its height and twice its length, as shown in the top right side of the figure. The dough is then cut in the center. The right segment is lifted and placed over the left segment to reform the initial square. This is the mathematical equivalent of the baker's motions. The operation is repeated again and we see that at the bottom, the cat's head is now embedded in four layers of dough. Even after only two of these transformations the cat's head is clearly decimated. After 20 such transformations, the head will be distributed across 2^{20} or approximately one million layers of dough. There is no way to identify the head from this distribution across the layers of dough.

The baker argument, originally due to Rössler, turns out to be generic. Two initially nearby orbits, represented by the labels A and B on the cat's head, cannot separate an infinite distance on a finite attractor At first the two points are adjacent with A to the left of B. After one iteration B is to the left of A and slightly above. After two iterations A is to the right of B and B is slightly below. The relative positions of the two points is not predictable from one iteration to the next. In general, the attractor structure (cat's head) must afford ample opportunity for trajectories to diverge and follow increasingly different paths (different layers of dough). The finite size of the attractor insures that these diverging trajectories will eventually pass close to one another again, albeit on different layers of the attractor which are not directly accessible. One can visualize these orbits on a chaotic attractor

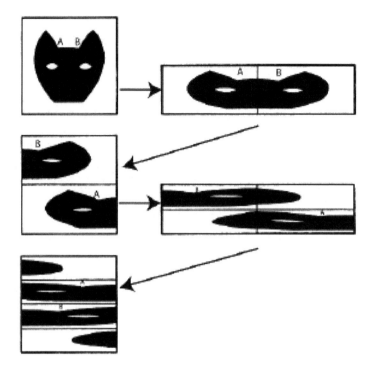

Figure 2.3   Arnold's cat being decimated the stretching and folding operation that accompanies the dynamics of a chaotic attractor. After only two operations the cat's head is unrecognizable.

being shuffled by this process, like a dealer in Las Vegas shuffles a deck of cards. This process of stretching and folding creates folds within folds ad infinitum, resulting in the attractor having a fractal structure in phase space.

One of the fascinating aspects of the chaos generated by such nonlinear discrete maps is that two such maps, when coupled, synchronize their dynamics. Even the chaos of the two coupled maps become synchronized after a very short time, as first observed by Pecora and Carroll [24]. We note that this remarkable property of chaos synchronization persists into the domain of fractional maps [34].

2.3.2   Carleman Embedding

Here we shift gears to demonstrate how linear methods can be used to obtain analytic solutions to certain classes of nonlinear dynamical equations. We sketch the Carleman embedding technique applied to an ordinary differential equation

$$\frac{dQ}{dt} = F(Q) \tag{2.39}$$

where $F(Q)$ is analytic in Q and in general is a polynomial. In this approach we introduce a function of the moments of Q:

$$\phi_n = Q^n, \tag{2.40}$$

where $n \in Z_+$ is contained in the set of non-negative integers. Consequently, Eq.(2.39) can be used to construct the set of linear differential-difference equations

$$\frac{d\phi_n}{dt} = \sum_{m \leq n} C_{nm} \phi_m. \tag{2.41}$$

Equation (2.41) is an infinite-order linear system of ordinary differential equations in which the dynamics of Eq.(2.39) are embedded. Note that each of the above steps can be generalized to multiple dimensions without changing the conclusions, but for the sake of clarity we do not present that level of generality here.

Let us consider how these ideas can be applied to obtaining the specific solution to a particular equation. Consider the quadratically nonlinear rate equation

$$\frac{dQ}{dt} = -aQ + bQ^2, \tag{2.42}$$

whose solution is simply obtained by the transformation of variables $z = 1/Q$:

$$Q(t) = \frac{a}{b} \frac{Q_0}{Q_0 + \left[\frac{a}{b} - Q_0\right] e^{at}}. \tag{2.43}$$

The solution has a sigmoidal shape and in the two limits for $a, b < 0$:

$$\lim_{t \to 0} Q(t) = Q_0, \tag{2.44}$$

and

$$\lim_{t \to \infty} Q(t) = \frac{a}{b} > 0, \tag{2.45}$$

asymptotically reaching a constant level as depicted in Figure 2.4.

Montroll [21] demonstrated how to solve the rate equation with a quadratic nonlinearity Eq.(2.42) using the Carleman embedding technique to obtain the solution to the quadratic rate equation from the infinite set of linear equations. It will become abundantly clear that this technique is much longer to implement than the straight forward substitution of variables used above. However what is lost in time is more than made up for in generality since most nonlinear rate equations cannot be solved in closed form. Consider the set of moments given by Eq.(2.40) to obtain the rate equation

$$\frac{d\phi_n}{dt} = \frac{dQ^n}{dt} = nQ^{n-1}\frac{dQ}{dt},$$

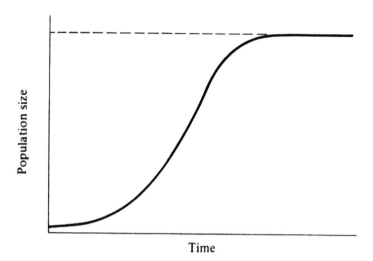

Figure 2.4 The solution to the quadratically nonlinear rate equation when $a, b < 0$ is the familiar logistic equation for population growth. The parameter a/b is the limiting value of the population, $r = -a$, and the initial population is taken to be small.

and from Eq.(2.43)

$$\phi_1(0) = Q_0 \equiv C \; ; \; \phi_n(0) = C^n,$$

and setting $at = \tau$ results in the linear system of equations

$$\frac{d\phi_n}{d\tau} = -n\phi_n + \frac{b}{a}n\phi_{n+1}. \tag{2.46}$$

This latter system is the infinite set of linear equations in which the quadratic nonlinear equation is embedded at $n = 1$.

Here we use Laplace transforms to solve Eq.(2.46). The Laplace transform of the solution is

$$\widehat{\phi}_n(s) = \int_0^\infty e^{-s\tau}\phi_n(\tau)\,d\tau \equiv LT\left\{\phi_n(\tau) ; s\right\}, \tag{2.47}$$

and using the initial conditions

$$LT\left\{\frac{d\phi_n(\tau)}{d\tau} ; s\right\} = s\widehat{\phi}_n(s) - C^n, \tag{2.48}$$

results in the system of linear equations for the Laplace variable

$$(s + n)\,\widehat{\phi}_n(s) - \frac{b}{a}n\widehat{\phi}_{n+1}(s) = C^n. \tag{2.49}$$

In matrix form we now have

$$\mathbf{A}\hat{\boldsymbol{\phi}} = \mathbf{C},\tag{2.50}$$

for the infinite dimensional linear system.

For pedagogic reasons we write the matrix equation explicitly as

$$
\begin{pmatrix}
(s+1) & -\frac{b}{a} & 0 & 0 & 0 \\
0 & (s+2) & -\frac{2b}{a} & 0 & 0 \\
0 & 0 & (s+3) & -\frac{3b}{a} & 0 \\
0 & 0 & 0 & (s+4) & -\frac{4b}{a} \\
\cdot & \cdot & 0 & 0 & \cdot
\end{pmatrix}
\begin{pmatrix}
\varphi_1 \\ \phi_2 \\ \phi_3 \\ \cdot \\ \cdot
\end{pmatrix}
=
\begin{pmatrix}
C \\ C^2 \\ C^3 \\ \cdot \\ \cdot
\end{pmatrix}.
\tag{2.51}
$$

Montroll gives us the inverse coefficient matrix, which is not difficult to obtain but does require some labor

$$
\mathbf{A}^{-1} =
\begin{pmatrix}
\frac{1}{(s+1)} & \frac{b}{a}\frac{1}{(s+1)(s+2)} & \left(\frac{b}{a}\right)^2\frac{2}{(s+1)(s+2)(s+3)} & \left(\frac{b}{a}\right)^3\frac{2\cdot3}{(s+1)(s+2)(s+3)(s+4)} & \cdot & \cdot \\
0 & \frac{1}{(s+2)} & \frac{b}{a}\frac{2}{(s+2)(s+3)} & \left(\frac{b}{a}\right)^2\frac{2\cdot3}{(s+2)(s+3)(s+4)} & \cdot & \cdot \\
0 & 0 & \frac{1}{(s+3)} & \frac{b}{a}\frac{3}{(s+3)(s+4)} & \cdot & \cdot \\
0 & 0 & 0 & \frac{1}{(s+4)} & \cdot & \cdot \\
0 & 0 & 0 & 0 & \cdot & \cdot \\
0 & 0 & 0 & 0 & 0 & \cdot
\end{pmatrix}.
\tag{2.52}
$$

Multiplying Eq.(2.51) on the left by the inverse matrix provides a formal solution for all the moments of the Laplace variable. Embedded in this set of solutions is the infinite series solution to our original equation

$$
\begin{aligned}
\hat{\phi}_1(s) &= \frac{C}{s+1} + \frac{b}{a}\frac{C^2}{(s+1)(s+2)} \\
&\quad + \left(\frac{b}{a}\right)^2\frac{2C^3}{(s+1)(s+2)(s+3)} + \left(\frac{b}{a}\right)^3\frac{3\cdot2C^4}{(s+1)(s+2)(s+3)(s+4)} + \cdots \\
&= \frac{a}{b}\sum_{n=1}^{\infty}\left(\frac{b}{a}C\right)^n\frac{\Gamma(n)\,\Gamma(s+1)}{\Gamma(n+s+1)} = \frac{a}{b}\sum_{n=1}^{\infty}\left(\frac{b}{a}C\right)^n B(n,s+1),
\end{aligned}
\tag{2.53}
$$

where we have replaced the ratio of gamma functions by the beta function

$$
B(n,m) = \int_0^1 z^{n-1}(1-z)^{m-1}dz = \frac{\Gamma(n)\,\Gamma(m)}{\Gamma(n+m)}.
\tag{2.54}
$$

Hence the Laplace variable solution can be re-expressed using this integral to obtain

$$
\hat{\phi}_1(s) = \frac{a}{b}\sum_{n=1}^{\infty}\left(\frac{b}{a}C\right)^n \int_0^1 z^{n-1}(1-z)^s dz,
$$

in which we can explicitly evaluate the sum to obtain

$$\widehat{\phi}_1(s) = C \int_0^1 \left(1 - \frac{b}{a}Cz\right)^{-1} (1-z)^s dz. \tag{2.55}$$

Introducing the final transformation of variables $(1-z) = e^{-\tau}$ and noting the change in integration limits allows us to write

$$\widehat{\phi}_1(s) = C \int_0^\infty \frac{e^{-s\tau} d\tau}{\frac{b}{a}C + \left(1 - \frac{b}{a}C\right)e^\tau}$$

and using the definition of the Laplace transform and $\tau = at$ we obtain from the argument in the integral using $C = Q_0$

$$Q(t) = \phi_1(t) = \frac{a}{b} \frac{Q_0}{Q_0 + \left(\frac{a}{b} - Q_0\right)e^{at}} \tag{2.56}$$

which is the same as Eq.(2.43).

Our goal of solving a nonlinear equation by explicitly solving an infinite system of linear equations has been realized. Moreover, as Montroll noted, the matrix $\mathbf{A}$ is a tridiagonal matrix with no non-zero elements below the central diagonal. He goes on to observe that if the right hand side of Eq.(2.39) were the polynomial

$$F(Q) = Q + a_2 Q^2 + \ldots + a_n Q^n \tag{2.57}$$

the matrix $\mathbf{A}$ would still be triangular but with $n-1$ filled diagonal lines above the central diagonal. The elements of the inverse of a tridiagonal matrix are known to have a simple form in terms of determinants so an explicit series expansion can be constructed for the solution to Eq.(2.39). Montroll [21] shows that a formal solution of the infinite set of linear equations resulting from sets of coupled nonlinear equations can be constructed from the inverse of tridiagonal matrices. We do not reproduce that formal solution here but refer the reader back to the original paper.

2.3.3 Infinite-order Linear Representation

It has been shown that the Carleman embedding technique is equivalent to the Koopman [11] linearization under the proper conditions [13]. Koopman demonstrate that a Hamiltonian generating a set of nonlinear equations of motion has an equivalent description in terms of an infinite set of linear dynamic equations.

Consider the multivariable rate equation

$$\frac{d\mathbf{V}(t)}{dt} = \mathcal{O}\mathbf{V}(t), \tag{2.58}$$

where $\mathcal{O}$ is an operator acting on the vector $\mathbf{V}(t) = \{V_1(t), V_2(t), ..., V_N(t)\}$ and having the general form

$$\mathcal{O} = \sum_{j=1}^{N} C_j(\mathbf{V}) \frac{\partial}{\partial V_j} \tag{2.59}$$

with the coefficients $C_j(\mathbf{V})$ being polynomials in the variables. With this general notation we can describe most dynamical systems of interest. The solution to Eq.(2.58) is formally given by

$$\mathbf{V}(t) = \exp(\mathcal{O}t)\mathbf{V}(0). \tag{2.60}$$

and the exponential operator is defined by

$$\exp(\mathcal{O}t) = \sum_{k=0}^{\infty} \frac{(\mathcal{O}t)^k}{\Gamma(k+1)}. \tag{2.61}$$

With this series representation it can be seen that there is no ambiguity in the action of the operator in the argument of the exponential function, since it requires the calculation of a series of operator moments of the form $\mathcal{O}^n \mathbf{V}(0)$.

Let us examine the condition defined by Eq.(2.58) more carefully. The generic system described by Eq.(2.60) involves a spectrum of eigenvalues γ_j of the operator $\mathcal{O}$. The eigenfunctions associated with this problem are given by the scalar product

$$\phi_j(\mathbf{V}) \equiv (\mathbf{V}, \chi_j) \tag{2.62}$$

where χ_j are the eigenfunctions of the adjoint operator $\mathcal{O}^\dagger$

$$\mathcal{O}^\dagger \chi_j = -\gamma_j \chi_j \tag{2.63}$$

normalized such that the inner product

$$(\mathbf{v}_j, \chi_k) = \delta_{j,k} \tag{2.64}$$

where $\mathbf{v}_j$ is the jth eigenvector of $\mathcal{O}$. The dynamics of the eigenvectors in time are given by

$$\frac{d\phi_j}{dt} = (\frac{d\mathbf{V}}{dt}, \chi_j) = (\mathcal{O}\mathbf{V}, \chi_j) = (\mathbf{V}, \mathcal{O}^\dagger \chi_j) = -\gamma_j(\mathbf{V}, \chi_j) = -\gamma_j \phi_j, \tag{2.65}$$

which has the exponential solution

$$\phi_j(\mathbf{V}, t) = e^{-\gamma_j t} \phi_j(\mathbf{V}_0). \tag{2.66}$$

Consequently, the state vector has the eigenfunction decomposition, as long as the operator $\mathcal{O}$ has a complete spectrum,

$$\mathbf{V}(t) = \sum_j (\mathbf{V}, \chi_j) \mathbf{v}_j = \sum_j \mathbf{v}_j \phi_j(\mathbf{V}, t) \tag{2.67}$$

and we have the exact series solution to the nonlinear dynamic equation expressed in terms of the infinite set of eigenvalues and eigenfunctions

$$\mathbf{V}(t) = \sum_j \mathbf{v}_j e^{-\gamma_j t} \phi_j(\mathbf{V}_0). \tag{2.68}$$

Thus, in principle, an arbitrary nonlinear dynamical system can be solved in terms of the eigenvalues and eigenfunctions of an infinite-order linear system.

2.4   After Thoughts

It occurs to me that much of what is fundamental about this chapter may have been obscured by the various formalisms. So let me attempt a summary based on thinking about phenomena as being linear and consequently as being expressible in terms of integer powers of basic units. In 1832 Gauss proposed that the dimensions of length, mass, and time are absolute and formed the basis for the development of the science of mechanics as formulated by Newton in 1687. These dimensions were thought to be irreducible in that they cannot be derived from one another, nor can they be resolved from anything more fundamental; they can only be directly experienced, which is to say operationally defined through experiment.

However, in addition to fundamental dimensions there are derived dimensions, for example, speed, the distance traveled divided by the time interval necessary to cover the distance.   Speed uses two of the basic dimensions: distance and time. Force, momentum and energy, on the other hand, use all three of the fundamental dimensions in various combinations. In other areas of physics there are additional fundamental dimensions, such at temperature in thermodynamics and charge in electricity and magnetism.

To demonstrate the power of this linear perspective I once did an experiment in teaching and wrote a book on biomechanics with a colleague [31] adopting the perspective that all the equations were derivable from dimensional analysis.   We learned a great deal from the effort and found that the biophysical processes involved in all the biodynamic activities from sprinting to swimming could be understood using this linear perspective. When the experimental data showed the phenomenon to be outside the linear regime, the notion of scaling could guide us to the next level of complexity.

So how does this elementary discussion of units orient our thinking about physical processes. For one thing we can determine the dynamics of simple systems without solving any equations of motion. We use the notation [L] as the unit of length, [T] the unit of time, and [M] the unit of mass. In this way the unit of speed is denoted $[V]=[L]/[T]$, momentum by $[P]=[M][L]/[T]$ and acceleration by $[a]=[L]/[T^2]$. In general we suppose that a measured quantity A has dimension that can be expressed in terms of the units of length, mass and time to the powers p, q and r:

$$[A] = [L^p M^q T^r] = [L^p][M^q][T^r] \tag{2.69}$$

where the units are seen to factor. Consequently, if speed is measured we obtain $p = 1$, $q = 0$ and $r = -1$, since speed is independent of mass and varies inversely with time. If A is the momentum then $p = 1$, $q = 1$ and $r = -1$ and so on.

So how do we use this reasoning to solve equations of motion? Consider the action of gravity g causing a stone to fall, after being dislodged from a cliff. How far does the stone fall in a time t? We express the distance of the fall in terms of the variables in the system, the acceleration of gravity g, the mass of the stone M and the time t:

$$\text{distance} = C \text{ acceleration}^p \times \text{ mass}^q \times \text{ time}^r$$

where p, q and r are unknown parameters to be determined and C is an overall constant that cannot be determined using this method. Inserting the dimensions for each observable in this equation we write the dimensional equation

$$[L] = [LT^{-2}]^p [M]^q [T]^r$$

and using the requirement that the dimension of each unit on both sides of the equation must be the same (*dimensional homogeneity*) we obtain the set of relations:

$$\text{exponent of } L \quad : \quad 1 = p,$$
$$\text{exponent of } T \quad : \quad 0 = -2p + r,$$
$$\text{exponent of } M \quad : \quad 0 = q.$$

The solutions to these equations yield the exponents $p = 1$, $q = 0$ and $r = 2$ resulting in the equation that the distance s traveled in a time t is

$$s = Cgt^2 \tag{2.70}$$

where $C = 1/2$ is found by other means. OK this was easy. Let's try something less obvious, say Newton's problem and try and find the speed of sound in air.

West and Griffin [31] recognized that given the lack of formalism Newton reasoned in a way very similar to the dimensional analysis used here. He first argued that sound, a disturbance of the air, must depend on the density of air ρ. Next that force is required to move the air molecules so he reasoned that a local force linearly binds a molecule to its local position, and to determine the effect of a macroscopic disturbance a large area must be considered. Therefore the pressure $\mathcal{P}$ (force/area) is a more reasonable quantity to consider than the force on a single molecule. Thus, the speed of sound should be determined by the pressure (force/area) and linear mass density ρ (mass/length) of the air. The dimensional equation expressing this physical reasoning is

$$\text{speed of sound} = C \, (\text{pressure})^p \times \, (\text{mass density})^q$$

or in dimensional terms

$$\frac{[L]}{[T]} = \left(\frac{[M]\,[LT^{-2}]}{[L^2]}\right)^{p}\left(\frac{[M]}{[L^3]}\right)^{q}. \tag{2.71}$$

Using dimensional homogeneity yields

$$\begin{aligned}
\text{exponent of } L &: \quad 1 = -p - 3q, \\
\text{exponent of } T &: \quad -1 = -2p, \\
\text{exponent of } M &: \quad 0 = -p - q,
\end{aligned}$$

which after some algebra yields the parameter values $p = 1/2$, $q = -p = -1/2$. Inserting these values of the exponents into the equation for the speed of sound in air and setting $C = 1$ produces

$$V = \sqrt{\frac{P}{\rho}}, \tag{2.72}$$

the value obtained by Newton. Note that inserting the values $p = -q = -1/2$ into the exponent of L equations yields a tautology.

At the moment this might seem like an esoteric but perhaps interesting exercise. However, we subsequently show the importance of this reasoning, once we introduce fractals and the fractional calculus and find ourselves without the traditional equations of motion to characterize the dynamics of a complex system.

2.5 Appendix Chapter 2

The classical equations of motion for the system variables in Section 2.2.2, described by a Hamiltonian H, have the form

$$\frac{dQ}{dt} = \frac{\partial H}{\partial P} = P, \tag{2.73}$$

and

$$\frac{dP}{dt} = -\frac{\partial H}{\partial Q} = -U'(Q) + \sum_{\nu}\Gamma_{\nu}q_{\nu}, \tag{2.74}$$

where the prime denotes the derivative of the potential with respect to Q. The force law given by Eq.(2.74) contains the unknown time-dependent bath coordinates $\mathbf{q}$. To eliminate these coordinates from the system equations of motion we consider Hamilton's equations for the heat bath:

$$\frac{dq_{\nu}}{dt} = \frac{\partial H}{\partial p_{\nu}} = \frac{p_{\nu}}{m_{\nu}}, \tag{2.75}$$

$$\frac{dp_{\nu}}{dt} = -\frac{\partial H}{\partial q_{\nu}} = -m_{\nu}\omega_{\nu}^2 q_{\nu} + \Gamma_{\nu}Q. \tag{2.76}$$

In a symmetrical manner we see that the force law for the bath depends on the unknown time-dependent system displacement Q. The linear nature of these equations allows us to write Eq.(2.76) as the second-order differential equation for a harmonic oscillator driven by $Q(t)$ and therefore its solution as an initial value problem may be reduced to quadratures

$$q_\nu(t) = q_\nu(0)\cos\omega_\nu t + \frac{p_\nu(0)}{m_\nu\omega_\nu}\sin\omega_\nu t + \frac{\Gamma_\nu}{m_\nu\omega_\nu}\int_0^t Q(\tau)\sin\omega_\nu(t-\tau)\,d\tau. \quad (2.77)$$

Still following Lindenberg and West [18, 29] we facilitate our discussion by integrating the integral term in this last equation by parts to obtain

$$q_\nu(t) - \frac{\Gamma_\nu}{m_\nu\omega_\nu^2}Q(t) = \left[q_\nu(0) - \frac{\Gamma_\nu}{m_\nu\omega_\nu^2}Q(0)\right]\cos\omega_\nu t + \frac{p_\nu(0)}{m_\nu\omega_\nu}\sin\omega_\nu t$$
$$- \frac{\Gamma_\nu}{m_\nu\omega_\nu^2}\int_0^t P(\tau)\cos\omega_\nu(t-\tau)\,d\tau. \quad (2.78)$$

The equation of motion for the system of interest may now be obtained by substituting the solution for the bath displacement given by Eq.(2.78) into the system force equation (2.74) to obtain

$$\frac{dP}{dt} + U_m'(Q) - \int_0^t P(\tau)K(t-\tau)\,d\tau = f(t), \quad (2.79)$$

which is the generalized Langevin equation introduced in the text.

In this model we have explicit expressions for the three new functions we have introduced. First there is the memory kernel given by

$$K(t-\tau) = \sum_\nu \frac{\Gamma_\nu^2}{m_\nu\omega_\nu^2}\cos\omega_\nu(t-\tau), \quad (2.80)$$

and the strength of the memory depends on the coupling coefficients $\{\Gamma_\nu\}$ and spectrum of the environmental oscillators $\{\omega_\nu\}$. Next there is the additive fluctuating force with the average value removed

$$f(t) = \sum_\nu \left\{\left[q_\nu(0) - \frac{\Gamma_\nu}{m_\nu\omega_\nu^2}Q(0)\right]\cos\omega_\nu t + \frac{p_\nu(0)}{m_\nu\omega_\nu}\sin\omega_\nu t\right\}, \quad (2.81)$$

that depends on the initial state of the environment $\{q_\nu(0), p_\nu(0)\}$ and system displacement $Q(0)$. Finally there is the "modified system potential"

$$U_m(Q) = U(Q) - Q^2\sum_\nu \frac{\Gamma_\nu^2}{2m_\nu\omega_\nu^2}, \quad (2.82)$$

in which the potential function has been shifted from its isolated form to one that is shifted due to the interaction with the environment.

We discuss the implications of the GLE subsequently.

References

[1] Arbesman, A. 2011. *Scientometrics* **86**, 245–250.

[2] Arnold, V.I. 1983. *Geometrical Methods in the Theory of Ordinary Differential Equations*, Springer-Verlag, New York.

[3] Bachelier, L. 1900. *Annal scientifiques de l'ecole normale supérieure Sup. (3)* No. 1017; English translation by A.J. Boness, in Cootners P., 1954, *The random character of the stock market*, MIT University Press, Cambridge, MA.

[4] Bernoulli, D. 1755. *Reflexions et Eclaircissements sur les Nouvelles Vibrations des Cordes Expones dans les Memoires de l'Academie,* Roy. Acad. Berlin 147.

[5] Carleman, T. 1932. *Acta Mathematica.* **59**, 63.

[6] Cohen, B.I., K.M. Watson and B.J. West. 1976. *Phys. Fluids* **19**, 345.

[7] Einstein, A. 1905. *Ann. Physik.* **17**, 549.

[8] Fermi, E., J. Pasta and S. Ulam. 1965. In *Collected Works of Enrico Fermi, Vol. II*, 978, Chicago.

[9] Ford, G.W., M. Kac and P. Mazur. 1965. *J. Math. Phys.* **6**, 504.

[10] Greenblatt, S. 2011. *The New Yorker*, August 8.

[11] Koopman, B.O. 1931. *PNAS* **17**, 315.

[12] Kowalski, K. and W.-H. Steeb. 1991. *Nonlinear Dynamical Systems and Carleman Linearization*, World Scientific, Singapore.

[13] Kowalski, K. 1997. *J. Math. Phys.* **38**, 2403.

[14] Kubo, R. 1957. *J. Phys. Soc.* **12**, 570.

[15] Langevin P. 1908, *Comptes Rendus Acad. Sci. Paris* **146**, 530.

[16] Lévy, P. 1925. *Calcul des probabilities*, Gauther-Villars, Paris.

[17] Lévy, P. 1937. *Théorie de Padditioon des variables aléatoires*, Guthie-Villars, Paris.

[18] Lindenberg, K. and B.J. West. 1990. *The Nonequilibrium Statistical Mechanics of Open and Closed Systems*, VCH Publishers, New York.

[19] Metzler, R. and J. Klafter. 2000. *Phys. Rept.* **339**, 1–77.

[20] Modis, T. and A. Debecker. 1992. *Tech. Forec. and Social Change* **41**, 111–120.

[21] Montroll, E.W. 1978. In *AIP Con. Proc.* **46**, 337.

[22] Montroll, E.W. and B.J. West. 1979. In *Fluctuation Phenomena*, Eds. E.W. Montroll and J.L. Lebowitz, *Studies in Statistical Mechanics, Vol. VII*, North-Holland, Amsterdam ; Second edition 1987.

[23] Ott, E. 1993. *Chaos in Dynamical Systems*, Cambridge University Press, New York.

[24] Pecora, L.M. and T.L. Carroll. 1990. *Phys. Rev. Lett.* **64**, 821.

[25] Perrin, J. 1913. *ATOMS*. French edition, Livrairie Felix Alcan; translated by D.Ll. Hammick, 1990, Ox Bow Press, Woodbridge, CO.

[26] Peterson, I. 1988. *The Mathematical Tourist*, W.H. Freeman and Co., New York.

[27] Price, D.J. de Solla. 1986. *Little Science, Big Science .... and Beyond*, Columbia University Press, NY; includes text of *Little science, big science*, originally published 1963.

[28] Stewart, I. 1989. *Does God Play Dice? The Mathematics of Chaos*, Basil Blackwell, Cambridge, MA.

[29] Tarasov, V.E. 2012. *Cent. Eur. J. Phys.* **10**, 382.

[30] West, B.J. 1985. *An Essay on the Importance of Being Nonlinear*, Lect. Notes in Biomath. **62**, Springer-Verlag, Berlin.

[31] West, B.J. and L. Griffin. 2004. *Biodynamics: Why the Wirewalker Doesn't Fall*, Wiley-LISS, Hoboken, NJ.

[32] West, B.J. 2013. *Fractal Physiology and Chaos in Medicine*, 2nd Ed., World Scientific, New Jersey.

[33] Wiggins, S. 1990. *Introduction to Applied Nonlinear Dynamical Systems and Chaos*, Springer-Verlag, New York.

[34] Wu, G.C. and D. Baleanu. 2014. *Signal Processing* **102**, 96–99.

[35] Zabusky, N.J. and M.D. Kruskal. 1965. *Phys. Rev. Lett.* **15**, 241.

CHAPTER 3

New Ways of Thinking

Up to this point we have seen that linear science, by which we mean the science of phenomena that have been described by linear equations of motion, can describe a wide range of complicated behavior. If this were not the case the industrial society of the last two centuries would not have been possible since the engineering on which it was built was predominantly linear. However such analyses are no longer adequate for controlling complex phenomena in the physical, social and life sciences of today's information society. Modernity requires us to think about the mathematics we use to describe complexity in large measure because it is mathematical reasoning that enables scientists and engineers to untangle the knots of the real world. We explore the more obvious reasons why the traditional calculus including differential equations are not sufficient to capture the full range of dynamics found in natural and artificially constructed processes and events.

In the previous chapter we found that finite dimensional nonlinear equations of motion may be replaced by infinite-order sets of linear rate equations. The formal equivalence of these two descriptions has been established [49], but we have barely begun to understand how the instabilities in nonlinear systems, as manifest in such phenomena as chaos, can be expressed by infinite sets of linear equations. Although such exact linearization methods are potentially fruitful for the exploration and development of new ways of thinking about and understanding complexity they are not the paths we have chosen. Our path lays in a different direction, although we do intersect such alternative paths multiple times in our travels.

The dynamics of complex nonlinear phenomena demands that we extend our horizons beyond analytic functions and classical analysis. These forays into the frontiers suggest that the functions necessary to describe complex phenomena may lack traditional equations of motion. Consequently, the replacement of ordinary differential equations by alternative descriptions, whether linear or not, need to be studied.

To explore this lack of traditional dynamic equations we introduce fractional thinking, which is a kind of in-between thinking; between the integer-order moments, such as the mean and variance, there are fractional moments required when empirical integer moments fail to converge; between the integer dimensions there are the fractal dimensions that are important when data have no characteristic scale length; and between the integer-valued operators that are local in space and time, are the non-integer operators necessary to describe dynamics that have long-time memory and spatial heterogeneity. Complex phenomena require new ways of thinking and the *fractional calculus* provides one framework for that thinking [118].

We encounter strange ideas in our odyssey, such as a dimension with a real and imaginary part; the real part is associated with the slope of an IPL and the imaginary part with a log-periodic modulation of that IPL. This behavior has a number of exemplars, including the average diameter of the bronchial airways of mammalian lungs. Another peculiar idea is how the size of a network (organ, group, city, etc.) nonlinearly influences its functionality, for example, the enhanced creativity of city dwellers over country folk. Finally, the connection between allometry and the fractional calculus was certainly an unexpected linkage.

3.1   Why Now?

The implementation of the fractional calculus in the physical, social and life sciences has languished in part because until recently the larger scientific community did not acknowledge a need for it. The calculus of Newton and Leibniz and the analytic functions that solve the differential equations resulting from Newton's force laws have historically been seen as necessary and sufficient to provide a proper and complete mechanical description of the physical world. On the other hand, experiment indicates that a broad range of physical, biological and social phenomena cannot be understood using the analytic functions we have come to rely on in physics. These functions do not capture the complex dynamics of such common physical phenomena as earthquakes and hurricanes [95]; everyday social phenomena including group consensus [102], economic unpredictability as in stock market crashes [58], high frequency finance [25] and healthcare networks [98]; or the familiar psychological activity of cognition and habituation [121]. The inherent complexity of these phenomena is beyond the scope of the familiar nineteenth century analysis that forms the mathematical foundation of physics and engineering since that time. Understanding complexity as an extended class of problems with common structural and mathematical properties requires a new way of modeling and consequently more innovative thinking.

Phenomena that require the notion of non-integer derivatives and/or integrals for their interpretation were believed to be interesting curiosities that lay outside mainstream science. However the increased sensitivity of experimental tools, the enhanced data processing techniques, the vast amounts of data made available by social media, and the ever increasing computational capabilities have all contributed to the expansion of science in such a way that those phenomena once thought to be outliers, are now center stage. These curious processes are now described as exotic scaling phenomena, but as we discuss here and elsewhere [123, 127], in order to form a basic understanding of such knotty processes requires a new mathematical perspective. Such a perspective might well be provided by the fractional calculus that is able to quantify the coupling of variations in phenomena across widely separated scales in both space and time.

An apparently different strategy, that of network science, has recently been used to model physical, social and life science phenomena in part because the complexity of phenomena in these discipline is manifest as emergent from the network dynamics. Complexity in networks may be broadly partitioned into topology that relates to the ways in which the elements of a network are interconnected in a scale-free distribution, and chronology that relates to the timing of significant events within network dynamics that also scale. The scaling topology of networks has been widely adopted by the network science community as a measure of complexity and is the topic of discussion in texts [68, 123]; whereas temporal complexity has only recently been identified as an important measure of complex network dynamics [127]. The scaling observed in numerical calculations of large complex dynamic networks is shown herein to interleave with the fractional calculus.

Exemplars of complex phenomena, used in subsequent chapters to highlight various aspects of the fractional calculus, are homogeneous turbulence and foraging, with their non-Normal statistics in time; the dynamics of viscoelastic materials, with their non-Newtonian dynamical properties; and the phase transitions in complex social networks. The non-Normal statistics, non-Newtonian dynamics and phase transitions are each in their different ways connected to the fractional calculus. So now let us explore how this calculus forces changes in our way of thinking.

3.2 Through the Looking Glass

Alice was a little girl who, one summer afternoon, chased a white rabbit down a hole and found herself in *Wonderland*. Recall that Alice was confused by the apparent lack of rules governing the world in which she found herself. Things that were relatively simple back home seemed unnecessarily complicated in Wonderland, as suggest by the modification of the classic woodcut in Figure 3.1. When Alice and her new friends discuss a cube they visualize quite different objects and which mental construct corresponds more closely to reality depends on how objects in the world are measured. In the clockwork universe from which she came a three-dimensional cube represents a volume. In our version of Wonderland objects do not smoothly fill space but erupt at all scales, filling space in an intermittent and heterogeneous way, as suggested in the figure by a Seripinski cube with a dimension between two and three. It is analogous to the realization that all our lives we have been speaking in prose, so too all our lives we have been living in Wonderland.

Figure 3.1 In this classic view of the tea party, when the Mad Hatter, White Rabbit and Door Mouse talk about a fractal dimensional Seripinski gasket Alice invisions a Euclidean cube. (Adapted from https://www.cs.cmu.edu/~rgs/alice25a.gif)

After some time Alice does come to understand that rules do exist; they are just different rules from those that determined how the world she had left behind operates. Much like Wonderland it is not that the quantitative reasoning discussed in this essay does not have rules; it is just that the rules of quantification are very different from those that most of us were taught in school. Our concern is with how the rules have changed and what those changes imply about the phenomena they describe and how we are to understand them.

Why is the fractional calculus entailed by complexity? The short answer is that the fractional calculus is probably not required to understand any specific instance of complexity. However we interpret this lack of uniqueness in the same way we do that of Newton not using his then newly formulated fluxions in his discussion of mechanical motion in the *Principia*. He confined his often remarkable arguments to geometry, the mathematical/scientific language of the day, but a close inspection of some of the geometrical constructions reveal the insight provided by the fluxion interpretation being in the background. Eventually it was accepted that the geometrical arguments of Newton and others lent themselves more readily to the differential calculus and modern theoretical physics was launched.

In an analogous way I believe that many of the complex phenomena of interest to science today have required ingenious but often tortuous explanations using traditional methods. Scientists are now showing that these processes are more naturally described using the insights provided by the fractional calculus. This is not just a statement of bias; more broadly it is a research vision that has only been partially realized [55, 118].

3.2.1 A Previous Paradigm Shift

In the nineteenth century it was believed that data sets gathered from social phenomena have a Normal distribution, such as those measured by Galton, a cousin of Charles Darwin. Galton was convinced that social phenomena could be understood by counting, so he counted the number of times people coughed at the opera, the number of illegitimate children conceived by the aristocracy and other such unlikely phenomena. He believed that the frequency with which an event occurs and the duration of its persistence are indicative of the nature of the underlying causes. In this he was correct, so let us further examine how complex phenomena were understood in the nineteenth century in order to better appreciate what is different about our present day understanding.

At a time when travel required horse-drawn carriages, reading involved candlelight or the fireplace, and gentlemen drank, gambled and whored, a new philosophy of the world was taking shape. In natural philosophy, which today we call the physical sciences, the predictability of the celestial mechanics of Newton became the paradigm of science. Consequently, the predictability of phenomena was the hallmark of science and anyone seeking to be thought a scientist or natural philosopher emphasized the predictability of phenomena. It eventually became apparent, however, that when the number of elements in a physical process became large and the interaction relatively week, the direct solution to the equations of motion was impossible. Consequently, the trajectory of the single particle was bundled into an ensemble of such trajectories, and the deterministic prediction of the single particle became the prediction of the average behavior of an ensemble to particles. This extension of scientific predictability spread into the social sciences.

A contemporary of the time, John Herschel, wrote in the Edinburgh Review in 1850:

> Men began to hear with surprise, not unmingled with some vague hope of ultimate benefit, that not only births, deaths, and marriages, but the decisions of tribunals, the result of popular elections, the influence of punishments in checking crime—the comparative value of medical remedies, and different modes of treatment of diseases—the probable limits of error in numerical results in every department of physical

inquiry—the detection of causes physical, social, and moral—nay, even the weight of evidence, and the validity of logical argument—might come to be surveyed with that lynx-eyed scrutiny of a dispassionate analysis, which, if not at once leading to the discovery of positive truth, would at least secure the detection and proscription of many mischievous and besetting fallacies.

Herschel was addressing the new statistical way of thinking about social phenomena that was being championed by social scientists such as Quetelet [77], among others. As Cohen [22] pointed out, one way of gauging whether the new statistical analysis of society was sufficiently profound to be considered a revolution was to consider the intensity of the opposition to this new way of thinking. One of the best-known opponents to statistical reasoning was the philosopher John Stuart Mill [63] who was not bashful about his skepticism:

It would indeed require strong evidence to persuade any rational person that by a system of operations upon numbers, our ignorance can be coined into science.

In other words Mill did not believe that it was possible to construct a mathematics that would take the uncertainty that prevails in the counting of social events and transform that uncertainty into the certainty of prediction required by science. Of course there are many that today share Mill's skepticism and view statistics as an inferior way of knowing.

The scientists of the eighteenth and nineteenth centuries would claim that science ought to be able to predict the exact outcome of experiment, given that the experiment is sufficiently well specified and the experimenter is sufficiently skillful. However, that is not what was or is observed. No matter how careful the experimenter, no matter how sophisticated the experimental equipment, each time an experiment is conducted a different result is obtained. The experimental results are not necessarily wildly different from one another, but different enough to question the wisdom of claiming exact predictability. This variability in the outcome of ostensibly identical experiments was an embarrassment and required careful thought, since the nineteenth century was the period when the quantitative measurement of phenomena became the same as knowledge. Consequently, if precision is knowledge, then variability must be ignorance or error; and this is what was believed.

It should be noted here that we have introduced an intellectual bias. This bias is the belief that there is a special kind of validity associated with being able to characterize a phenomenon with a number or set of numbers. This particular bias is one that is shared by most physical scientists, such as physicists or chemists. I point this out to alert the reader that numbers are representations of facts and are not facts in themselves, what da Vinci called experience. It is the underlying facts that are of importance and not necessarily the numbers associated with them. However, it is also true that it is usually easier to logically manipulate the numbers than to perform equivalent operations on the underlying facts, demonstrating why mathematical models are so important. Mathematical models of physical and social phenomena replace the often torturous arguments made by lawyers and philosophers concerning the facts of the world, with the deductive manipulations of mathematics applied to the representations of those facts. Consequently, a scientist can proceed from an initial configuration and predict with confidence the final configuration of a network, because the mathematical reasoning is error-free, even if the experimental realization of the phenomenon is not.

The central fact regarding experiments is the variability in outcome and, the polymath Gauss, at the tender age of nineteen, recognized that the average value was the best representation of the ensemble of outputs of a particular set of experiments. The general perspective that arose in subsequent decades is that any particular measurement has little or no meaning in itself, it is only the collection of measurements, the ensemble, that has a scientific interpretation and this meaning is manifest through the distribution function. The distribution function, also called the PDF, associates a probability with the occurrence of an error in the neighborhood of a given magnitude. If we consider the relative number of times an error of a given magnitude occurs in a population of a given (large) size, that is, the frequency of occurrence, we obtain an estimate of the probability an error of this order will occur. Gauss [34] and Adrian [1] were the first scientists to systematically investigate the properties of measurement errors and in so doing set the course of experimental science for subsequent centuries.

But what does it mean?

Consider a slightly different example from that of measuring a physical or biological phenomenon. Suppose I wanted to estimate the quality of teaching by a university professor. I might ask a number of students to estimate the teaching capability of a given professor on a scale of 1 to 10 in order to obtain a quantitative characterization of the complex process of teaching. Suppose this professor gets an average grade of 8.5 from her/his students. How reliable is this estimate of teaching capability? Two distinct cases are of interest. In the first case the lowest grade s/he received is 7.5 and the highest is 9.5. From this I could conclude that the students' opinions are in close agreement. In the second case I hypothesize that the students' grades range over the full interval 1 to 10. There would be more of the higher grades in order for the professor to receive the average of 8.5, but there must also be a significant minority that strongly objected to the way the course is taught. Without making a judgment as to which teaching method is preferable, the professor in the second case is certainly more controversial since the students' responses are so varied.

Thus, these two situations in which the student population thinks the professor is a good teacher are really quite different. The single number (average) is not adequate to give a clear picture of the professor's teaching and distinguish between these two cases. A second number, the degree of variability in the students' responses (standard deviation) gives more insight. In fact the standard deviation provides a measure of the width of the bell-shaped curve, so that together with the average value we know all we can about this phenomenon when the Normal distribution is applicable. This shift from strict determinism was the manner in which the effects of complexity were estimated in the nineteenth and twentieth centuries.

However, the Normal PDF is not the only one available to us today and as was found in the latter half of the twentieth century it usually does not apply to complex phenomena. Consequently, our path to understanding complexity shall also lead us into the realm of uncertainty that is quite different from the rather simple picture presented by Normal stochastic processes.

3.2.2   A Contemporary Shift

We have left and continue to leave complexity explicitly undefined in our discussion because like nonlinearity it is often defined by what it is not. In this way we consider phenomena or structures to be complex when traditional analytic functions are not able to capture their full richness in space and/or time. It was believed for a long time that physical theories such as classical mechanics could be used to describe with absolute certainty the dynamics of highly idealized systems using limiting concepts abstracted from the real world. However, that particular bubble was burst by Poincaré [75], even though it took over half a century for his discovery of the significance of nonlinear dynamics and the non-predictability of non-integrable Hamiltonian systems to penetrate mainstream physics. Historically it was left to statistical physics to restore to the mechanical description of complex systems the uncertainty observed in actual measurements and to construct the associated PDF as a measure of that uncertainty. Probability theory provided the first universally accepted systematic treatment of physical complexity and was the mathematical foundation of kinetic theory.

Ludwig Boltzmann, one of the architects of the kinetic theory of gases, spent a great deal of time thinking about the discontinuous changes in particle velocity that occur in collisions and wondering about their proper mathematical representation [14]. He believed that such microscopic dynamics should be described by continuous but non-differentiable functions such as the one developed by Weierstrass and which is discussed subsequently herein. This theme of using continuous non-analytic functions to describe complex physical processes was taken up by Jean Perrin who received the Noble Prize in Physics for his diffusion experiments determining Avogadro's number [74]. Perrin stated that curves without tangents (derivatives) are more common in the physical world than those special ones like the circle that have tangents.

He was adamant in his arguments emphasizing the importance of non-analytic functions to describe complex physical phenomena such as molecular diffusion [74].

The backbone of equilibrium statistical mechanics is the Poisson distribution

$$P(t) = \exp\left[-\beta t\right] \qquad (3.1)$$

where t is the time between events, such as in the decay of a radioactive particle, that occur at an average rate β. All the moments of a Poisson process are finite as are those of the Gaussian distribution. Eq.(3.1) is referred to as the Boltzmann distribution when t is interpreted as the system's energy.

However, empirical PDFs for complex processes are non-Poisson and non-Gaussian. Very often they have an IPL form with diverging first and second moments. For example, in complex networks the time between events is often given by the waiting-time PDF:

$$P(t) \propto 1/t^{\mu}, \qquad (3.2)$$

rather than the more familiar exponential form for the Poisson process. Empirical exemplars of such complex phenomena are the time intervals between earthquakes of a given magnitude [71], the waiting time between solar flares [35], the time from one breath to another [100], the inter-event times in electroencephalograms (EEGs) [31] and the decrease in human memory with time [4]. In many, if not all, such empirical distributions the power-law index is $\mu < 2$ and consequently the average time between events, the first moment, diverges. For these latter distributions there is no characteristic time scale for the process and as we mentioned earlier the traditional assumption that time averages and ensemble averages coincide is violated, that is, the statistics in such complex phenomena are non-ergodic.

Figure 3.2 displays the PDFs for the size of neuronal quakes (number of neurons involved in a discharge event) as first observed by Beggs and Plenz [9] and the time interval between neuronal spikes. These are exemplars of topological complexity and temporal complexity, respectively, associated with the same physiologic phenomenon of multiple neuron discharge. However they are also representative of the two types of complexity evident in complex networks in other contexts, such as the distribution in the sizes of earthquakes and the distribution of time intervals between earthquakes of a given size [123]; the number of emails sent and the time interval between the receipt of and reply to an email [70].

It is important that the reader have some perspective as to the generality of such laws. They appear to be independent of context, occurring in geophysics, economics, sociology, medicine, astrophysics, urban growth, and on and on. West and Grigolini [123] complied a list of empirical laws that are presented in Table 3.1 and Table 3.2. Each of these empirical laws has an IPL form. These lists are not intended to be exhaustive, but they are none the less impressive, because they cover over five centuries of data collection and fall into the domain of ten distinct disciplines. The year 1883 on the list for da Vinci is obviously not the date that he recorded his observations; it is the date of the first compilation and translation of his notebooks [82]. As we pointed out in this earlier work the recorded phenomena are representative of the interconnectedness of the phenomena that have historically been partitioned into the various disciplines of science. The intent here is to demonstrate that the distribution of Gauss is no where in evidence when the data for complex phenomena are examined. Once complexity enters by the door, Normal statistics leaves by the window.

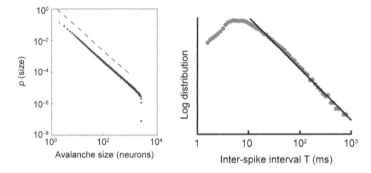

Figure 3.2 Left: probability of a given number of neurons involved in a neuronal avalanche (adapted from [64] with permission). Right: the time interval between neuronal spikes (adapted from [52] with permission). Both are inverse power laws. Note that such "laws" are valid over a limited domain and the mechanism by which they are truncated is often of interest. The numerical value of the slopes, being the IPL index of the PDFs, are not important here except to note they are less than two.

Scientists often rediscover things that have been known for a long time outside their particular discipline and it is not unusual for two or more scientists to uncover an effect or give the same explanation for a phenomenon at essentially the same time. This is no where more evident than with the explanation of IPL phenomena over the past century, where the same mechanism has been postulated to explain their bizarre statistics over and over again. In the remarkable paper by Simkin and Roychowdhury [91] the authors trace the theoretical understanding of most of the effects recorded in Table 3.1 and Table 3.2, as well as some that are not included, to a mechanism originally proposed in 1922 by the mathematician Yule [136, 132] to explain the experimental data on the sizes of biological genera collected by the biologist Willis [133]. This mechanism is closely related to what today is called 'preferential attachment'. They also review, as they say, the history of reinvention of closely related branching processes, random graphs and coagulation models. All interesting topics that we mention in passing to provide context.

It is perhaps worthwhile to indicate that the centuries old scaling rule of da Vinci is still being used in a contemporary setting. The self-similarity arising from the da Vinci scaling rule shown in Table 3.1 is still being critiqued using biomechanical models [65, 123]. The rule has also been used to understand the wind induced stresses in the branching of botanical trees [26]. It is argued that da Vinci's rule is a consequence of the branch diameters being adjusted to resist wind-induced loads, a phenomenon awkwardly called thigmomorphogenesis [26].

Table 3.1 A list of empirical laws for complex phenomena is given according to their accepted names in the disciplines in which they were first developed. The power-law index α is different for each of the laws and is intended to be generic; the value listed is from the original reference and need not be the accepted value today.

Discipline	Law's Name	Form of Law
Anthropology		
1913 [5]	Auerbach	$\Pr(\ city\ size\ rank\ r) \propto 1/r$
1998 [83]	war	$\Pr(intensity{>}I) \propto 1/I^\alpha$
1978 [108]	1/f-music	$\mathrm{Spectrum}(f) \propto 1/f$
Biology		
1992 [109]	DNA sequence	symbol spectrum(frequency f) $\propto 1/f^\alpha$
2000 [66]	ecological web	$\Pr(k\ species\ connections) \propto 1/k^{1.1}$
2001 [46]	protein	$\Pr(k\ connections) \propto 1/k^{2.4}$
2000 [45]	metabolism	$\Pr(k\ connections) \propto 1/k^{2.2}$
2001 [51]	sexual relations	$\Pr(k\ relations) \propto 1/k^\alpha$
Botany		
1883 [82]	da Vinci	*branching*; $d_0^\alpha = d_1^\alpha + d_2^\alpha$
1922 [132]	Willis	*# of genra(# of species N)* $\propto 1/N^\alpha$
1927 [67]	Murray	$d_0^{2.5} = d_1^{2.5} + d_2^{2.5}$
Economics		
1897 [72]	Pareto	$\Pr(\ income\ x) \propto 1/x^{1.5}$
1998 [32]	price variations	$\Pr(stock\ price\ variations\ x) \propto 1/x^3$
Geophysics		
1894 [71]	Omori	$\Pr(aftershocks\ in\ time\ t) \propto 1/t$
1933 [84]	Rosen-Rammler	$\Pr(\#\ ore\ fragments < size\ r) \propto r^\alpha$
1938 [56]	Korčak	$\Pr(\ island\ area\ A{>}a) \propto 1/a^\alpha$
1945 [41]	Horton	$\frac{\#\ segments\ at\ n}{\#\ segments\ at\ n{+}1} = $ constant
1954 [36]	Gutenberg-Richter	$\Pr(earthquake\ magnitude < x) \propto 1/x^\alpha$
1957 [37]	Hack	*river length* $\propto$ *(basin* area$)^\alpha$
1977 [56]	Richardson	*length of coastline* $\propto 1/(ruler\ size)^\alpha$
2004 [107]	forest fires	frequency density(*burned area A*) $\propto 1/A^{1.38}$
Information Theory		
1999 [42]	World Wide Web	$\Pr(\ k\ connections) \propto 1/k^{1.94}$
1999 [27]	Internet	$\Pr(k\ connections) \propto 1/k^\alpha$

Table 3.2. A list of empirical laws for complex phenomena is given according to their accepted names in the disciplines in which they were first developed. The power-law index α is different for each of the laws and is intended to be generic; the value listed is from the original reference and need not be the accepted value today.

Discipline	Law's Name	Form of Law
Physics		
1918 [86]	1/f-noise	Spectrum$(f) \propto 1/f$
2002 [35]	solar flares	$\Pr(\textit{time between flares } t) \propto 1/t^{2.14}$
2003 [85]	temperature anomalies	$\Pr(\textit{time between events } t) \propto 1/t^{2.14}$
Physiology		
1959 [80]	Rall	$\textit{neurons; } d_0^{1.5} = d_1^{1.5} + d_2^{1.5}$
1963 [99]	mammalian vascular network	$\textit{veins \& arterial; } d_0^{2.7} = d_1^{2.7} + d_2^{2.7}$
1963 [112]	bronchial tree	$d_0^3 = d_1^3 + d_2^3$
1973 [59]	McMahon	metabolic rate$(\textit{body mass } M) \propto M^{0.75}$
1975 [134]	radioactive clearance	$\Pr(\textit{isotope expelled in time } t) \propto 1/t^{\alpha}$
1987 [114]	West-Goldberger	$\textit{airway diameter(generation } n) \propto 1/n^{1.25}$
1991[40]	mammalian brain	surface area $\propto$ volume$^{0.90}$
1992[100]	interbreath variability	# of breaths $(\textit{interbreath time } t) \propto 1/t^{2.16}$
1993 [73]	heart beat variability	power spectrum$(\textit{frequency } f) \propto f$
2007 [31]	EEG	$\Pr(\textit{time between EEG events}) \propto 1/t^{1.61}$
2007 [20]	motivation and addiction	$\Pr(k \textit{ behavior connections}) \propto 1/k^{\alpha}$
Psychology		
1957 [97]	psychophysics	perceived response $(\textit{stimulus intensity } x) \propto x^{\alpha}$
1963 [88]	trial and error	reaction time$(\textit{trial } N) \propto 1/N^{0.91}$
1961 [39]	decision making	utility$(\textit{delay time } t) \propto 1/t^{\alpha}$
1991 [4]	forgetting	percent correct recall$(\textit{time } t) \propto 1/t^{\alpha}$
2001 [29]	cognition	response spectrum$(\textit{frequency } f) \propto 1/f^{\alpha}$
2009 [47]	neurophysiology	$\Pr($phase-locked interval $<\tau) \propto 1/\tau^{\alpha}$
Sociology		
1926 [54]	Lotka	$\Pr(\# \textit{ papers published rank } r) \propto 1/r^2$
1949 [137]	Zipf	$\Pr(\textit{ word has rank } r) \propto 1/r$
1963 [76]	Price	$\Pr(\textit{ citation rank } r) \propto 1/r^3$
1994 [7]	urban growth	population density $(\textit{radius } R) \propto 1/R^{\alpha}$
1998 [111]	actors	$\Pr(k \textit{ connections}) \propto 1/k^{2.3}$

It is perhaps worthwhile to quote ourselves rather than attempting to paraphrase a summary of the new paradigm shift [123]:

Mandelbrot presented many examples of physical, social and biological phenomena that cannot be properly described using the traditional tenets of dynamics from physics. The functions required to explain these complex phenomena have properties that for a hundred years were thought to be mathematically pathological. He argued that, rather than being pathological, these functions capture essential properties of reality and therefore are better descriptors of the world than are the traditional analytic functions of theoretical physics. The fractal concept shall be an integral part of our discussion, sometimes in the background and sometimes in the foreground, but always there...fractals can be taken as one of a number of working definitions of complexity...

3.3 Non-differentiability can be Physical

The first scientific application of the fractional calculus facilitated the understanding of the dynamics of viscoelastic materials such as taffy, mud and rubber. The equations of motion for such materials fall outside traditional analytic dynamics and fluid dynamics because the material properties are neither those of solids or fluids [79], but somewhere between the two. Historically the viscoelastic equations of motion are of an integro-differential form that were shown to have an equivalent interpretation in terms of the fractional calculus [30, 87]. To assist in understanding the reason for this equivalence we examine the notion of dimension.

Over two millennia ago Euclid organized the understanding of structure within the physical world into classical geometry, giving us the metrics of points, lines, planes and other surfaces. Fifteen hundred years later da Vinci considered these concepts in the context of painting and extended them beyond the rigid constraints of Euclid. He used geometry to bound the artistic forms of natural events such as churning water and birds in flight, as well as to determine the most pleasing proportions of the human body. Five centuries after da Vinci, Mandelbrot [56] pointed out "the emperor had no clothes", which is to say that lightening does not move in straight lines, clouds are not spheres and most physical phenomena violate the underlying assumptions of Euclidean geometry, giving mathematical form to da Vinci's observations. Pursuing his observation to their logical conclusions Mandelbrot introduced the idea of fractional or fractal dimension into the scientific lexicon and proceeded to catalogue the myriad of physical, social and biological phenomena that ought to be described by his fractal geometry and fractal statistics.

So what is a fractal dimension? Technically the dimension of an object is determined by how it is measured, or more pedantically how it is covered. For example, given a ruler of unit length η the length of a curve L can be determined by laying the ruler end to end N times to obtain $L = N\eta$. If $L = L_0$, a constant, then the number of units of measure necessary to cover the line varies inversely with the unit $N = \{L_0/\eta\}$, where $\{B\}$ is the largest integer value of B. Proceeding upward in dimension the area A of a surface can be determined by placing a unit of area η^2 over the surface N times until it is completely covered to obtain $A = N\eta^2$. If $A = A_0$ then the number of units of measure necessary to cover the area varies inversely with the unit area $N = \{A_0/\eta^2\}$. The procedure is clear: to determine the "covering" of an object of dimension D, N unit intervals η^D are required. Consequently, the dimension D of an object is defined by

$$D = \frac{\log N}{\log (1/\eta)} \tag{3.3}$$

in the limit $\eta \to 0$. The limit is taken so that concepts like the length of a curve or the area of a surface are independent of the measuring instrument and are therefore objective. Note however that this definition does not require the dimension be integer. In fact when the dimension D is not integer, more and more structure emerges as $\eta \to 0$ and the traditional measure of length and area cease to have meaning. The existence of non-integer or fractal dimensions led to the development of fractal geometry and fractal statistics.

The decades of the 1980s and 1990s witnessed an explosion of applications of fractal geometry in disconnected fields of study from statistical physics [60], to the convoluted foldings of the surface of the brain and the branching network of the mammalian lung [130], to the growth patterns of cities [7], and to intermittent search strategies [10]. In these and many other studies it became apparent that not just static structures are fractal, but the dynamics of complex phenomena are fractal as well, including their statistical fluctuations [123]. These fractal descriptions suggest the need for a calculus capable of systematically treating fractal dimensions and to do this requires an understanding of the scaling properties of fractals.

3.3.1 Fractal Evolution

Fractals give rise to a number of mathematical as well as scientific curiosities. For example it is possible to construct a continuous line connecting two points a finite Euclidean distance apart and show that line to be infinitely long. Such a fractal curve has the property of being self-similar at all scales, that is, in the neighborhood of any point along the curve there is variability. In fact a fractal curve has such a high degree of variability that it is not possible to draw a tangent to the curve at any point. As a limit is taken in the vicinity of any point more and more structure of the curve is revealed and the derivative of the curve becomes ill-defined as anticipated by Perrin [74] in his study of diffusion, see Feder [28] and Mandelbrot [56] for a complete inventory of fractal properties.

The nineteenth century mathematician Weierstrass was able to construct a Fourier series that had the property of being continuous everywhere, but being nowhere differentiable. A century later Mandelbrot [56] generalized the Weierstrass function (GWF) to the form :

$$W(t) \equiv \sum_{n=-\infty}^{\infty} \frac{1}{a^n} \left[1 - \cos (b^n \omega_0 t) \right], \tag{3.4}$$

with the parameter values $b > a > 1$. The continuous curve depicted in Figure 3.3 is generated by the GWF and is seen to have the property of self-similarity. The tip of one of the excursions of the curve is magnified within the box where it is seen that the tip reproduces the entire geometric structure of the curve on a smaller scale. This process of selection and magnification can be repeated indefinitely as suggested by the change in scales indicated in the three levels of magnification shown in the figure.

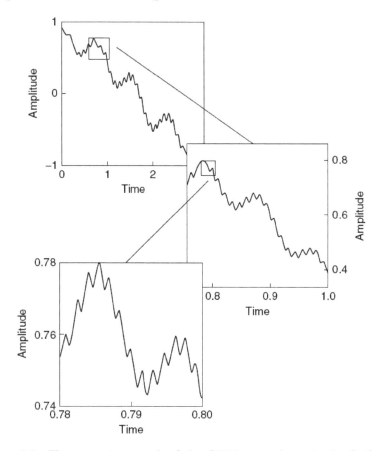

Figure 3.3 The curve is a graph of the GWF using the series Eq.(3.4). The function depicted by the curve obeys a scaling relation Eq.(3.6) characterized by a fractional dimension.

It is straight forward to show that multiplying time in the GWF by the parameter b results in the renormalization group equation

$$W(bt) = aW(t), \tag{3.5}$$

obtained by re-indexing terms in the infinite series. We shall have more to say about renormalization group theory in the sequel. The solution to the renormalization group equation is of the scaling form [60, 96]:

$$W(t) = A(t)t^{\mu}, \tag{3.6}$$

which when inserted into Eq.(3.5) yields

$$A(bt)(bt)^{\mu} = aA(t)t^{\mu},$$

resulting in the separate equations

$$A(t) = A(bt), \tag{3.7}$$

and

$$a = b^{\mu}.$$

Thus, the power-law index is related to the two parameters in the series expansion by

$$\mu = \frac{\ln a}{\ln b}, \tag{3.8}$$

which is the ratio of the influence of the relative scale size to the scale size itself in the GWF. The function $A(t)$ must be equal to $A(bt)$ so that it is periodic in the logarithm of the time interval with period $\ln b$. The renormalization group scaling in Eq.(3.5) is the analytic manifestation of the self-similarity property of the GWF. The general form of the solution to the modulation function Eq.(3.7) is given by

$$A(t) = \sum_{n=-\infty}^{\infty} A_n \exp\left[i2\pi n \frac{\ln t}{\ln b}\right], \tag{3.9}$$

where the set of time-independent coefficients $\{A_n\}$ are determined by the elements of the series Eq.(3.4). In general the coefficient can be used to fit experimental data, as we subsequently show.

It is clear from Figure 3.3 that if any segment of the curve is magnified the entire curve is again revealed due to the property of self-similarity. Finally the time derivative of the GWF can be shown to diverge for $b > a$ [56]. Consequently, this function is not the solution to any traditional equation of motion and before Mandelbrot's work was considered by most scientists to be a mathematical curiosity and not to describe any 'real' time-dependent process. Of course there were those few that could see farther than others.

The potential physical significance of the Weierstrass function was first recognized by Richardson [81], who had measured the increasing span of plumes of smoke ejected from chimneys and driven by fluctuating wind fields, see Figure 3.4. From his observations he speculated that the turbulent air speed, which was known to be non-differentiable, could be characterized by a

Weierstrass function. This was motivated in part by the observation that the span of the plume increased in time t as t^β with $\beta \geq 3$, a value inconsistent with molecular diffusion for which $\beta = 1$. Half a century after Richardson's speculation Mandelbrot established that turbulent velocity fields are fractal statistical processes and that the eddy cascade model of turbulence invented by Kolmogorov [48], was in fact a dynamic fractal so that turbulence has no characteristic space/time scale.

Figure 3.4 The smoke plumes from chimneys are seen to expand with distance from the source. In 1926 Richardson [81] speculated that this could be described by a Weierstrass function, since the turbulent velocity field of the wind was not differentiable.

The details of the fractal description of turbulence is of interest because this story has been repeated again and again in multiple disciplines over the past half century. Not the specifics, but the notion that complex phenomenon are self-similar and therefore do not possess characteristic scales in space and/or time. This recognition of self-similarity in turn led to the reinterpretation of existing data sets in terms of geometric, dynamic, or statistical fractal processes. The crucial feature of the process is the scaling of the data, which appears in multiple forms; reaching back five hundred years to da Vinci's branching relation for the limbs of trees and the tributaries of streams [82], jumping forward four hundred years to the distribution of income by Pareto

[72], up to the contemporary probability of occurrence of wars of a given size [83] and the scale-free character of complex networks [2].

The oscillatory coefficient in the scaling solution of the GWF given in Eq.(3.9) has been shown to be a consequence of the underlying process having a complex dimension [90, 92]. The Weierstrass function can be expressed as

$$W(t) = \sum_{n=-\infty}^{\infty} A_n t^{D_n}; \quad D_n \equiv \frac{\ln a}{\ln b} + i2\pi \frac{n}{\ln b}, \tag{3.10}$$

and D_n is a fractal dimension with a real and an imaginary part. Such complex dimensions have been observed in the architecture of the human lung [113] as well as in earthquakes [95], turbulence [89], and financial crashes [123].

The most important point for us here is that once a scale-free process has been identified as being described by a fractal function we know that the traditional calculus will not be available to determine its dynamics, because the integer-derivatives of fractal functions diverge. This is where fractional operators enter the story. When a process is continuous, but no where differentiable, its evolution must be described by the fractional calculus. We return to this observation repeatedly and with each repetition we hopefully answer one or two of the many questions you now have regarding the behavior of non-differentiable functions. However before we examine fractal functions and the operators that determine how they change in space and time, let us look a bit more carefully at the utility of the notion of a complex fractal dimension.

3.3.2 Complex Fractal Dimension

Over forty years ago Novikov [69] in his investigation of the properties of turbulent fluid flow discovered periodically modulated scaling properties of intermittent stochastic processes. He considered a general Poisson process supplemented by "nested" pulses of activity. The power spectral density function was consequently determined to satisfy a scaling relation of the renormalization group form and the predicted spectrum was a modulated IPL. Thus, whether the functional form of Eq.(3.9) arises in the study of the moments of a processes or in its spectrum, it indicates a process that is void of a characteristic scale and has long-range correlations induced by nested bursts of activity, as also pointed out by West and Fan [116].

The periodic variability of an observable average was first discussed in a physiologic context by West *et al.* [113] and subsequently by Shlesinger and West [90] to describe the scaling behavior of the bronchial airway of the mammalian lung. In this application to physiology the renormalization group solution takes the heuristic form

$$\langle d; z \rangle = \frac{1}{z^\delta} \left[A_0 + A_1 \cos\left(2\pi \frac{\ln z}{\ln b} \right) \right], \tag{3.11}$$

where $\langle d; z \rangle$ is the average diameter of the bronchial airway at generation z, A_0 and A_1 are empirical constants, and the generation z denotes the number of branchings of the bronchial tree starting from the trachea $z = 1$. The bronchial tree was characterized as having the complex fractal dimension:

$$D = \delta + i\frac{2\pi}{\ln b} \quad ; \quad \delta = \frac{\ln(1/a)}{\ln b} \tag{3.12}$$

whose imaginary part produces the modulation of the dominant IPL behavior of the average diameter. This periodic modulation is clearly seen in Figure 3.5 for four distinct species, human, dog, rat and hamster for all generations and is apparently a general property of mammalian species. The modulation was first observed using the so-called fractal lung model [113].

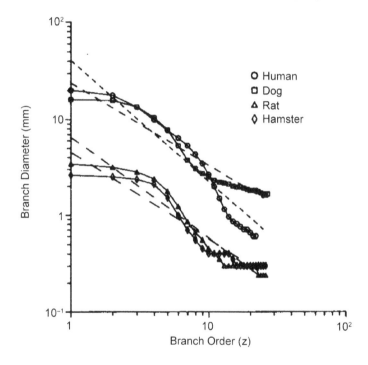

Figure 3.5 The data from [78] for the average diameter of the mammalian lung for four distinct species is compared with the predictions of the fractal model of the lung [113]. The symbols are the data points and the solid curves the results of the fractal model. The parameter values yield the slopes $\delta = 1.26$ (humans), 1.05 (rats), 0.86 (dog) and 0.90 (hamster), determined by the dashed lines; periods $lnb = 2.20$ for humans and rats; $lnb = 2.40$ for dogs and hamsters, as determined by the solid curves (From [90] with permission.)

The periodic modulation of the IPL behavior of the average diameter in the bronchial airway is only one of many such phenomenological regularities

observed in physiology.   Another is cerebral blood flow (CBF) velocity
measured using transcranial Doppler ultrasonography, which is not strictly
constant. West *et al.* [117] use the dimensionless relative dispersion, the ratio
of the standard deviation to the mean, to show by systematically aggregating
the data that the correlations in the beat-to-beat CBF time series data is the
modulated IPL depicted in Figure 3.6. This scaling of the CBF time series
indicates the existence of long-time memory in the underlying control process.
They argued that the control system has allometric properties that enable it
to maintain a relatively constant perfusion.

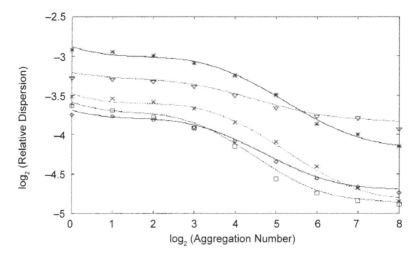

Figure 3.6 The logarithm base 2 of the relative dispersion is plotted versus the
exponent of 2 in the numbers of aggregated data points for each of six subjects.
Each time series is two hours long and consists of between seven and eight thousand
data points. The lines are the best fits to the data and it is clear that the curves
describe modulated inverse power laws. (From [117] with permission.)

Correlations across time scales, for example, are revealed in the structure
of the variance of the time series $X(t)$ as a function of the scale size

$$Var\left[X(bt)\right] = aVar\left[X(t)\right]. \tag{3.13}$$

Where a social or natural system exhibits such a form one should look
for a dynamical basis for this behavior. The dynamical instabilities in the
underlying process would be revealed in the scaling behavior of the statistical
measure. This scaling behavior is observed directly in the time series from
many phenomena, such as, for example, in intervals between beats of the
heart, the stride interval in human gait, and in DNA sequence data, see for
example [119]. The relative dispersion is a more reliable statistical measure
of the scaling in time series data than many other measures that have been
used.

The CBF time series for two hours of data from each of six subjects is processed to obtain the relative dispersions depicted in Figure 3.6. The processing procedure is to first calculate the relative dispersion using all the data as indicated by the zero point on the horizontal axis in the figure. Next each of the neighboring data points are added together to obtain half the original number of data points and the relative dispersion is calculated again. This is the 1 point on the horizontal axis in the figure. These data points are aggregated again in the same way and the relative dispersion is calculated a third time to obtain point 2. This process is repeated six more times with the data and plotted as shown. For each subject the relative dispersion is successively aggregated in this way and plotted against the size of the aggregation and is seen to yield a modulated IPL. The underlying theory on which this processing technique is based is given by Bassingthwaighte *et al.* [6] for fractal time series where the relative dispersion is shown to satisfy a renormalization group relation.

These are only two examples of physiologic phenomena with complex fractal dimensions that are revealed through modulation of the processed time series data. The technique for determining the scaling behavior of a data set is based on the calculation of the relative dispersion, which outside physiology provides a probe into the interactions between and among individuals and social institutions. The method is to coarse grain the data set at successively larger scales and to compare the level of dispersion in the resulting data set at each of the resultant scales.

Sornette and Johansen [93] also implemented the renormalization group idea to postdict stock market crashes, that is to 'predict' historical stock market crashes, fitting the log-periodic modulation of the solution to the renormalization group to historical data. This distinctive modulation is obtained using the Dow Jones time series financial data for the United States stock market crash of 1929 depicted in Figure 3.7. The solid line segment is the RG solution fit to the data. Predicting the occurrence of a crash from historical stock market time series subsequently became equivalent to predicting the arrival of an extrema that Sornette calls a 'dragon king' [94].

The interpretation of time series extrema in medical pathologies as dragon kings has not yet been done systematically. However given the recent progress by Cavalcante *et al.* [19] in real time forecasting of an impending extreme event (dragon king), and the fact that it is demonstrably possible to perturb the system to suppress the onset of the dragon king for certain chaotic mechanisms may become extremely important in a medical context and elsewhere.

3.4   The Size Effect

We have discussed fractals and fractal dimensions, but it probably has not yet been made clear how the measurable properties of systems depend on such dimensions. The connection between form and function was clearly observed

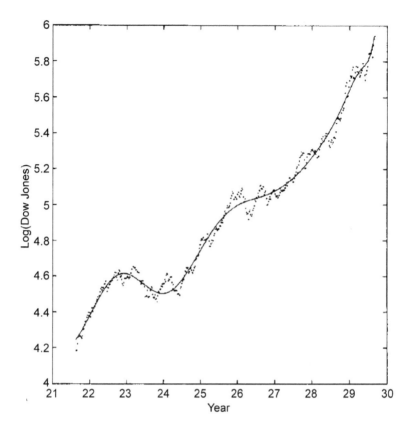

Figure 3.7 The time dependence of the logarithm of the Dow Jones stock exchange index from June 1921 to September 1929 and the best fit of the parameters using the log-periodic modulated solution to the renormalization group relation. (From [93] with permission.)

by D'Arcy Thompson [101] early in the last century, when he explored problems of scale, size, and shape in the biological sciences. Complex systems both inside and outside the biological sciences are replete with size-dependent relations in which a function of the system changes with the size of the system. Such linkages are known as allometry relations (ARs). If we denote some function of a system by Y and the size of the system by X the allometry relation is usually written as

$$Y = aX^b, \tag{3.14}$$

where a is the allometry coefficient and b the allometry exponent. Equation(3.14) does not however represent the empirical AR, which is strictly a relation between average quantities, that is,

$$\langle Y \rangle = a \langle X \rangle^b, \tag{3.15}$$

and the brackets denote an average over the data set.

The difference between Eqs.(3.14) and (3.15) immediately brings to our attention one of the major problems in constructing a theoretical understanding of the origins of ARs. The two variables X and Y are random and from the Jensen inequality [44] we have for the nonlinear function $Y = G(X)$ that

$$\langle Y \rangle = \langle G(X) \rangle \neq G(\langle X \rangle), \tag{3.16}$$

where the brackets denote an ensemble average. Consequently, the empirical AR given by Eq.(3.15) cannot be realized by a direct average of Eq.(3.14) over the data set since $G(X) = X^b$ is not linear for $b \neq 1$. However there is a theory that explains how an AR could originate from the scaling of the underlying statistical fluctuations [124, 126].

Scaling is a nearly ubiquitous property of complex networks indicating that the observables simultaneously fluctuate over many time and/or space scales. The existence of ARs has been closely tied to fractal geometry by some investigators [128], but herein we follow the argument that the origin of AR resides in the scaling of fractal statistics and not necessarily in fractal geometry [125]. Mandelbrot [56] identified a number of ARs masquerading under a variety of empirical 'laws' and argued that they were a consequence of complex phenomena not having a characteristic scale.

Fractal statistics are inhomogeneous in space and intermittent in time and it is the statistical scaling that is evident at increasing levels of resolution. The phase space description of the dynamics of statistical fractals are shown to be given by fractional equations in Chapter 7. In anticipation of that discussion we note here that the PDF $P(q, t)$ for a fractal random variable $Q(t)$ satisfies the scaling relation

$$P(q, \lambda t) = \lambda^{-\mu} P(q, t). \tag{3.17}$$

Time series with such statistical properties are found in multiple disciplines including finance [57], economics [58], neuroscience [3, 103], geophysics [106], physiology [121] and general complex networks [123]. A complete discussion of PDFs with such scaling behavior is given by Beran [11] in terms of the long-term memory captured by the scaling exponent. An example of a scaling PDF is given by

$$P(q, t) = \frac{1}{t^\mu} F_q \left(\frac{q}{t^\mu} \right), \tag{3.18}$$

and is found in Section 7.5.3 to be the general solution to a fractional phase space equation.

Of course the stochastic variables of interest in the context of ARs are not physical space and time, they are the measures of functionality and size of the allometry phenomena being investigated. We use Eq.(3.18) to write

$$\langle q \rangle = \int q P(q, t) dq = a t^\mu, \tag{3.19}$$

and the allometry coefficient is given by

$$a = \int y F_q(y) \, dy, \quad y \equiv \frac{q}{t^\mu}. \tag{3.20}$$

If we identify the function variable as $q = Y$, the average measure of size as $t = \langle X \rangle$ and finally the scaling exponent with the allometry exponent $\mu = b$ we can interpret Eq.(3.19) as the empirical AR given by Eq.(3.15). Therefore the scaling properties of the PDF solution to the fractional phase space equation entail the allometry relation [124, 125], as we show in detail subsequently.

3.4.1 Sleep-wake Scaling

In a typical night, we have intermittent intervals of wakefulness that interrupt our sleep. The distributions of intervals of wakefulness and that of sleep are quite different and are determined by the complex interactions between neurons in many areas of the brain [53]. The sleep patterns of humans, cats, rats and mice have been recorded and found to have a number of similarities. On the one hand, the distribution of arousal times is found to be IPL, with an IPL index of 2.2, independent of species. On the other hand, the distribution of sleep intervals is found to be exponential with a relaxation time that is species specific.

Lo *et al.* [53] argue that having an IPL for the distribution of wake intervals suggests a scale-free dynamics typical of fractal phenomena. They make the further connection of such dynamics to critical phase transitions in systems, where fluctuations are important over a broad range of scales. In addition, they determine that the relaxation time for the exponential distribution of sleep intervals satisfies an allometry relation. The phenomenological relaxation time is found to increase as a non-integer power of the average mass of the species (0.2 ± 0.03) or with the average brain mass (0.18 ± 0.03). They reach the tentative conclusion that the rate of transition from sleep to wake is positively correlated with the metabolic rate per unit body mass, but caution that more species need to be examined before one can have confidence in this allometry relation for sleep patterns.

3.4.2 Natural Science

Natural science encapsulates the information contained in vast amounts of biological and botanical data through allometry relations of the form Eq.(3.14), which were introduced into science in the early nineteenth century. The allometry laws in the natural sciences fall into two groups: 1) group 1 interrelates variables within a specific species such as the average total body mass (TBM) of an animal to the average mass of a specific organ within that animal, for example, the average TBM to the average mass of a deer's antlers or to the average mass of a crab's claws [43], the relation between the mass of white matter to grey matter in the brain [23]; 2) group 2 interrelates variables across species such as the average time interval between heart beats or breaths among mammals to the average TBM.

The most famous allometry relation is perhaps that between the average basal metabolic rate and the average TBM of multiple mammalian species as depicted in Figure 3.8. The metabolic rate refers to the total utilization of chemical energy for the generation of heat by the body of an animal. In Figure 3.8 the "mouse-to-elephant" curve depicts the average metabolic rate for mammals and birds $\langle Y \rangle$ plotted versus the average TBM $\langle X \rangle$ and the straight line segment is the linear regression fit to data by Eq.(3.14) with the slope given by the power-law index $b \approx 0.75$ and the intercept of the vertical axis yields a.

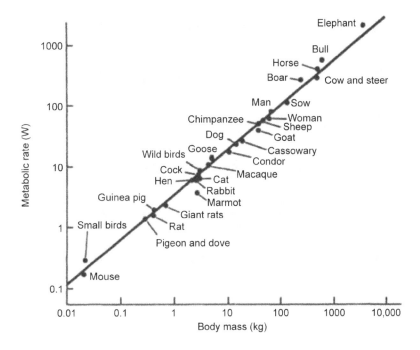

Figure 3.8 The mouse-to-elephant curve of metabolic rates (MR) of mammals and birds plotted versus the TBM on log-log graph paper. The solid line segment is best fit by linear regression to the data of $MR = a\,(TBM)^b$ with $b = 0.74$. (from Schmidt-Nielson [110] with permission).

Over the past decade or so there has been a significant controversy regarding the theoretical basis for the allometry relation between the metabolic rate and TBM. At the heart of the controversy is the notion of supplying nutrients to an organ by means of a fractal delivery network [128], that being a network of tubes with diameters that decreases with successive branchings in a manner similar to that anticipated by da Vinci and shown in

Figure 3.9. The controversy, as is so often the case in science, revolves around the details, such as how precisely self-similarity enters into the modeling. But this level of detail need not concern us here except to laud the introduction of fractal geometry at a fundamental level in natural science. My own take on the controversy is that the introduction of fractals is admirable, but needs to be done at the level of statistics and not geometry in order that the allometry relations can be expressed in terms of average quantities. This is not just a matter of personal taste, but concerns the treatment of data and how one interprets the ARs. Consequently, the power-law index is still theoretically uncertain as are the allometry coefficients, in fact, it is only very recently that statistical theory has been used to predict a co-variation of these two parameters [126].

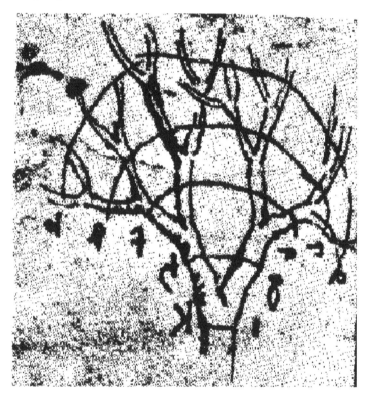

Figure   3.9 Leonardo   da   Vinci's   sketch   of   relationship between   the   branches   of   a   tree   of   equal   generation   number. (http://www.bonsaibasho.com/micromarket/#/library/library/a135).

It is not out of place here to note that Brown *et al.* [15] observed that the physical relations observed in data are not in themselves physical laws, but are rather the emergent outcomes of physical laws operating in complex systems containing many interacting components. They go on to hypothesize that at least some ecological scaling relations reflect the outcome of underlying laws and that the empirical ARs may be used to suggest how universal principles of ecology arise from the laws of physics, chemistry and biology. They [129] have even proposed that fractal geometry and allometric scaling may be seen as the "fourth dimension" of life. It has even been suggested that the relationship between cardiovascular structure and function with body size may play a critical role in medicine [24, 119].

3.4.3 Information Transfer Hypothesis

A mechanism to explain how information on the size of an organism is communicated to the organs within the organism was hypothesized by Hempleman *et al.* [38]. Their hypothesis involved modifying the neural spike code according to body size to convey this information. They point out that action potential spike trains are the mechanisms for long distance information transmission in the nervous system and that the modification requires mass-dependent scaling of neural coding in order to preserve information transmission with decreasing body size. The hypothesis is that some phasic physiological traits are sufficiently slow in large animals to be neural rate coded, but are rapid enough in small animals to require neural time coding. These traits include such activities as breathing rates that scale with an allometry exponent of $-1/4$.

Hempleman *et al.* tested for this allometry scaling of neural coding by measuring action potential spike trains from sensory neurons that detect lung CO_2 oscillations linked to breathing rate in birds ranging in body mass over two decades of scale size. While it is well known that spike rate codes occur in the sensing of low frequency signals and spike timing codes occur in the sensing of high frequency signals, their experiment was the first designed to test the transition between these two coding schemes in a single sensory network due to variation in TBM. The results of their experiments on breathing rate was an allometry exponent in the interval $-0.26 \leq b \leq -0.23$. The implications of these experiments strongly suggest the need to continue such investigations.

The human body consists of a network of complex physiologic networks over which information is shuttled back and forth. The control system governing these interconnections is the neural network and we realize that there exists an overlap between information theory, the empirical laws governing neuronal control and $1/f$ noise [61]. The last is a consequence of the remarkable property that $1/f$ signals are encoded and transmitted by sensory neurons more efficiently than are white noise signals [135]. The phenomenon of $1/f$ noise was discovered by Schottky [86] at the opening of the last century in his study of electrical conductivity. Between then and now this spectral form has been found in biological, economic, linguistic, medical, neurological and social phenomena as well as in physics [121]. The spectra of such complex phenomena are given by

$$P(f) \propto 1/f^\alpha,$$

and the spectral index α, falls within the interval $0.5 < \alpha < 1.5$. Complex phenomena span the dynamic range from the macroscopic behavioral level down to the microscopic dynamical level. West and Grigolini [122] pointed out that $1/f$ variability appears in body movements such as walking, postural sway, and movement in synchrony with external stimulation such as a metronome; also in physiologic networks as manifest in heart rate variability, human vision, the dynamics of the human brain and in human cognition; also $1/f$ noise is measured at the level of single-ion channels and in single neuron adaptation to various stimuli. Each of these psychophysical phenomena manifests $1/f$ variability and is reviewed and referenced by West *et al.* [120].

Such information-dominated phenomena in the social and life sciences behave differently from the physical phenomena in thermodynamic equilibrium. Consequently the procedures borrowed from the physical sciences to describe their dynamics must be generalized to be applicable in these more complex domains. An intuitive understanding of this difference was first articulated in 1948 by the mathematician N. Wiener [131]:

> We have a system of high energy coupled to a message low in energy, but extremely high in amount of information, i.e., of great negative entropy. This is unlike the usual situation in thermodynamics, where all the coupled systems enjoy high entropy. But it may happen in the development of such a system that the internal coupling causes the information, or negative entropy, to pass from the part at low energy to the part at high energy, so as to organize a system of vastly greater energy than that of the present instantaneous input.

In this way the network of lower energy has the stronger influence because it has the greater information. The mathematical/computational verification of Wiener's intuition required over half a century and uses the properties of complex networks. His conjecture became the Principle of Complexity Management (PCM) [120] in which the level of information exchange between two complex networks is determined by their relative complexity. A complete discussion of PCM has been presented in the context of complex webs [123, 127].

The principle has subsequently been generalized to include non-ergodic phenomena where the average over an ensemble distribution function differs from a time series average. The original proof of the PCM relied on taking ensemble averages and therefore it was not clear whether the PCM applied to phenomena for which only a single non-ergodic time series was available. Recent results indicate that the cross correlation between two complex non-ergodic networks is maximum when they have the same degree of complexity. Consequently we have the principle of complexity matching, where the maximum amount of information is exchanged when the degree of complexity of the interacting networks are the same [105].

3.4.4 Social Allometry

Our interest in sociology in this section is limited to an allometry context, in which the patterns that emerge in the data are dependent on the size of the social group. In large urban centers size dependencies are found in both a city's physical structure [7] and in its functionality. For example the behavior of people [62] changes with the size of the city to which they belong. In addition among the most compelling aspects of urban life, for example, income level and innovation, are observed to be allometric phenomena.

Urban allometry was studied by Batty *et al.* [8], who characterized urban spatial structure in large cities through the distribution of buildings in terms of their volume, height and area. They point out that the allometry hypothesis suggests the existence of critical ratios between geometric attributes that are fixed by the functioning elements just as in living organisms. An exemplar is the dependence of natural light on the surface area of a building, so that to maintain a given ratio of natural light to building volume the shape of the building must change with increasing size. Consequently, the volume is not given by the traditional geometrical argument in which the surface area would be raised to the $3/2$ power, but is found empirically to have an allometry index $b \approx 1.3$. They interpret this value of the index to mean that the volume does not increase as rapidly with increasing surface area as it would for strict geometric scaling or strict rationality on the part of the builder. A number of such ARs are found between the volume, area, height and perimeter of buildings indicating the strong influence of allometry on human design.

Cities have throughout history promoted the extremes in human activity generating creativity, wealth as well as crime. These measurable extremes do not escape the quantification provided by allometry relations. The ARs for wealth and innovation in urban centers are concave with an allometry exponent for the population greater than one, $b > 1$, whereas the ARs accounting for infrastructure are convex with $b < 1$ [12, 13]. The convex urban ARs share the economy of scale that is enjoyed by biological networks since $\langle Y \rangle / \langle X \rangle \propto \langle X \rangle^{b-1}$ decreases with network size and as pointed out by Bettencourt *et al.* [12] this economy of scale facilitates the optimized delivery of social services, such as health care and education. They go on to contrast the convex with the concave urban ARs that focus on the growth of occupations oriented toward innovation and wealth creation. Of particular interest is their discussion of the scaling of rates of resource consumption with city size in direct correspondence to physiologic time. They [12] emphasize the concave situation for processes driven by innovation and wealth creation having $b > 1$, manifest in an increasing pace of urban life within larger cities [62], which they [12] quantitatively confirm for urban crime rates, spread of infectious diseases and pedestrian walking.

3.4.5   Strength of Materials

Let us now shift gears to consider the strength of a solid material and consider how the strength of material varies with size. This was a problem identified by Galileo who observed [33]:

> From what has already been demonstrated, you can plainly see the impossibility of increasing the size of structures to vast dimensions either in art or in nature...; so also it would be impossible to build up the body structures of men, horses and other animals so as to hold together and perform their normal functions if these animals were to be increased enormously in height; for this increase in height can be accomplished only by employing a material which is harder and stronger than usual, or by enlarging the size of the bones, thus changing their shape until the form and appearance of the animals suggests a monstrosity.

Carpinteri *et al.* [16, 17] extend the observation of Galileo to the size effects entailed by the fractal features of the microstructure making up a heterogeneous material. As they discuss [17], once the fractal geometry of the microstructure is set the quantities characterizing the failure process of a disordered material can be determined. For example, the nonlinearity in the constitutive equations result from the concentration of stress on sets that are self-similar over a broad spectrum of spatial scales. They explicitly work out cases including stress distributions on spatial Cantor sets to develop intuition about the properties of real material with fractal microstructure such as concrete. In the case of concrete the stress concentration on a fractal set is a consequence of the heterogeneity of the aggregates within the concrete and Carpinteri *et al.* [16] have outlined the topologic framework for the mechanics of such deformable fractal media using local fractional derivatives and obtain predictions consistent with experiment. The constitutive equations in this situation are fractional.

3.5   After Thoughts

What can we learn about complex phenomena, such as the beating heart, breathing, or walking; the stock market; the growth of cities, or any other complex phenomena for that matter using erratic time series? The scaling observed in the data encourages us to think about these phenomena in ways that would probably not have suggested themselves without the analysis. In other words, processing erratic time series using scaling ideas provides a measure that allows us to understand these phenomena in new and different ways. This is not so different from the nineteenth century scientist who predicted exact results, but continued to obtain scatters of data that only approximated the predictions. In the earlier situation science was forced to introduce statistical uncertainty into her predictions in order to describe the influence of the complexity that was not being explicitly accounted for in the theoretical prediction. The change in perspective from a single certain outcome to an ensemble of outcomes, with the average value corresponding to the prediction, constituted a true paradigm shift. This introduced a fundamentally different criterion for what constituted knowledge and which subsequently found its way into all the corners of science.

There have been other paradigm shifts in science, such as the realization that the equations of classical mechanics used to calculate the orbits of celestial bodies could not be solved perturbatively, as astronomers had been doing for two centuries. But none has been so pervasive as the shift from the determinism of classical mechanics to the randomness in statistical mechanics in order to understand the complexity of many-body systems. The techniques developed to understand the physical mechanisms necessary to explain such collective behavior as phase transitions, for example, renormalization group (RG) theory, were the forerunners of today's paradigm shift.

We subsequently explore some of the implications of RG theory in the larger context of modeling complex phenomena in general. We draw some analogies between the phenomenon of physical phase transitions and the multiple time correlations observed in social and physiological data sets. The formalism of RG theory was developed to describe the physical phenomenon of phase transitions, which is a qualitative change in the equilibrium state of matter. Liquids become solid as the temperature is lowered sufficiently. In a gas the interactions among the molecules are relatively short-range, so the particles fly about only occasionally colliding with one another. As the temperature is lowered the particles slow down, the forces become longer-range, the particles begin to coordinate their behavior and liquid is formed. As the temperature is further lowered the forces become very long-range, keeping each molecule in the vicinity of lattice points in space and a solid transitions from the liquid.

Not until RG theory was developed was there a formal mathematical context in which the singular behavior of phase transitions could be explained without recourse to a specific mechanistic model. It appears that the long-range spatial forces in observed phase transitions may have a correspondence with the long-range temporal forces seen in physiological phenomena, suggesting an analogy between aggregate physical spatial structures and multiple time scales in physiological structures.

We demonstrated heuristically that certain physiologic time series satisfy scaling relations and at least one measure of statistics, the relative dispersion, satisfies a RG relation. We also demonstrated that the simplest solution to the RG relation $F(\lambda x) = \gamma F(x)$ has the form of an IPL with power-law index $\alpha = -\ln\gamma/\ln\lambda$, which encompasses all the empirical IPLs in Tables 3.1 and 3.2. The more general solution has a complex power-law index that captures the log-periodic modulation of a number of scaling phenomena whose dominant behavior is IPL. This led to the notion of a complex fractal dimension [115], which we examine in more detail in subsequent chapters.

References

[1] Adrian, R. 1809. *The Analyst: or Mathematical Museum* **1**, 93–109.

[2] Albert, R. and A.-L. Barabási. 2002. *Rev. Mod. Phys.* **74**, 47–97.

[3] Allegrini, P., P. Paradissi, D. Menicci and A. Geminani. 2010. *Front. Physiol.* **1**, 28. doi: *10.3389/fphys.2010.00128*.

[4] Anderson, J.R. and L.J. Schooler. 1991. *Psych. Sci.* **2**, 396–408.

[5] Auerbach, F. 1913. *Petermanns Mitteilungen*.

[6] Bassingthwaighte, J.B., L.S. Liebovtch and B.J. West. 1994. *Fractal Physiology*, Oxford University Press, Oxford.

[7] Batty, M. and P. Longley. 1994. *Fractal Cities*, Academic Press, San Diego.

[8] Batty, M., R. Carvalho, A. Hudson-Smith, R. Milton, D. Smith and P. Steadman. 2007. *Proc. 6th Int. Space Syntax Symp.*, Istanbul.

[9] Beggs, J. and I. Plenz. 2003. *J. Neurosci.* **23**, 11167-783 11177.

[10] Bénichou, O., C. Loverdo, M. Moreau and R. Voituriez. 2011. *Rev. Mod. Phys.* **83**, 81.

[11] Beran, J. 1994, *Statistics for Long-Memory Processes*, Chapman & Hall, New York.

[12] Bettencourt, L.M.A., J. Lobo, D. Helbing, C. Kuhnert and G.B. West. 2007. *PNAS* **104**, 7301.

[13] Bettencourt, L.M.A., J. Lobo, D. Strumsky and G.B. West. 2010. *Plos one* **5**, e1354-1.

[14] Boltzmann, L. 1987. *Lectures on the Principles of Mechanics, Vol. 1*, 66, Leipzig: Barth; originally published 1904.

[15] Brown, J.H., V.K. Gupta, B. Li, B.T. Milne, C. Restrepo and G.B. West. 2002. *Phil. Trans. R. Soc. Lond. B* **357**, 619–626.

[16] Carpinteri, A., B. Chiaia and P. Cornetti. 2000. *Rend. Sem. Mat. Univ. Pol. Torino* **58**, 57–68.

[17] Carpinteri, A., P. Cornetti and K.M. Kolvankar. 2004. *Chaos, Solitons & Fractals* **21**, 623–632.

[18] Carroll, L. 1960. *The Annotated Alice, Alices Adventures in Wonderland & Through the Looking Glass*, illus. J. Tenniel, Introduction and Notes by M. Gardner, Clarkson N. Potter, NY.

[19] Cavalcante, H.L.D. de S., M. Oria, D. Sornette, E. Ott and D.J. Gauthier. 2013. *Phys. Rev. Lett.* **111**, 198701-1.

[20] Chambers, R.A., W.K. Bickel and M.N. Potenza. 2007. *Neurosci. Biobehav. Rev.* **31**, 1017–1045.

[21] Chialvo, D.R. 2010. *Nature Phys.* **6**, 744.

[22] Cohen, I.B. 1987. In *The Probabilistic Revolution: Vol. 1 Ideas in History*, MIT Press, Cambridge MA.

[23] Cuvier, G. 1812. *Recherchces sur les ossemens fossils*, Paris.

[24] Dewey, F.E., E. Rosentha, D.J. Murphy, V.F. Froelicher and E.A. Ashley. 2008. *Circulation* **117**, 2279–87.

[25] Dacorogna, M.M., R. Gencoy, U. Müller, R.B. Olsen and O.V. Pictet. 2001. *An Introduction to High Frequency Finance*, Academic Press, San Diego, CA.

[26] Eloy, C. 2011. *Phys. Rev. Lett.***107**, 258101-1.

[27] Faloutsos, M., P. Faloutsos and C. Faloutsos. 1999. *Comput. Commun. Rev.* **29**, 251.

[28] Feder, J. 1980. *Fractals*, Plenum Press, New York.

[29] Gilden, D.L. 2001. *Psych. Rev.* **108**, 33–56.

[30] Glöckle, W.G. and T.F. Nonnenmacher. 1991. *Macromolecules* **24**, 6426.

[31] Gong, P., A.R. Nikolaev and C. van Leeuwen. 2007. *Phys. Rev. E* **76**, 011904.

[32] Gopikrishnan, P., M. Meyer, L.A.N. Amaral and H.E. Stanley. 1998. *Eur. Phys. J. B.* **3**, 139–140.

[33] Galilei, G. 1954. *Two New Sciences*, translated by H. Crew and A. de Salvio, Dover, NY; first published in 1665.

[34] Gauss, F. 1809. *Theoria motus corposrum coelestrium*, Hamburg.

[35] Grigolini, P., D. Leddon and N. Scafetta. 2002. *Phys. Rev. E* **65**, 046203.

[36] Gutenberg, B. and C.F. Richter. 1954. In *Seismicity of the Earth and Associated Phenomena*, 2nd Ed., 17–19, Princeton University Press, Princeton, N.J.

[37] Hack, J.T. 1957. *U.S. Geol. Surv. Prof. Paper* **294-B**, 45.

[38] Hempleman, S.C., D.L. Kilgore, C. Colby, R.W. Bavis and F.L. Powell. 2005. *J. Exp. Biol.* **208**, 3065–3073.

[39] Herrnstein, R.J. 1961. *J. Exp. Anal. Behav.* **29**, 267–272.

[40] Hofman, M.A. 1991. *J. Hirnforsch* **32**, 103–111.

[41] Horton, R.E. 1945. *Bull. Geol. Soc. Am.* **56**, 275–370.

[42] Huberman, B.A. and L.A. Adamic. 1999. *Nature* **401**, 131.

[43] Huxley, J.S. 1931. *Problems of Relative Growth*, NY, The Dial Press.

[44] Jensen, J.L.W.V. 1906. *Acta Mathematica* **30** (1): 175.

[45] Jeong, H., B. Tombor, R. Albert, Z.N. Oltvai and A.-L. Barabási. 2000. *Nature* **407**, 651.

[46] Jeong, H., S.P. Mason, A.-L. Barabási and Z.N. Oltvai. 2001. *Nature* **411**, 41.

[47] Kitzbichler, M.G., M.L. Smith, S.R. Christensen and E. Bullmore. 2009. *PLoS Comp. Biol.* **5**, 1–13, www.ploscombiol.org.

[48] Kolmogorov, A.N. 1941. *Coptes Rendus (Dokl.) Akad. Sci. URSS* **26**, 115.

[49] Kowalski, K. 1997. *J. Math. Phys.* **38**, 2403.

[50] Li, W. 1992. *IEEE Transactions on Information Theory* **38** (6): 18421845.

[51] Lilerjos, F., C.R. Edling, L.A.N. Amaral, H.E. Stanley and Y. Aberg. 2001. *Nature* **411**, 907.

[52] Liu, W., R. Yan, W. Jing, H. Gong and Pl. Liang. 2011. *Protein & Cell* **2**, 764.

[53] Lo, C.-C., T. Chou, T. Penzel, R.E. Scammell, R.E. Strecker, H.E. Stanley and P.Ch. Ivanov. 2004. *PNAS* **101**, 17545–17548.

[54] Lotka, A.J. 1926. *J. Wash. Sci.* **16**, 317.

[55] Magin, R.L. 2006. *Fractional Calculus in Bioengineering*, begell house inc., New York.

[56] Mandelbrot, B.B. 1977. *Fractals, Form and Chance*, W.F. Freeman, San Francisco, CA.

[57] Mandelbrot, B.B. 1997. *Fractals and Scaling in Finance*, Springer, New York.

[58] Mantegna, R.N. and H.E. Stanley. 2000. *An Introduction to Econophysics*, Cambridge University Press, Cambridge, UK.

[59] McMahon, T. 1973. *Science* **179**, 1201–1204.

[60] Meakin, P. 1998. *Fractals, scaling and growth far from equilibrium*, Cambridge Nonlinear Science Series 5, Cambridge University Press, Cambridge, MA.

[61] Medina, J.M. 2009. *Phys. Rev. E* **79**, 011902.

[62] Milgrim, S. 1970. *Science* **167**, 1461.

[63] Mill, J.S. 2002. *System of Logic*, 8th Edition, University Press of the Pacific, Honolulu; first published in 1882.

[64] Millman, D., S. Mihalas, A. Kirkwood and E. Niebur. 2010. *Nature Physics* **6**. 801.

[65] Minamino, R. and M. Tateno. 2014. *PLOS ONE* **9**, e9353S.

[66] Montoya, J.M. and R.V. Solé. 2000. arXiv; cond-mat/0011195

[67] Murray C.D. 1927, *J.Gen. Physiology* **10**, 725–729.

[68] Newman, M.E.J. 2010. *Networks, An Introduction*, Oxford University Press, Oxford, New York.

[69] Novikov, E.A. 1966. *Soviet Phys.-Dokl.* **11** (6), 497.

[70] Oliveria, J.G. and A.-L. Barabási. 2005. *Nature* **437**, 1241.

[71] Omori, F. 1894. *J. College of Science, Imperial University of Tokyo* **7**, 111–200.

[72] Pareto, V. 1897. *Cours d'Economie Politique*, Lausanne and Paris.

[73] Peng, C.-K., J. Mietus, J.M. Hausdorff, S. Havlin, H.E. Stanley and A.L. Goldberger. 1993. *Phys. Rev. Lett.* **70**, 1343–1346.

[74] Perrin, J. 1990. *ATOMS*, translated by D.Ll. Hammick, Ox Bow Press, Woodbridge, CO; French edition, Livrairie Felix Alcan, 1913.

[75] Poincaré, H. 1888. *Mémoire sur les courves définies parles equations différentielles, I–IV, Oevre 1*, Gauthier-Villars, Paris.

[76] Price, D.J. de Solla. 1963. *Little Science, Big Science*, Columbia University Press, NY.

[77] Quetelet, A. 1842. *A Treatise on Man and the Development of His Faculties*, William and Robert Chambers, Edenburgh.

[78] Raabe, O.H., H.C. Yeh, G.M. Scham and R.F. Phalan. 1976. *Traceobronchial Geometry: Human, Dog, Rat, Hamster*, Lovelace Foundation for Medical Education and Research, Albuquerque.

[79] Rabotnov Yu. N. 1977. *Elements of Hereditary Solid Mechanics*, MIR Pub., Moscow.

[80] Rall, W. 1959. *Annals of New York Academy of Science* **96**, 1071.

[81] Richardson, L.F. 1926. *Proc. Roy. Soc. Lond. Ser. A* **110**, 709–725.

[82] Richter, J.P. 1970. *The Notebooks of Leonardo da Vinci*, Vol. 1, Dover, New York; unabridged edition of the work first published in London in 1883.

[83] Roberts, D.C. and D.L. Turcotte. 1998. *Fractals* **6**, 351–357.

[84] Rosen, P. and E. Rammler. 1933. *J. Inst. Fuel.* **7**, 29–36.

[85] Scafetta, N. and B.J. West. 2004. *Phys. Rev. Lett.* **92**, 138501.

[86] Schottky, W. 1918. *Annalen der Physik* **362**, 541–567.

[87] Scott Blair, S.G., B.C. Veinoglou and J.E. Caffyn. 1947. *Proc. Roy; Soc. Ser. A* **187**, 69.

[88] Seibel, R. 1963. *J. Exp. Psych.* **66**, 215–226.

[89] Shlesinger, M.F., B.J. West and J. Klafter. 1987. *Phys. Rev. Lett.***58**, 1100.

[90] Shlesinger, M.F. and B.J. West. 1991. *Phys. Rev. Lett.* **67**, 3200.

[91] Simkin, M.V. and V.P. Roychowdhury. 2011. *Phys. Rept.* **502**, 1.

[92] Sornette, D. 1994. *Phys. Rept.* **297**, 239.

[93] Sornette, D. and A. Johansen. 1997. *Physica A* **245**, 411.

[94] Sornette, D. 1998. *Phys. Rept.* **297**, 239.

[95] Sornette, D. 2003. *Phys. Rept.* **378**, 1.

[96] Stanley, H.E. 1979. *Introduction to Phase Transitions and Critical Phenomena*, Oxford University Press, Oxford, UK.

[97] Stevens, S.S. 1957. *Psychol. Rev.* **64**, 153–181.

[98] Sturmberg, J.P. and C.M. Martin. 2013. *Handbook of Systems and Complexity in Health*, Springer, New York.

[99] Suwa, N., T. Nirva, H. Fukusawa and Y. Saski. 1963. *Tokoku J. Exp. Med.* **79**, 168–198.

[100] Szeto, H.H., P.Y. Cheng, J.A. Decena, Y. Cheng, D. Wu and G. Dwyer. 1992. *Am. J. Physiol.* **262** (*Regulatory Integrative Comp. Physiol.* **32**) R141–R147.

[101] Thompson, D.W. 1961. *On Growth and Form* (1915); unabridged ed., Cambridge University Press, Cambridge, UK.

[102] Turalska, M., M. Lukovic, B.J. West and P. Grigolini. 2009. *Phys. Rev. E* **80**, 021110-1.

[103] Turalska, M., E. Geneston, B.J. West, P. Allegrini and P. Grigolini. 2012. *Front. Physio.* **3**. doi:10.3389/fphys.2012.00052.

[104] Turalska, M., P. Grigolini and B.J. West. 2013. *Sci. Rept.* **3**, 1–8.

[105] Turalska, M., A. Svenkeson and B.J. West. "Synchronization and anti-synchronization of dynamically coupled networks", under review.

[106] Turcotte, D.L. 1992. *Fractals and chaos in geology and geophysics*, Cambridge University Press, Cambridge.

[107] Turcotte, D.L. and B.D. Malamud. 2004. *Physica A* **340**, 580–589.

[108] Voss, R.V. and J. Clark. 1978. *J. Acoust. Soc. Am.* **63**, 258–263.

[109] Voss, R.V. 1992. *Phys. Rev. Lett.* **68**, 3805.

[110] Schmidt-Nielsen, K. 1984. *Scaling: Why is Animal Size So Important ?*, Cambridge University Press, Cambridge.

[111] Watts, D.J. and S.H. Strogatz. 1998. *Nature* **393**, 440.

[112] Weibel, E.R. 1963. *Morphometry of the Human Lung*, Academic Press, New York.

[113] West, B.J., V. Bhargava and A.L. Goldberger. 1986. *J. Appl. Physiol.* **60**, 189.

[114] West, B.J. and A. Goldberger. 1987. *American Scientist* **75**, 354.

[115] West, B.J. and M.F. Shlesinger. 1989. *Int. J. of Mod. Phys. B* **395**-819; B.J. West and M.F. Shlesinger 1990, *Am. Sci.* **78**, 40–45.

[116] West, B.J. and X. Fan. 1993. *Fractals* **1**, 21.

[117] West, B.J., R. Zhang, A.W. Sander, S. Mimiyer, J.H. Zuckerman and B.D. Levine. 1999. *Phys. Rev. E* **59**, 3492.

[118] West, B.J., M. Bologna and P. Grigolini. 2003. *Physics of Fractal Operators*, Springer, Berlin.

[119] West, B.J. 2006. *Where medicine went wrong*, Studies of Nonlinear Phenomena in Life Science vol. 11, World Scientific, Singapore.

[120] West, B.J., E. Geneston and P. Grigolini. 2008. *Physics Reports* **468**, 1–99.

[121] West, B.J. and P. Grigolini. 2010. *Physica A* **389**, 5706.

[122] West, B.J. and P. Grigolini. 2010. *Med. Hyp.* **75**, 475.

[123] West, B.J. and P. Grigolini. 2011. *Complex Webs; Anticipating the Improbable*, Cambridge University Press, UK.

[124] West, B.J. and D. West. 2012. *Frac. Calc. & App. Analysis* **15**, 1.

[125] West, D. and B.J. West. 2012. *Int. J. Mod. Phys. B* **26**, 1230013-1.

[126] West, D. and B.J. West. 2013. *Phys. of Life* **10**, 210.

[127] West, B.J., M. Turalska and P. Grigolini. 2014. *Network of Echoes; Immitation, Innovation, and Invisible Leaders*, Springer, New York.

[128] West, G.B., J.H. Brown and B.J. Enquist. 1997. *Science* **276**, 122.

[129] West, G.B., J.H. Brown and B.J. Enquist. 1999. *Science* **284**, 1677–79.

[130] Weibel, E.R. 2000. *Symmorphosis: On form and function in shaping life*, Harvard University Press, Cambridge, MA.

[131] Wiener, N. 1948. *Ann. NY Acad. Sci.* **50**, 197–220.

[132] Willis, J.C. and G.U. Yule. 1922. *Nature* **109**, 177.

[133] Willis, J.C. 1922. *Age and area: a study in geographical distribution and origin of species*, Cambridge University Press, Cambridge, IUK.

[134] Wise, M.E. 1975. In *Statistical Distributions in Scientific Work, Vol. 2*, Eds. G.P. Patil et al., D. Reidel, Dordrecht-Holland, 241–262.

[135] Yu, Y., R. Romero and T.S. Lee. 2005. *Phys. Rev. Lett.* **94**, 108103.

[136] Yule, G.U. 1925. *Phil. Trans. Roy. Soc. London B* **213**, 21.

[137] Zipf, G.K. 1949. *Human Behavior and the Principle of Least Effort: An Introduction to Human Ecology*, Addison-Wesley, Cambridge, MA.

CHAPTER 4

Simple Fractional Operators

Calculus is the formal method by which scientists carry out quantitative reasoning. In the analysis of complex phenomena, particularly those that change over time, the traditional guideposts such as identifying causality and making verifiable predictions become problematic. One way to anticipate the improbable and prepare for unintended consequences is through the systematic handling of fractal operators, using the fractional calculus and fractional differential equations. But we also need to incorporate the more traditional notions of complexity, such as randomness into the discussion. Therefore we start our tour of the tools required to think newly about complexity with a brief introduction to random walks using fractional difference equations. Such fractional random walks have been used to model complex phenomena from fluctuations in the financial market [24] to climate change [17].

In the continuum limit fractional difference equations become fractional differential equations. There are many books that develop the mathematics that takes one from the discrete to the continuous fractional operators, for example, Samko *et al.* [56], but that path is not followed herein. Instead fractional operators are introduced by means of prescriptions that highlight the departure from the standard differential calculus. The intent is to underscore the properties of the phenomena whose dynamics require a fractional dynamic description and consequently are described by non-differentiable functions.

Examples of fractional equations of motion discussed in some detail are for viscoelastic materials that lie between solids and liquids, and whose dynamics are given by fractional rate equations. The solution to such fractional equations of motion are shown to compare remarkably well with a library of experiments testing the relaxation properties of 'exotic' materials. A different kind of example is drawn from the history of the back-reaction of a fluid to the motion of a heavy particle through it. Over a century ago careful analysis revealed a fractional differential viscosity modifying the fluid dynamic description of a passive scalar in a fluid. This analysis is applied to a generalization of Brownian motion and the comparison of analytic results with experiment are quite good.

It is not only physical scientists that are interested in the behavior of in-between material, however. It is also the physician studying traumatic brain injury (TBI), the bioengineer examining the stress on the joints of athletes, and the natural scientist seeking to understand the dependence of a given functionality on the size of an animal, who also recognize that living matter, such as cells, tissue and organs are viscoelastic. This is part of what makes bioengineering, not to say medicine, so difficult. They are, in large part, the study of the exceptional. The viscoelastic behavior of physiologic networks, such as the lungs, have been successfully modeled using factional stress-strain relations [26].

Ionescu *et al.* [26] explain how the stress relaxation of the lungs is better described by an IPL than by an exponential, and show how this relaxation can be expressed as a function of the pressure using lumped parameters. Structural changes in the lungs with disease are quantified using the parameters of the fractional-order impedance model of the lungs. Another application of these ideas to living tissue is the generalization to fractional-order of the Bloch equations describing the precession of nuclear spins. The solution to these fractional rate equations are expressed in terms of MLFs and lay the ground work for the further extension to the subsequent discussion of MRIs.

4.1   Random Walks

Most physical scientists are introduced to statistical processes through the study of diffusion or radioactive decay. Their intuition of such unpredictable processes is developed by way of the physics and not the mathematics, although a certain level of mathematical intuition does come in time. Here we consider how we might generalize the physical, social and biological statistical processes of interest and along with them the underlying mathematics. One way to think differently about processes that cannot be characterized by a specific scale, such as a fractal process, is to replace integer difference equations with fractional difference equations. We do this for discrete random walks and determine the meaning of discrete fractional differences. This exercise provides the first insight into the potential utility of fractional thinking.

4.1.1 Rayleigh Type

A familiar derivation of the bell-shaped distribution of Gauss starts from the simple random walk process that was first recognized by Lord Rayleigh [53] the same year Einstein explained molecular diffusion [13] that being 1905. Lord Rayleigh had considered the acoustic field resulting from a large but finite number of sources all with the same frequency, but with independent random phases and had published the results of his investigation [52]. Some years later he read an article in *Nature* by Pearson [49] who had posed the following problem:

> A man starts form point O and walks a distance in a straight line, he then turns through any angle whatever and walks a distance a in a second straight line. He repeats this process n times. Required, the probability that after these n stretches he is a distance between r and $r + \delta r$ from his starting point.

Lord Rayleigh [53] published an asymptotic solution to the random walk problem for a large number of steps n in the same issue of *Nature*, that being the Rayleigh PDF, or equivalently a two-dimensional Gaussian PDF. His solution was based on the analysis he had done twenty-five years earlier on acoustic waves with random phases. The two-paragraph solution to Pearson's question was titled "The problem of the random walk". The name stuck; but in all fairness it should have been named the Rayleigh walk, or at least the Rayleigh random walk (RRW).

In one dimension the RRW argument involves updating the displacement of a walker at step n given by Q_n by adding a discrete random number ξ_n for the step length to obtain

$$Q_{n+1} = Q_n + \xi_{n+1}. \tag{4.1}$$

Introducing the down-shift operator B into the random walk equation enables us to write Eq.(4.1) as

$$(1 - B) Q_{n+1} = \xi_{n+1}, \tag{4.2}$$

since $BQ_{n+1} = Q_n$. The total displacement of the walker along the line after N steps is

$$Q_N = \xi_1 + \xi_2 + \cdots + \xi_N, \tag{4.3}$$

which becomes the continuous variable $Q(t)$ in the limit of the number of steps N becoming infinitely large and the step size Δt becomes infinitesimally small, such that the continuous time is $t = N\Delta t$. In this limit the statistical properties of the random walk are given by the Gaussian PDF in terms of the phase space variable q rather than the dynamic variable Q :

$$P(q,t) = \frac{1}{\sqrt{2\pi Dt}} \exp\left[-\frac{q^2}{2Dt}\right]. \tag{4.4}$$

The variance in the PDF increases linearly in time t and D is the strength of the fluctuations [13, 53], that is, denoting by brackets an average over an ensemble of realization of the random steps the strength of the noise is given by

$$\langle \xi_n \xi_k \rangle = 2D\delta_{nk} \tag{4.5}$$

where δ_{nk} is the Kronecker delta $= 0$ for $n \neq k$ and $= 1$ for $n = k$. In the analysis of classical diffusion the constant D is the diffusion coefficient and is a property of the embedding fluid, called the environment in Chapter 2.

Figure 4.1 Cartoon of a two-dimensional random walk or drunkard's walk as given by Gamow [16].

This model of diffusion has been used to describe what was accepted as complex phenomena in the nineteenth and early twentieth centuries. It enabled an understanding of how stirred cream mixes with your morning coffee, how rumors spread through social groups [47], how oil spills spread in the deep ocean [46], and a myriad of other phenomena involving the dynamics of a variable interacting with an unknown and often unknowable environment. Classical diffusion is one of those remarkably successful theories that is thoroughly understood and which has provided deep insight into a wide variety of phenomena. However the formal equation rests on two assumptions: 1) the dynamics are linear and 2) space is homogeneous and time is isotropic. When the phenomenon being described is truly complex it is probably not linear and often lacks the smoothness properties in space and time.

4.1.2 Fractional Type

We begin our deconstruction of the Rayleigh random walk by no longer assuming that the time of interest in the phenomenon being studied is isotropic. This assumption of the isotropy of time dates back to Newton's *Principia* along with the infinite extent of homogeneous space. However it is not the time measured by the clock that we introduce in our reformulation, but time as experienced by the individual engaged in a complex stochastic process. There are a number of ways to incorporate this new view of time into the dynamics and we shall have occasion to examine a number of them in due course. For the moment we restrict our concern to the first major modification of the RRW model. Hosking [24], who was interested in economic processes and their associated time series, generalized RRWs using the fractional difference equation

$$(1 - B)^{\alpha} Q_n = \xi_n \tag{4.6}$$

where the index α is non-integer. Here the discrete index n is the time variable and the fact that the power index is not an integer introduces some peculiar features into how a random impulse at time n influences the displacement of the walker at that time. This new random walk is still a linear additive process, but it is no longer simple.

Hosking was able to establish that the operator on the left side of Eq.(4.6) has an inverse that can be expressed by a binomial expansion. The total displacement of the walker after n steps using this expansion can be written as [69]:

$$\begin{aligned} Q_n &= \sum_{k=0}^{\infty} \binom{\alpha}{k} (-B)^k \xi_n \\ &= \sum_{k=0}^{\infty} \frac{\Gamma(k+\alpha)}{\Gamma(k+1)\Gamma(\alpha)} \xi_{n-k}. \end{aligned} \tag{4.7}$$

Consequently, Q_n is influenced by random impulses stretching infinitely far back in time, as indexed by k with the relative impact of each impulse on the displacement being determined by the ratio of Gamma functions. As the step index (discrete time) $k \to \infty$ the ratio of Gamma functions becomes, using Stirling's approximation, proportional to

$$\frac{\Gamma(k+\alpha)}{\Gamma(k+1)\Gamma(\alpha)} \propto k^{\alpha-1} \text{ for } k \gg \alpha, \tag{4.8}$$

which is an IPL since $|\alpha| < 1/2$ in the analysis [69]. The statistics of the displacement remain Gaussian because the relation of the total displacement to the random impulses is linear and additive, and therefore the central limit theorem insures that the displacement has Gaussian statistics. Note that the variance no longer increases linearly in time. In general the time dependence of the variance is quite complicated but in the asymptotic limit it behaves as an IPL

$$Var\,[Q_k] \propto k^{2\alpha-1}. \tag{4.9}$$

This scaling in terms of the parameter α is our first indication of a possible connection between fractional dynamics and temporal complexity, albeit the dynamics in this random walk are still discrete. We refer to this process as a fractional random walk (FRW).

The FRW generates a random process with memory. One way to understand the introduction of memory is to consider the simple random walk of Pearson and Rayleigh:

$$(1 - B)\,X_n = Q_n \tag{4.10}$$

and use the FRW process Q_n to generate the random impulses driving the displacements, that is, step sizes in a new random walk. In this case the walk has a solution that in the continuum limit has the stationary autocorrelation function [69]

$$C(\tau) = \langle X(t+\tau)X(t)\rangle \propto \tau^{2H-2}\,. \tag{4.11}$$

The power-law index, the Hurst exponent H, is given by $H = \alpha+1/2$, so that since it is assumed that $0 \le H \le 1$, the fractional index lies in the interval $-1/2 \le \alpha \le 1/2$. This new process has an IPL spectrum given by the Fourier transform of the autocorrelation function

$$S(\omega) = \mathcal{FT}\,\{C(\tau),\omega\} \propto \frac{1}{\omega^{2H-1}}. \tag{4.12}$$

Equation (4.12) results from Eq.(4.11) using a Tauberian theorem in which the Fourier transform of the monomial in time t^{β} yields $1/\omega^{\beta+1}$ [72]. The scaling exponent H is called the Hurst exponent, following Mandelbrot [39], who named it after the Civil Engineer who discovered this scaling in his study of 1500 years of time series data on the floods and droughts of the Nile River.

An extended discussion of the Hurst exponent and the properties of the underlying processes are found in the excellent book by Feder [14] on fractals. In particular, random walks with $H > 1/2$ have a positive correlation, with long (short) intervals more likely to be followed by long (short) intervals. Random walks with $H < 1/2$ have a negative correlation, with long (short) intervals more likely to be followed by short (long) intervals. We stress that when the Hurst exponent $H \ne 1/2$ the random walk has long-time memory. Such memory has been observed consistently in financial time series, since Hosking first introduced fractional difference equations into the study of economic time series in 1981.

In the physical sciences the generalization of RRWs to complex phenomena was made by Montroll and Weiss [44]. They argued that in complex materials individual lattice sites have complicated structures, resulting in a walker waiting at such sites for time intervals of random duration, before taking steps of random length. Their approach was the first systematic generalization of random walks that abandoned both historical assumptions of spatial homogeneity and temporal isotropy. The transition PDF for taking fluctuating steps in space and time was incorporated into an integral equation for the PDF for the total displacement of the particle in a given time resulting in the continuous time random walk (CTRW) model. This model is important for the present discussion because it provided the first systematic theory that connected the local statistics of the random walker to an integral equation that under certain general conditions reduces to fractional derivatives in both space and time. The resulting fractional diffusion equation provided the first explanation of anomalous diffusion [45, 62], that is, diffusion in which the variance of the processes does not increase linearly with time.

The RRW process has a continuous analogue given by a Langevin equation [31] that provides a stochastic model of particle dynamics. Analogously the discrete FRW has a continuous fractional Langevin equation (FLE). In the latter case the connection between temporal complexity and the fractional derivative in time should be evident. Before exploring the fractional generalization of the dynamics we review a few of the simpler formal properties of the fractional calculus and introduce some of the phenomena that entail its implementation.

4.1.3   Climate Change

In keeping with the intent of this book I feel obliged to remark on a timely topic of political and scientific debate, that being climate change. Such discussions were earlier conducted under the umbrella of global warming, but the average global temperature of the earth has stopped increasing, consistent with a preliminary prediction we made in 2008 [58] in keeping with the variable periodicity of the total solar irradiance reaching the Earth. As anticipated, global warming has reached a plateau, thereby forcing a change in what to call the political discussion and someone made the decision to replace 'global warming' with the more neutral phrase 'climate change'. The latter now seems to be the accepted label.

In my own research we analyzed the variability in the Earth's average global temperature and concluded that it was strongly tied to solar variability [57, 67]. This connection was not universally embraced and fueled an already existing controversy. The most recent contribution to the debate uses fractional difference equations to analyze both average global temperature and sunspot numbers and the potential relationship between the two [17]. These latter authors conclude that there is no statistically significant influence of the variation in the number of sunspots on the variation in average global temperature. We sketch how they arrived at their conclusion and introduce a little of the physics necessary to understand the debate over whether climate change is natural, anthropogenic or a combination of the two.

It is well known that the Earth's short-term temperature anomalies share the same complexity index as solar flares; where the complexity index is the index of the IPL of the time intervals between events in each of the two time series. West and Grigolini [67] showed that this index equality is not accidental and argued that it is a consequence of the principle of complexity management (PCM), which is the information transfer between the sun and the earth, based on the crucial role of non-Poisson renewal events in complex networks. We applied the PCM to the linking of Earth's climate to total solar irradiance (TSI), and used as a surrogate for the variability in TSI the time intervals between solar flares. The average global temperature is a consequence of the TSI being absorbed and redistributed by the Earth's atmosphere and oceans by means of nonlinear hydrothermal dynamic processes [34].

Gil-Alana *et al.* [17] approach the physics problem of the linking of the Earth and Sun as a question in data processing, using the fractional differencing techniques to incorporate long-time dependence into the analysis. For average global temperature data they use an auto-regressive integrated moving average (ARMA) generalized to fractional differences [69]. They adopt Eq.(4.6) in order to model the hyperbolic decay of temperature fluctuations, which is to say, IPL decay of the autocorrelation function with the resulting fractionally integrated ARMA called ARIFMA. A careful study of the global temperature data yields a robust value of $\alpha = 0.46$ to account for the estimated global average temperature increase of $0.57\ °C$ over the last one hundred years. They further generalized the ARIFMA to include the periodic variation in sunspot data and used the fractional cyclical model:

$$\left(1 - 2B\cos\omega_r + B^2\right)^{\alpha/2} Q_n = \xi_n, \qquad (4.13)$$

for the analysis of sunspot data. They define $\omega_r = 2\pi r/T$, the frequency of sunspot oscillations, with $r = T/s$, so that s indicates the number of time periods per cycle, while r refers to the frequency that has a pole or singularity in the spectrum of Q_n, the number of sunspots at time n. Fitting the data they estimate $r = 11$ and $\alpha = 0.40$. Note that for $r = 0$ the fractional polynomial in Eq.(4.13) becomes $(1 - B)^\alpha$ and the equation reduces to the FRW.

They conclude that since the two time series have different fractional integration orders, 0.46 and 0.40, for temperature and sunspots, respectively, that they cannot have a common stochastic trend. In addition, they maintain that, the two series must display a pole or singularity in the spectrum at the same zero frequency, which did not occur in their results.

The story does not end here however. Scafetta [60] responded to Gil-Alana *et al.* [17], showing that the two time series are related when the nonlinear nature of the two physical processes are properly taken into account and they are subsequently characterized by cyclical fractional models. The emphasis on the nonlinear nature of the physics is central to Scafetta's argument as he and I stressed in *Disrupted Networks: From Physics to Climate Change* [68]. As is often the case, herein I will not be able to do justice to the physics in the limited space available and strongly urge the interested reader to consult the literature for an in depth understanding of the argument. Scafetta points out that showing that a particular statistical analysis does not highlight the existence of a relationship between two physical processes does not imply that the two processes are necessarily physically or statistically unrelated.

The "pole" observed in the average global temperature record by Gil-Alana *et al.* [17] was found to be an artifact of their discrete Fourier representation of time series [60]. Scafetta argues that when the temperature record is examined more systematically a number of periodic modulations are uncovered in addition to the solar periodicity, including lunar cycles and astronomical oscillations of the Heliosphere by Jupiter and Saturn and the Hale solar magnetic cycle. Thus, the temperature record like that of sunspot time series is a member of the same statistical family; a cyclical fractional process, but with a non-cyclic anthropogenic component emerging in the last few decades.

One must now recognize that sunspot data are also a surrogate for TSI and that Gil-Alana *et al.* [17] implicitly assumed that the number of sunspots is linearly related to other solar phenomena, thereby making it a good surrogate for TSI. However this assumption is physically flawed due to the complexity of solar dynamics and the coupling among multiple solar phenomena as Scafetta itemized [60]. It is difficult to escape the conclusion that sunspot time series are not a suitable surrogate for TSI.

We applaud the use of fractional difference equations for data processing, but sophisticated data processing does not relieve the analyst of the obligation to properly interpret the data in the context from which it is extracted. A final cautionary note regarding the assertions made in scientific publications is also in order. My collaborators and I never use sunspot number data in our analyses, specifically not in any of the dozen articles Gil-Alana *et al.* [17] conspicuously cite in their article.

4.2 Fractional Derivatives

The question of fractional derivatives was first raised by de l'Hôpital in a letter to Leibniz in 1695. He asked what one would obtain from the nth order derivative of a function when n is not an integer, in particular, when $n = 1/2$. Before giving the algebraic result of such an operation Leibniz responded with what has become one of those most remarkable understatements that is recalled with a smile: "It will lead to a paradox from which one day useful consequences will be drawn."

Starting from the properties of Gamma functions extended into the complex plane it is possible to write the arbitrary derivative of a monomial as

$$\frac{d^\alpha}{dt^\alpha}\left[t^\beta\right] = \frac{\Gamma\left(\beta+1\right)}{\Gamma\left(\beta+1-\alpha\right)}t^{\beta-\alpha}. \tag{4.14}$$

This is not a particularly surprising equation until we stipulate that the order of the derivative α is not an integer. Consider the case of interest to de l'Hôpital $\alpha = 1/2$ for various values of β :

$$\frac{d^{1/2}}{dt^{1/2}}\left[t^{-1/2}\right] = 0 \tag{4.15}$$

$$\frac{d^{1/2}}{dt^{1/2}}\left[1\right] = \frac{1}{\sqrt{\pi t}} \tag{4.16}$$

$$\frac{d^{1/2}}{dt^{1/2}}\left[t\right] = \sqrt{\frac{t}{\pi}}. \tag{4.17}$$

The top result is strange and is a consequence of the divergence of the Gamma function in the denominator in Eq.(4.14). The middle result is perhaps even stranger in that the fractional derivative of a constant yields a time-dependent function. Of course, neither of these curious findings is consistent with the ordinary calculus and has to do with the non-local nature of fractional derivatives as we shall learn. Finally, the last example expresses Leibniz's answer to de l'Hôpital's question.

These three fractional derivatives alert us to the fact that we have entered into a world in which the rules for quantitative analysis are different from what we have always believed, but they are not arbitrary. It remains to be seen if this mathematical world can explain the complexity of the physical, biological and social worlds in which we live. For a mathematician such a question is not always of interest and the development of the fractional calculus proceeded in the mathematical diaspore for over three hundred years, independently of the mathematics being developed for the physical sciences. Only recently have fractional operators attracted the attention of physical scientists, see for example, Magin [36]; Podlubny [50]; and West *et al.* [66].

4.2.1 Fractional Differentials and Limits

A critical reader would probably still be suspicious of the definition given by Eq.(4.14) for non-integer α. The fact that it coincides with the traditional definition when α is integer does not guarantee that it uniquely defines the non-integer case. To provide a demonstration of this expression we consider differentials and limits. The mathematician/physicist Mark Kac used to say; "A demonstration is what is required to convince a reasonable person, but it takes a proof to convince a mathematician." As the reader has undoubtedly concluded this essay is written for reasonable people and not for mathematicians.

Consider the generalization of the discrete shift operator introduced in the discussion of FRWs to the continuous case. The definition of a continuous time forward shift operator acting on the function $Q(t)$ is

$$B_\tau Q(t) = Q(t + \tau). \tag{4.18}$$

A Taylor expansion of $Q(t + \tau)$ in the vicinity of t allows us to identify the forward shift operator with the exponential

$$B_\tau Q(t) = e^{\tau D_t} Q(t), \tag{4.19}$$

where we identify the operator D_t with the ordinary time derivative. Consequently, the time derivative of the function can be expressed by the tautology

$$\frac{dQ(t)}{dt} = \lim_{\tau \to 0} \frac{B_\tau - 1}{\tau} Q(t) = D_t [Q(t)], \tag{4.20}$$

all of which is very familiar. Now let us extend these ideas to the non-integer case.

The fractional-order derivative can be determined by considering the generalization of Eq.(4.20)

$$\lim_{\tau \to 0} \frac{(B_\tau - 1)^\alpha}{\tau^\alpha} Q(t) = D_t^\alpha [Q(t)], \tag{4.21}$$

and it remains to provide a proper interpretation of the operator $D_t^\alpha [\cdot]$. The fractional derivative defined by Eq.(4.21) can be expressed by the integral equation

$$D_t^\alpha [Q(t)] = \frac{1}{\Gamma(1 - \alpha)} \frac{d}{dt} \int_0^t \frac{Q(t')dt'}{(t - t')^\alpha}, \tag{4.22}$$

which is the Riemann-Liouville (RL) fractional derivative [50, 66]. We emphasize that this is only one of the may forms that have been constructed for a fractional derivative [56].

If we introduce the monomial

$$Q(t) = t^\beta$$

into Eq.(4.22) we obtain

$$D_t^\alpha [t^\beta] = \frac{1}{\Gamma(1 - \alpha)} \frac{d}{dt} \int_0^t \frac{t'^\beta dt'}{(t - t')^\alpha}. \tag{4.23}$$

Inserting the variable $z = t'/t$ into Eq.(4.23) yields

$$D_t^\alpha \left[t^\beta \right] = \frac{1}{\Gamma(1-\alpha)} \frac{d}{dt} \left[t^{\beta+1-\alpha} \int_0^1 \frac{z^\beta dz}{(1-z)^\alpha} \right],$$

which readily integrates to the Beta function

$$\int_0^1 \frac{z^\beta dz}{(1-z)^\alpha} = \frac{\Gamma(\beta+1)\Gamma(1-\alpha)}{\Gamma(\beta+2-\alpha)}.$$

After some algebra involving gamma functions we obtain

$$D_t^\alpha \left[t^\beta \right] = \frac{\Gamma(\beta+1)}{\Gamma(\beta+1-\alpha)} t^{\beta-\alpha} \qquad (4.24)$$

in complete agreement with Eq.(4.14).

Thus, we see that the simple expression for an arbitrary derivative given by Eq.(4.14) is completely consistent with the limiting procedure given in Eq.(4.21), the latter having the integral form of the RL fractional derivative. It is also evident that with $\beta = 0$ we have for the RL fractional derivative of a constant the non-zero value

$$D_t^\alpha [1] = \frac{\Gamma(1)}{\Gamma(1-\alpha)} t^{-\alpha}$$

and when $\alpha = 1/2$ reduces to the value given by Eq.(4.16).

4.2.2 Differentiating Fractal Functions

It is worth re-emphasizing that there is no single fractional calculus, just as there is no single geometry. Different definitions of fractional operators, differentials and integrals, have been constructed to satisfy various needs, desires and constraints. There are a number of excellent texts that review the mathematics of the fractional calculi [42, 56], others that emphasize the engineering applications of the fractional operators [36, 50], and still others that provide physical interpretations of those operators [41, 66]. In other words the literature is much too vast to cover here. However many insights into complexity can be made by judiciously choosing various forms of the fractional operators that have been used in specific applications.

To facilitate subsequent discussion we denote the Laplace transform of a time-dependent function $Q(t)$ by $\widehat{Q}(s)$:

$$\widehat{Q}(s) \equiv \int_0^\infty e^{-st} Q(t) dt \equiv \mathcal{LT}\{Q(t); s\} \qquad (4.25)$$

and write the Laplace transform of the Caputo fractional derivative [7] for $\alpha < 1$:

$$\mathcal{LT}\left[\partial_t^\alpha \left[Q(t)\right]; s\right] \equiv s^\alpha \widehat{Q}(s) - s^{\alpha-1} Q(0), \tag{4.26}$$

where $Q(0)$ is the initial value of the dynamic variable $Q(t)$. In the more general case where $\alpha > 1$ the dependence of the Laplace transform on initial values yields

$$\mathcal{LT}\left[\partial_t^\alpha \left[Q(t)\right]; s\right] \equiv s^\alpha \widehat{Q}(s) - \sum_{j=0}^{m} s^{\alpha-1-j} Q^{(j)}(0) \tag{4.27}$$

where $Q^{(j)}(0)$ is the initial condition for the j-th derivative of the function and the size of the index for the fractional derivative determines the number of initial derivatives required $m \le \alpha \le m+1$ in the sum.

There are, as we said, a number of definitions of fractional integrals and derivations, each depending on a given set of assumptions. The Riemann-Liouville (RL) fractional operators are particularly useful [36, 50, 66]. The RL fractional integral is defined

$$D_t^{-\alpha}\left[Q(t)\right] \equiv \frac{1}{\Gamma(\alpha)} \int_0^t \frac{Q(t')dt'}{(t-t')^{1-\alpha}}, \tag{4.28}$$

the RL fractional derivative is defined

$$D_t^\alpha\left[Q(t)\right] \equiv D_t^n D_t^{\alpha-n}\left[Q(t)\right], \tag{4.29}$$

and the operator index is in the range $n-1 \le \alpha \le n$ for integer n. The Laplace transform of the RL fractional derivative for $\alpha < 1$ is

$$\mathcal{LT}\left[D_t^\alpha \left[Q(t)\right]; s\right] \equiv s^\alpha \widehat{Q}(s) - D_t^{\alpha-1}\left[Q(t)\right]\big|_{t=0}. \tag{4.30}$$

The choice of the fractional operator to be used is therefore dependent on the process being investigated, which requires evaluation of a fractional integral for the initial condition. This term can in some cases be evaluated using knowledge of $Q(0)$ and $\dot{Q}(0)$ or by placing restrictions on $Q(t)$ such as its being a bounded function [42].

It is now possible to return to the GWF and examine its RL fractional derivative using Eq.(4.30) in which we replace $Q(t)$ with $W(t)$ from Eq.(3.4). The inverse of the Laplace transform of the RL fractional derivative of the generalized Weierstrass function we denote as

$$W^\alpha(t) = \mathcal{LT}^{-1}\left[s^\alpha \widehat{W}(s); t\right]. \tag{4.31}$$

The fractional derivative of the GWF therefore satisfies a RG relation [40, 63]:

$$W^\alpha(bt) \equiv D_t^\alpha\left[W(bt)\right] = \frac{a}{b^\alpha} W^\alpha(t), \tag{4.32}$$

which can be solved just as we did previously to obtain

$$W^{\alpha}(t) = A(t)t^{\gamma} \;\; ; \;\; \gamma = \frac{\log a}{\log b} - \alpha. \tag{4.33}$$

Comparing the scaling index in Eq.(3.8) with that in Eq.(4.33) we see that taking the fractional derivative of the GWF changes the scaling parameter by a factor α and the resulting function can be shown to converge [54].

A factional operator of order α acting on a fractal function of fractal dimension D consequently yields another fractal function with fractal dimension $D \pm |\alpha|$. The fractional operator can be either a derivative or an integral; the fractional derivative increases the fractal dimension, thereby making the function more erratic; whereas the latter decreases the fractal dimension, thereby making the function smoother. The fact that the fractional operator acting on a fractal function converges supports the conjecture that the fractional calculus can provide an appropriate description of the dynamics for fractal phenomena made by Rocco and West [54].

To be a bit more mathematical the Hölder condition for the GWF is given by

$$|W(t) - W(t')| \le C \, |t - t'|^{\beta} \tag{4.34}$$

and the Hölder exponent lies in the interval $0 \le \beta \le 1$. As Carpinteri *et al.* [8] point out it is possible to prove that the fractal dimension of the graph of this function is $2 - \beta$ and therefore greater than one. They go on to say that the Weierstrass function admits continuous fractional derivatives of order lower than β, indicating a direct relation between the fractal dimension and the maximum order of differentiability, with the greater fractal dimension having the lower differentiability. To make use of this property in the physical domain they make use of the *local fractional derivative* (LFD) concept that was developed to examine the local properties of fractal structures such as the strength of materials discussed in Section 3.4.

In the same notation as the RL fractional operators the Caputo fractional derivative can be written for $\alpha < 1$ as

$$\partial_t^{\alpha} [Q(t)] = \frac{1}{\Gamma(-\alpha)} \int_0^t \frac{dt'}{(t - t')^{\alpha}} \dot{Q}(t') \tag{4.35}$$

where $\dot{Q}(t)$ is the derivative of $Q(t)$ with respect to t. The Laplace transform of Eq.(4.35) is given by

$$\mathcal{LT} \{\partial_t^{\alpha} [Q(t)] ; s\} = s^{\alpha-1} \mathcal{LT} \left\{ \dot{Q}(t); s \right\} = s^{\alpha-1} \left[s\widehat{Q}(s) - Q(0) \right], \tag{4.36}$$

which simplifies to Eq.(4.26). An advantage of using the Caputo rather than the RL fractional derivative is that the fractional derivative of a constant is now zero as we were brought up to believe it ought to be and we do not need the initial condition for a fractional derivative.

There is however a draw back to using the Caputo fractional derivative. Examining Eq.(4.35) we see that taking the derivative of order $\alpha < 1$ requires the function to have a first derivative. Therefore the Caputo fractional derivative is not applicable to a function describing a non-differentiable process such as turbulence or a fractal function such as the GWF. But it was precisely to gain understanding of complex phenomena that investigators introduced the fractional calculus into the physical, social and life sciences. This failure to address the non-differentiability of functions without returning to the RL fractional operators has been overcome, but we postpone that discussion and examine what has been learned using these two kinds of fractional operators with all their limitations.

4.3 Fractional Rate Equations

As we saw in Chapter 3 the simplest dynamic process is described by the relaxation rate equation for the dynamic variable $Q(t)$:

$$\frac{dQ(t)}{dt} = -\lambda Q(t), \tag{4.37}$$

whose solution is, of course, given by the exponential relaxation from the initial condition $Q(0)$ to zero

$$Q(t) = Q(0)e^{-\lambda t}. \tag{4.38}$$

This is the unique solution to the rate equation and provides everything we can know about the system. We can also interpret $Q(t)$ as the probability of the occurrence of an event, such as the decay of a radioactive particle where we interpret the initial condition as $Q(0) = \lambda$. Consequently, the rate equation describes the generation of events by a Poisson process where new events are generated at a rate λ.

4.3.1 Distribution of Rates

Of course the relaxation of disturbances in complex materials such as taffy or tar are not described by simple rate equations such as Eq.(4.37) and have been shown to require a fractional relaxation equation to describe their dynamics. One derivation of the fractional calculus representation of relaxation is based on the notion of self-similar dynamics as manifest through renormalization behavior. Glöckle and Nonnenmacher [18] argue that the renormalization concept may be applied to the rate equation by assuming the existence of many conformational substates separated by energy barriers. They assume a dichotomous stochastic process in which the relaxation between two states is not given by a single rate λ but by a distribution of rates such that the relaxation function is

$$Q(t) = Q(0) \int_0^\infty \rho(\lambda) \exp\left[-\lambda t\right] d\lambda \qquad (4.39)$$

where $\rho(\lambda)$ is the distribution of rates that represent the reaction kinetics and relaxation. However there are many other phenomena that can be modeled in this way, including thermally activated escape processes [10], intermittent fluorescence of single molecules [21] and nanocrystals [6], stochastic resonance [15] and blinking quantum dots [28], to name a few.

Nonnemacher and Metzler [48] introduced a fractal scaling model for the distribution of reaction rates from which they were able to derive a fractional differential relaxation equation. Here we write the equivalent of their equation using a Caputo derivative in time

$$\partial_t^\alpha \left[Q(t)\right] = -\lambda_0 Q(t). \qquad (4.40)$$

and we have used a zero subscript on the rate to indicate that it is a constant with units given by $1/time^\alpha$. The units of the relaxation rate are chosen for dimensional consistency. Subsequently we introduce a scaling parameter to make the time dimensionless, which is essentially the same thing as choosing these units for the relaxation constant. Using the Laplace transform of the Caputo fractional derivative we can replace Eq.(4.40) with

$$s^\alpha \widehat{Q}(s) - s^{\alpha-1} Q(0) = -\lambda_0 \widehat{Q}(s),$$

which after some rearrangement yields

$$\widehat{Q}(s) = \frac{s^{\alpha-1}}{s^\alpha + \lambda_0} Q(0). \qquad (4.41)$$

The time-dependent solution to Eq.(4.40) is obtained by inverse Laplace transforming Eq.(4.41).

The solution to the fractional rate equation Eq.(4.40) was first obtained by the mathematician Mittag-Leffler [43] at the opening of the twentieth century:

$$Q(t) = Q(0) E_\alpha \left(-\lambda_0 t^\alpha\right), \qquad (4.42)$$

in terms of the infinite series that now bears his name

$$E_\alpha \left(-\lambda_0 t^\alpha\right) = \sum_{k=0}^\infty \frac{(-\lambda_0 t^\alpha)^k}{\Gamma(k\alpha + 1)}. \qquad (4.43)$$

Note that with the units chosen for the relaxation rate the argument of the Mittag-Leffler function (MLF) is dimensionless. It is clear that the exponential simplicity of radioactive decay is here replaced by a more complex decay process, but the exponential simplicity is regained when $\alpha = 1$. We subsequently have a long and compelling story to tell about the MLF.

4.3.2 Viscoelastic Material Experiments

The time dependence of the MLF is depicted in Figure 4.2. At early times the MLF has the analytic form of a stretched exponential obtained as an approximation to the lowest-order terms in the series expression

$$\lim_{t \to 0} E_\alpha \left(-\lambda_0 t^\alpha \right) = 1 - \frac{\lambda_0 t^\alpha}{\Gamma \left(1 + \alpha \right)} + \cdots = \exp \left[-\frac{\lambda_0}{\Gamma \left(1 + \alpha \right)} t^\alpha \right]. \qquad (4.44)$$

The quality of this approximation is indicated by the long-dashed line segment in the figure, which deviates from the exact MLF solution only at long times. In rheology the dashed curve is the Kohlrausch-Williams-Watts law for stress relaxation.

Asymptotically in time the MLF yields an IPL

$$\lim_{t \to \infty} E_\alpha \left(-\lambda_0 t^\alpha \right) = \frac{1}{\lambda_0 \Gamma \left(1 - \alpha \right) t^\alpha}, \qquad (4.45)$$

as shown by the short-dashed line segment in Figure 4.2, which deviates from the exact MLF solution at early times. The short-dashed line corresponds to the Nutting law of stress relaxation.

The relation of the fractional relaxation equation and its solution, the MLF, to these two empirical laws for stress relaxation was explored by Glöckle and Nonnenmacher [19]. Note that the MLF smoothly joins these two empirical regions with a single analytic function. This fitting of the MLF parameters to data and the excellent fit across the entire region suggests that analytic functions are still useful for modeling complex physical phenomena, they just do not arise as solutions to the more familiar ordinary differential equations.

Of course this fractional generalization of the stress relaxation equations is phenomenological rather than fundamental. The empirical law that stress is proportional to strain for solids was provided by Hooke. For fluids Newton proposed that stress is proportional to the first derivative of strain. We emphasize that these two forms of stress relaxation are also empirical and not given by fundamental theory. Scott Blair *et al.* [61] suggested that a material with properties intermediate to that of a solid and a fluid, for example, a polymer, should be modeled by an integral over the strain. This was not an unreasonable assumption given that the integral equation they suggested was equivalent to a fractional derivative with an index between that of a constant $\alpha = 0$ and an ordinary derivative $\alpha = 1$.

Their conjecture is vindicated in Figure 4.2 where data from stress relaxation experiments using polyisobutylene are shown to be well fit by the MLF. Glöckle and Nonnenmacher [18] have also compared the theoretical results with experimental data sets obtained by stress-strain experiments carried out on natural rubber and have found agreement over more than 10 orders of magnitude. They have also successfully modeled self-similar protein dynamics in Myoglobin [20] and to a formulation of slow diffusion processes in biological tissue [30].

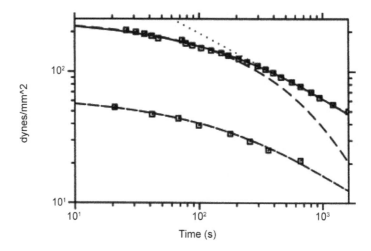

Figure 4.2 Stress relaxation at constant strain for two different initial conditions are indicated by the discrete data points. Upper: The solid curve is the MLF. The short-dashed curve is the inverse power law and the long-dashed curve is the stretched exponential. Lower: The dashed curve is the MLF and the boxes are data. The two fits use the MLF with different parameter values. The figure is from Glöckle and Nonnenmacher [18] with permission.

The asymptotic form of the MLF is an IPL as noted above. This asymptotic behavior suggests that perhaps the data sets that have been modeled strictly in terms of IPLs may in fact be more faithfully modeled using a MLF when examined more carefully. We explore this more fully in subsequent discussions involving PDF's.

The MLF is a direct generalization of the exponential function since the MLF series

$$E_\alpha\left(z\right) = \sum_{k=0}^{\infty}\frac{z^k}{\Gamma\left(k\alpha + 1\right)}, \quad \mathrm{Re}\left(\alpha > 0\right); \ \alpha, z \in \mathcal{C}, \qquad (4.46)$$

with $\mathcal{C}$ being the set of complex numbers, becomes the series for the exponential function when $\alpha = 1$. A straight forward generalization of the MLF carries a second index

$$E_{\alpha\beta}\left(z\right) = \sum_{k=0}^{\infty}\frac{z^k}{\Gamma\left(k\alpha + \beta\right)}, \quad \mathrm{Re}\left(\alpha > 0\right), \mathrm{Re}\left(\beta > 0\right); \ \alpha, \beta, z \in \mathcal{C}. \qquad (4.47)$$

These functions first arose in Mittag-Leffler's study of the summation of certain divergent series. Haubold *et al.* [22] provide a unified and detailed, if brief, account of the many guises of the MLF and their properties.

4.3.3 Fractional-Order Bloch Equations

The dynamic description of nuclear spins precessing in a static external magnetic field B_0 is given by the phenomenological Bloch equations for the magnetization vector $\mathbf{M} = (M_x, M_y, M_z)$ and the direction of orientation of the static magnetic field defines the z direction. Introducing the complex field $M_\pm = M_x \pm iM_y$ enables us to write the magnetization rate equations

$$\frac{dM_\pm(t)}{dt} = \left(\mp i\omega_0 - \frac{1}{T_2}\right) M_\pm(t),\tag{4.48}$$

$$\frac{dM_z(t)}{dt} = -\frac{1}{T_1}(M_z(t) - M_0),\tag{4.49}$$

with M_0 the equilibrium magnetization, $\omega_0 = \gamma B_0$ is the Larmor resonance frequency, with $\gamma/2\pi$ the gyromagnetic ratio for spin $1/2$ protons, T_1 the spin-lattice relaxation time and T_2 the spin-spin relaxation time. These equations are, as pointed out by Baleanu *et al.* [3], among many other things, the basis for image reconstruction and tissue contrast in MRI. Magin *et al.* [37] emphasize that there are a number of ways to generalize the Bloch equations to fractional differential form and the choice should be made based on which is the best suited for a given experimental situation. After a brief discussion concerning the interpretation of fractional derivatives in this context they opt to leave the interpretation open. We attempt to rectify this subsequently by associating the fractional derivative with memory in the material.

The formal generalization of the Bloch equations to fractional differential form is done using the Caputo derivative. The fractional equation for the rotating magnetization field components are

$$\partial_t^\alpha [M_\pm(t)] = \left(\mp i\omega_0' - \frac{1}{T_2'}\right) M_\pm(t),\tag{4.50}$$

and for the $z-$component of the magnetization

$$\partial_t^\alpha [M_z(t)] = -\frac{1}{T_1'}(M_z(t) - M_0),\tag{4.51}$$

where we have introduced the time scales τ_1 and τ_2 such that the primed parameters are related to the traditional parameters by

$$T_1' = T_1\tau_1^{\alpha-1}; \quad T_2' = T_2\tau_2^{\alpha-1}; \quad \omega_0' = \omega_0\tau_2^{1-\alpha}.\tag{4.52}$$

The primed parameters have been adjusted such that ω_0', $1/T_1'$ and $1/T_2'$ all have the units $(\text{sec})^{-\alpha}$, thereby retaining the appropriate units for Eqs.(4.50) and (4.51). Physically these scaling parameters are selected to view particular time time scales, nano-, micro- or millisecond; such that an experimenter can view a range of decay analogous to setting the time scale on an oscilloscope to view the decay in an RC circuit.

The solutions to the fractional-order Bloch equations are obtained using the methods introduced in Chapter 5 for solving linear fractional differential equations:

$$M_z(t) = M_z(0) E_\alpha\left(-t^\alpha/T_1'\right) + \frac{M_0}{T_1'} t^\alpha E_{\alpha,\alpha+1}\left(-t^\alpha/T_1'\right), \qquad (4.53)$$

involving the two-parameter MLF. Note that the solution to the linear fractional rate equation Eq.(4.53) was obtained in the previous section. The x and y components of the magnetization field are expressed in terms of

$$M_\pm(t) = M_\pm(0) E_\alpha\left(-\lambda_\pm t^\alpha\right); \quad \lambda_\pm = \frac{1}{T_2'} \pm i\omega_0', \qquad (4.54)$$

involving the MLF with a complex argument. A detailed discussion of these solutions is given by Magin *et al.* [37], which we do not reproduce here, but it is interesting to show their three-dimensional plot of the solution for one set of parameter values in Figure 4.3.

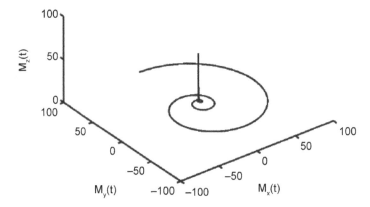

Figure 4.3 A plot of fractional order solution to the Bloch equations with $\alpha = 0.9$ (fractional model). For this plot, Eqs.(4.53) and (4.54) were used with $M_x(0) = 0$, $M_y(0) = 100$, $M_z(0) = 0$, $T_1' = 1$ (sec)$^\alpha$, $T_2' = 20$ (ms)$^\alpha$, and $f = 160$ Hz. (with permission of [37])

In addition to the fractional order two more parameters have been introduced into the above discussion, the times τ_1 and τ_2. As Magin *et al.* [37] state, the next step in the research is to determine reasonable values for these parameters. They go one to observe that such parameterization in viscoelastic materials, that undergo similar transient behavior, suggest that the relaxation parameters only differ from their classical values in special situations where the spins undergo multiscale relaxation.

It is left as an exercise for the student to show that the above solutions coincide with the classical ones when $\alpha = 1$. These latter solutions are depicted in three dimensions in Figure 4.4 and should be compared and contrasted with the fractional solutions in Figure 4.3 being guided by the discussion of the fractional harmonic oscillator. We return to this process in Chapter 6 where fractional diffusion is introduced in addition to fractional time.

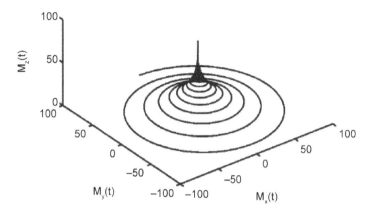

Figure 4.4 A plot of fractional order solution to the Bloch equations with $\alpha = 1.0$ (fractional model). For this plot, Eqs.(4.53) and (4.54) were used with $M_x(0) = 0$, $M_y(0) = 100$, $M_z(0) = 0$, $T_1' = 1$ (sec)$^\alpha$, $T_2' = 20$ (ms)$^\alpha$, and $f = 160$ Hz. (with permission of [37])

4.3.4 Three-scale Brownian Motion

As discussed previously one of the first successful phenomenological treatment of the dynamics of complex physical systems was given by Langevin [31]. In the stochastic differential equation of Langevin the time scale separation between the microscopic and macroscopic worlds enabled smoothing over the microscopic degrees of freedom, resulting in a differential equation of motion of average quantities over macroscopic time scales, with an additive random force. This is how the force equation for a heavy particle in a fluid of lighter particles is usually constructed to obtain the rate equation for a Brownian particle, one version of which was presented earlier.

In physics the concept of randomness and its formal explanation in terms of the central limit theorem are often introduced through the elegant phenomenon of molecular diffusion. This physical process describes uncertainty in the position of the diffusing particle that is buffeted by the lighter particles of the fluid in which it is embedded. Einstein's theory of diffusion, also known as Brownian motion, has a limited domain of validity. It relates the mean-square displacement (MSD) of a freely moving Brownian particle to the time t and the diffusion coefficient D by

$$\left\langle [Q(t) - \langle Q(t)\rangle]^2 \right\rangle = 2Dt. \tag{4.55}$$

This expression for the MSD is well known to be unphysical at early times where the average mean square velocity

$$\frac{\sqrt{\left\langle [Q(t) - \langle Q(t) \rangle]^2 \right\rangle}}{t} = \sqrt{\frac{2D}{t}} \qquad (4.56)$$

diverges as $t \to 0$. Consequently, the velocity of the Brownian particle becomes ill-defined for times shorter than the characteristic time scale for diffusion $\tau_D = M/\gamma$ for a particle of mass M and the Stoke's friction coefficient γ. Pussey [51] points out that Einstein concluded from his estimates of scale sizes that only the larger-scale diffusive random walks could be observed in practice.

On the other hand, Langevin's theory of Brownian motion is ostensibly valid for all times and on the time scale $t \ll \tau_D$ predicts ballistic motion. Consequently the MSD is not dependent on diffusion parameters and is given directly in terms of the temperature

$$\left\langle [Q(t) - \langle Q(t) \rangle]^2 \right\rangle = \frac{k_B T}{M} t^2 \qquad (4.57)$$

and the average velocity is

$$\frac{\sqrt{\left\langle [Q(t) - \langle Q(t) \rangle]^2 \right\rangle}}{t} = \sqrt{\frac{k_B T}{M}} \qquad (4.58)$$

a well-defined constant. At the turn of the twentieth century there was no way to measure this ballistic behavior and this partitioning of effects for Brownian motion fell into relative obscurity. However there were a few intrepid investigators that did return to these questions in the intervening years and finally technology enabled measurement.

Li *et al.* [33] were able to achieve the high resolution in space and time necessary to measure the ballistic regime of a Brownian particle using optical trapping interferometry. They determined the position and velocity of a $3\mu m$ diameter silicon sphere trapped in air. The optical trap was configured in a vacuum chamber and the experiment done at two distinct pressures. Without going into the details of the experiment it is sufficient for our purposes to note that the optical trap harmonically confines the particle in physical space, where it is subject to thermal collisions with the air particles in the chamber. It is evident from Figure 4.5 that the measured MSD deviates markedly from Einstein's theory of Brownian motion in the diffusive regime.

By way of contrast, the Langevin theory is expected to give the proper ballistic behavior. The solution to the Langevin equation for a harmonic oscillator driven by random noise was first obtained in 1930 by Uhlenbeck and Ornstein [64] and can be obtained using the theory given in Chapter 2 with $V(Q) = \frac{1}{2}\omega_0^2 Q^2$. The resulting MSD is given by [65]

$$\left\langle [Q(t) - \langle Q(t)\rangle]^2 \right\rangle = \frac{2k_B T}{M\omega_0^2} \left\{ 1 - e^{-t/2\tau_D} \left[\cos\omega_1 t + \frac{\sin\omega_1 t}{2\omega_1 \tau_D} \right] \right\} \qquad (4.59)$$

where the strength of the fluctuations driving the Langevin equation are proportional to the temperature of the ambient air and not the diffusion coefficient. The resonant frequency of the optical trap is ω_0 and $\omega_1 = \sqrt{\omega_0^2 - \left(\frac{1}{2\tau_D}\right)^2}$. This solution with the parameter values determined in the experiment is given by the solid curve in Figure 4.5 where it is evident that the MSD data essentially falls on the solution curve given by Eq.(4.59). The difference between the measured MSD and that predicted by Einstein's theory of diffusive motion is evident. The early time ballistic behavior predicted by the Langevin theory is therefore apparently vindicated.

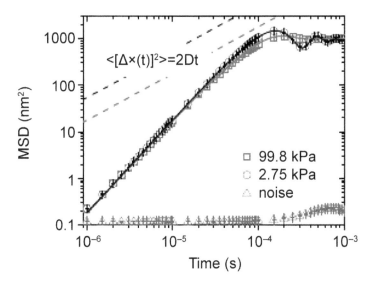

Figure 4.5 The MSDs of a 3 μm silica bead trapped in air at 99.98 kPa (red squares) and 2.75 kPa (black circle). They are calculated from 4×10^7 measurements for each pressure. The "noise" signal (blue triangles) is recorded when there is no particle in the optical trap. The solid lines are the theoretical predictions of Eq.(4.59). The prediction of Eistein's theory of free Brownian motion in the diffusive regime is shown in dashed lines for comparison. (from [33] with permission)

Like the other stories related herein this one does not end here, but continues as the experimental conditions are expanded. Even though Langevin's theory of diffusion appears to be exact over all time scales the validating experiments were done in air and not in water. It turns out that the ambient fluid makes a tremendous difference in the applicable theory, since the time scales for momentum relaxation in the gas and liquid phases differ by a factor of 50 due to the difference in the ambient fluid mass density. We have discussed two regimes, the microscopic and the macroscopic, but there is an intermediate regime, the mesoscopic, that can be important when the ambient fluid is liquid. The dynamics in the mesoscopic domain for a freely moving particle in a viscous liquid are described by an integro-differential equation. The force equation used by Langevin to describe Brownian motion is not given by the direct application of Newton's Third Law to a spherical particle in water. It is more subtle than that and requires that we take into account the inertia of the ambient fluid. The derivation of the equation for the motion of a heavy spherical particle in a fluid, taking into account the backflow of the ambient fluid around the Brownian particle, was first derived in 1885 by Boussinesq [5] and independently three years later by Basset [4].

For a spherical particle of radius R in a fluid with a viscosity η the force law is given by [11]

$$M\frac{dV(t)}{dt} = -6\pi\eta R V(t) - \frac{1}{2}M_0\frac{dV(t)}{dt} - U'(Q) + f(t)$$

$$-6R^2\sqrt{\pi\rho\eta}\int_0^t \frac{d\tau}{\sqrt{t-\tau}}\frac{dV(\tau)}{d\tau}, \tag{4.60}$$

where the first term on the right hand side of Eq.(4.60) is the ordinary Stokes' friction with coefficient $\gamma = 6\pi\eta R/M$, the second term is connected with the virtual mass of the sphere in an incompressible fluid, the third is a mechanical force modeled by the potential function $U(Q)$, the fourth is the random force generated by the ambient fluid, and the final term is the memory associated with the hydrodynamic retardation effects. Clercx and Schram [11] solve this equation using Laplace transforms and the time-dependent solution fits experimental data over the entire time domain [25].

A recent discussion of the Brownian particle's dynamics including the Basset force, that is, a retarded viscous force, was given by Mainardi and Pironi [38], which we express here in terms of the Caputo fractional derivative:

$$M\frac{dV(t)}{dt} + \lambda_0\partial_t^{1/2}[V(t)] = -\gamma_0 V(t) - U'(Q) + f(t), \tag{4.61}$$

where the parameters λ_0 and γ_0 are known functions of the fluid viscosity coefficient, the particle masses, and the radius of the Brownian particle. We refer the reader to [38] for details. The optical trap previously used by the scientists in Raizen's lab to measure the instantaneous velocity of a harmonically bound particle in air [33] and a bead in fluid [25] was again employed to test the theoretical predictions and interpretations of the Langevin equation modified to include the Basset force. The fit of theory [11] to experiment is excellent [29].

An earlier solution to this fractional Langevin equation was analyzed in terms of the generalized Langevin equation with fractional noise by Case [9].

He, and independently Widom [70], determined the asymptotic behavior of the velocity autocorrelation function. They showed that it decreases as an IPL in time

$$C_v(t) \propto t^{-3/2}. \tag{4.62}$$

Their explanation of this long-time tail was motivated by the observation in computer simulations of velocity correlations by Alder and Wainwright [1] in the previous year. The detailed hydrodynamic solution to this problem in the Laplace domain was given by Hinch [23] and eventually the exact solution was given in terms of MLFs by Mainardi and Pironi [38] using the fractional calculus.

The fractional derivative in Eq.(4.61) is the result of the back reaction onto the spherical particle by the ambient fluid flowing around it, resulting in the retarded viscous force. Of course, the solution to this equation is dominated by Stokes' dissipation due to viscosity at long times and the Brownian particle being 'heavy', which accounts for the success of the usual description of Brownian motion, without the inclusion of the fractional derivative term. However when the background fluid is not homogeneous or the Brownian particle is not 'heavy' the derivation of the rate equations needs to be re-examined.

Leptos *et al.* [32] conducted experiments on the motion of Brownian particles (tracers) suspended in a fluid of swimming Eukaryotic microorganisms of varying concentrations. The interplay between the inanimate tracer particles and the advection by flows from the swimming microorganisms results in their displacement having a self-similar PDF with a Gaussian core and exponential tails. Eckhardt and Zammert [12] re-analyzed these data and obtained an excellent fit to a MLF PDF based on the CTRW model.

A theoretical study of a simplified tracer-swimmer interaction by Zaid *et al.* [71] show that the non-Gaussian effect of the tails of the PDF arise from a combination of truncated Lévy statistics for the velocity field and the IPL decay of correlations in the ambient fluid. They further show that the dynamics of the PDF leading to the truncated Lévy statistics is given by a fractional diffusion equation, which we discuss subsequently.

It is evident that rigorous modeling of Brownian motion in heterogeneous fluids such as microbial suspensions in marine ecologies would potentially benefit from applications of the fractional calculus.

4.4 After Thoughts

In this chapter complexity, as understood in the late nineteenth and early twentieth centuries, is captured by the RRW and the resulting Gaussian statistics. This simple random walk picture was generalized to include long-time memory and spatial inhomogeneity. The effect of memory is modeled by means of a fractional difference equation that in the continuous limit yields a fractional time derivative resulting in the fractional Langevin equation. Spatial inhomogeneity and temporal anisotropy could be included in the dynamics through the CTRW introduced by Montroll and Weiss [44]. Thus, complexity is modeled by the kinetic equations for the PDF with IPL transition probabilities in both space and time, resulting in fractional diffusion equations. The stage is now set for their subsequent introduction and discussion as done in Chapter 7.

Simple fractional derivatives of algebraic functions were introduced and shown to give the same results as the integral expressions for both fractional derivatives and integrals in the appropriate limits. This was done, at least in part, to convince the reader that the intuition developed over years of applying the ordinary calculus to the interpretation of dynamic phenomena is still of value. That intuition continues to be useful in the applications of the fractional calculus. One of the most significant results presented in this chapter is the realization that fractal functions, whose integer derivatives diverge, do not have ordinary equations of motion. However, fractional derivatives of fractal functions can converge as in the case of the GWF, suggesting that the equations of motion for fractal phenomena are provided by the fractional calculus. We shall discuss this more fully in subsequent chapters.

The solution to the fractional rate equation was shown to be given by the MLF, which as mentioned is a generalization of the exponential function. Hook's law for simple materials assumes that stress is proportional to strain. However when the material under stress is neither solid nor liquid the relation between stress and strain can be a fractional differential equation, whose solution is a MLF. This analytic function, which may be new to the reader, smoothly joins the empirical stretched exponential form of stress relaxation at early times with the empirical IPL at late times. Consequently, complex materials have fractional equations of motion, as does the more familiar Brownian motion, when done properly and not when considered in the continuum limit of a RRW. Even in Brownian motion, memory and spatial inhomogeneity entail fractional derivatives in the equations of motion.

We mentioned that one of the peculiarities of fractional calculus was the fact that the RL fractional derivative of a constant is not zero. Caputo subsequently introduced a definition of a fractional derivative where this was no longer the case, that is, the Caputo derivative of a constant is zero. However his retrieval of the familiar property of derivatives required that the function be differentiable to at least the next integer order above the fractional one. Consequently, the Caputo derivative of order $\alpha < 1$ is only defined for functions that have first order derivatives. Almeida and Torres [2] point out that a function for which the Caputo derivative is not defined can still have a continuum of RL fractional derivatives with index less than one [55]. Consequently, the Caputo fractional derivative cannot be used to describe the dynamics of a non-differentiable process. Such truly complex processes require something more.

A simple alteration of the RL fractional derivative, which can be used to describe the dynamics of a non-differentiable process, was introduced by Guy Jumarie, who argued: "...the very reason for introducing and using fractional derivatives is to deal with non-differentiable functions" [27]. He suggested the RL fractional derivative be modified to include the initial value:

$$\mathcal{D}_t^\alpha [Q(t)] \equiv \frac{1}{\Gamma(1-\alpha)} \frac{d}{dt} \int_0^t \frac{dt'}{(t-t')^\alpha} [Q(t') - Q(0)] \quad, 0 < \alpha \leq 1, \quad (4.63)$$

which has the virtues of both the Caputo and RL fractional derivatives, without the counterintuitive limitations. The modified RL (MRL) fractional derivative is defined for arbitrary non-differentiable functions (there are no requirements on the derivatives of the function) and when applied to a constant yields zero. Moreover, the Laplace transform of the MRL derivative of a function coincide with that of the Caputo derivative.

It is reasonable to demonstrate that the MRL fractional derivative reproduces the results presented in this chapter. To do this we follow Jumarie [27] and consider the FRE for $\alpha < 1$:

$$\mathcal{D}_t^\alpha [Q(t)] = -\lambda_0 Q(t), \quad (4.64)$$

which using the definition of the MRL fractional derivative is

$$\frac{1}{\Gamma(1-\alpha)} \frac{d}{dt} \int_0^t \frac{dt'}{(t-t')^\alpha} [Q(t') - Q(0)] = -\lambda_0 Q(t). \quad (4.65)$$

Introducing the new variable $z = t'/t$ into the equation and integrating over time allows us to write

$$t^{1-\alpha} \int_0^1 \frac{dz}{(1-z)^\alpha} [Q(zt) - Q(0)] = -\lambda_0 \Gamma(1-\alpha) t \int_0^1 dz Q(zt). \quad (4.66)$$

We assume a solution of the form

$$Q(t) = \sum_{k=0}^\infty Q_k t^{k\alpha}, \quad (4.67)$$

as suggested by the factor t^α multiplying the integral. Inserting this series into Eq.(4.66), integrating and equating coefficients of equal powers of the time yields,

$$Q_{k+1} = -\lambda_0 \frac{\Gamma(k\alpha + 1)}{\Gamma(k\alpha + \alpha + 1)} Q_k,$$

which when iterated to the initial value $k = 0$ provides the expansion coefficients in terms of the initial value

$$Q_k = \frac{(-\lambda_0)^k}{\Gamma(k\alpha + 1)} Q_0.$$

Inserting this last expression into Eq.(4.67) yields

$$Q(t) = \sum_{k=0}^{\infty} \frac{(-\lambda_0)^k}{\Gamma(k\alpha+1)} t^{k\alpha} Q_0$$
$$= E_\alpha(-\lambda_0 t^\alpha) Q(0),$$

the series expression for the MLF.

The Caputo fractional derivative assumes a function is differentiable, whereas the MRL fractional derivative does not. Since the Laplace transform of the two operators is the same, the algebra for the solution to the equations of motion are the same, however the assumptions about the properties of the underlying phenomena are vastly different. This difference manifests itself in additional properties of the two operators and we shall focus on those different properties subsequently.

References

[1] Alder, B.J. and T.E. Wainwright. 1970. *Phys. Rev. A* **1**, 18–21.

[2] Almeida, R. and D.F.M. Torres. 2011. *Computers and Math. with Appl.* **61**, 3097–3104.

[3] Baleanu, D., R. Magin, S. Bhalekar and V. Daftardar-Gejji. 2015. *Com. Non. Sci.& Num. Sim.* **25**, 41–49.

[4] Basset, A.B. 1888. *A Treatise on Hydrodynamics, Vol.* **2**, Chapt. 22, pp. 285–297, Deighton Bell, Cambridge, MA.

[5] Boussinesq, K. 1885. *C.R. Acad. Sci. Paris* **100**, 935.

[6] Brokmann, X., J.P. Hermiere, G. Messin, P. Desbiolles, J.P. Bouchaud and M. Dahan. 2003. *Phys. Rev. Lett.* **90**, 120601-1.

[7] Caputo, M. 2001. *Fract. Calc. Appl. Anal.* **4**, 421.

[8] Carpinteri, A., B. Chiaia and P. Cornetti. 2000. *Rend. Sem. Mat. Univ. Pol. Torino* **58**, 57–68.

[9] Case, K.M. 1971. *Phys. Fluids* **14**, 2091–2095.

[10] Chvosta, P. and P. Reineker. 1997. *J. Phys. A* **30**, L307.

[11] Clercx, H.J.H. and P.P.J.M. Schram. 1992. *Phys. Rev. A* **46**, 1942.

[12] Eckhardt, B. and S. Zammert. 2012. *Eur. Phys. J. E* **35**, 96.

[13] Einstein, A. 1905. *Ann. Physik* **17**, 549.

[14] Feder, J. 1988. *Fractals*, Plenum Press, New York.

[15] Fraser, S.J. and R. Kapral. 1992. *Phys. Rev. A* **45**, 3412.

[16] Gamow, G. 1955. *One, Two, Three...Infinity*, The Viking Press, New York.

[17] Gil-Alana, L., O.S. Yaya and O.Il. Shittu. 2014. *Physica A* **396**, 42.

[18] Glöckle, W.G. and T.F. Nonnenmacher. 1991. *Macromolecules* **24**, 6426.

[19] Glöckle, W.G. and T.F. Nonnenmacher. 1993. *J. Stat. Phys.* **71**, 741.

[20] Glöckle, W.G. and T.F. Nonnenmacher. 1995. *Biophys. J.* **68**, 46.

[21] Haase, M., C.G. Hubneer, E. Reuther, A. Herrmann, K. Mullen and Th. Basche. 2004. *J. Phys. Chem. B* **108**. 10445.

[22] Haubold, H.J., A.M. Mathai and R.K. Saxema. 2011. *J. Appl. Math.* **2011**, 1–51.

[23] Hinch, E.J. 1975. *J. Fluid Mech.* **72**, 499.

[24] Hosking, J.T.M. 1981. *Biometrika* **68**, 165.

[25] Huang, R., I. Chavez, K.M. Taute, B. Lukic, S. Jeney, M.G. Raizen and E. Florin. 2011. *Nature Physics* **7**, 576.

[26] Ionescu, C.M., D. Copot and R. De Keyser. 2014. *19th World Cogress, The Int. Fed. Auto. Cont.*, Capetown, SA.

[27] Jumarie, G. 2006. *Computers and Math. with Appl.* **51**, 1367–1376.

[28] Jung, Y., E. Barkai and R.J. Silbey. 2002. *Chem. Phys.* **284**, 181.

[29] Kheifets, S., A. Simha, K. Melin, T. Li and M.G. Raizen. 2014. *Science* **343**, 1493.

[30] Köpf, M., R. Metzler, O. Haferkamp and T.F. Nonnenmacher. 1998. In *Fractals in Biology and Medicine*, Vol. II, Eds. G.A. Losa, D. Merlini, TF. Nonnenmacher and E.R. Weibel, Birkhäuser, Basel.

[31] Langevin, P. 1908. *Comptes Rendus Acad. Sci. Paris* **146**, 530.

[32] Leptos, K.C., J.S. Guasto, J.P. Gollub, A.I. Pesei and R.E. Goldstein. 2009. *Phys. Rev. Lett.* **103**, 198103.

[33] Li, T., S. Kheifets, D. Medellin and M.G. Raizen 2010, *Science* **328**, 1673.

[34] Lindenberg, K. and B.J. West. 1984. *J. Atmos. Sci.* **41**, 3021–3031.

[35] Lindenberg, K. and B.J. West. 1990. *The Nonequilibrium Statistical Mechanics of Open and Closed Systems*, VCH Publishers, New York.

[36] Magin, R.L. 2006. *Fractional Calculus in Bioengineering*, begell house inc., New York.

[37] Magin, R., X. Feng and D. Baleanu. 2009. *Magnetic Resonance A* **34** 16.

[38] Mainardi, F. and P. Pironi. 1996. *Extracta Mathematicae* **11**, 140.

[39] Mandelbrot, B.B. 1977. *Fractals; Form, Chance and Dimension*, W.H. Freeman & Co., San Francisco.

[40] Meakin, P. 1998. *Fractals, scaling and growth far from equilibrium*, Cambridge Nonlinear Science Series 5, Cambridge University Press, Cambridge, MA.

[41] Meerschaert, M. and A. Sikorskii. 2012. *Stochastic Models for Fractional Calculus*, DeGrugter. Berlin, Germany.

[42] Miller, K.S. and B. Ross. 1993. *An Introduction to the Fractional Calculus and Fractional Differential Equations*, John Wiley & Sons, New York.

[43] Mittag-Leffler, G.M. 1903. *C.R. Acad. Sci. Paris* **137**, 554.

[44] Montroll, E.W. and G. Weiss. 1965. *J. Math. Phys.* **6**, 167–181.

[45] Montroll, E.W. and B.J. West. 1979. In *Fluctuation Phenomena*, Eds. E.W. Montroll and J.L. Lebowitz, *Studies in Statistical Mechanics, Vol. VII*, North-Holland, Amsterdam; Second edition 1987.

[46] Murray, S.P. 1972. *Lumnology and Oceanography* **17**, 651.

[47] Newman, M.E.J. 2010. *Networks: An Introduction*, Oxford University Press, Oxford, New York.

[48] Nonnenmacher T.F. and R. Metzler. 1995. *Fractals* **3**, 557–566.

[49] Pearson, K. 1905. *Nature LXXII*, 294.

[50] Podlubny, I. 1999. *Fractional Differential Equations, Mathematics in Science and Engineering Vol.* **198**, Academic Press, San Diego.

[51] Pussey, P.N. 2011. *Science* **332**, 802.

[52] Rayleigh Lord 1880. *Philos. Mag. X*, 73–78.

[53] Rayleigh Lord 1905, *Nature* **72**, 318.

[54] Rocco, A. and B.J. West. 1999. *Physica A* **265**, 535.

[55] Ross, B., S.G. Samko and E.R. Love. 1994. *Real Anal. Exchange* **20**, 140–157.

[56] Samko, S.G., A.A. Kilbas and O.I. Marichev. 1993. *Fractional Integrals and Derivatives*, Gordon and Breach Science Pub., USA.

[57] Scafetta, N. and B.J. West. 2003. *Phys. Rev. Lett.* **90**, 248701.

[58] Scafetta, N. and B.J. West 2008, Physics Today 3, 50-51.

[59] Scafetta, N. 2013. *Solar Trends and Global Warming, Pattern Recognition in Physics* **1**, 37.

[60] Scafetta, N. 2014. *Physica A* **413**, 329.

[61] Scott Blair, S.G., B.C. Veinoglou and J.E. Caffyn. 1947. *Proc. Roy; Soc. Ser. A* **187**, 69.

[62] Seshadri, V. and B.J. West. 1982. *Proc. Nat. Acad. Sci. USA* **79**, 4501–4505.

[63] Stanley, H.E. 1979. *Introduction to Phase Transitions and Critical Phenomena*, Oxford University Press, Oxford, UK.

[64] Uhlenbeck, G.E and L.S. Ornstein. 1930. *Phys. Rev.* **36**, 823.

[65] Wang, M.C. and G.E. Uhlenbeck. 1945. *Rev. Mod. Phys.* **17**, 323.

[66] West, B.J., M. Bologna and P. Grigolini. 2003. *Physics of Fractal Operators*, Springer, Berlin.

[67] West, B.J. and P. Grigolini. 2008. *Phys. Rev. Lett.* **100**, 088501.

[68] West, B.J. and N. Scafetta. 2010. *Disrupted Networks: From Physics to Climate Change*, World Scientific, NJ.

[69] West, B.J. 2013. *Fractal Physiology and Chaos in Medicine*, 2nd Edition, World Scientific, NJ.

[70] Widom, A. 1971. *Phys. Rev. A* **3**, 1394–1396.

[71] Zaid, I.W., J. Dunkel and J.M. Yeomans. 2011. *J. R. Soc. Interface* **8**, 1314.

[72] Zygmund, A. 1977. *Trignometric Series, Vol. 1 & 2*, Cambridge University Press, Cambridge, UK; first published in 1935.

CHAPTER 5

Tomorrow's Dynamics

It is generally accepted that we completely understand the formal solutions to sets of coupled linear rate equations. However as with many things in science when we examine something that is well understood more closely we find unexpected problems. Therefore it should not come as a surprise that we encounter difficulties in determining the explicit properties of the formal matrix solution to the linear equations in all but the simplest cases. The existence of these problems are discussed in Section 5.1, because they also arise in obtaining explicit solutions to sets of coupled linear fractional rate equations. We mention this so as to distinguish between the problems associated with understanding the simplest dynamic equations in science and those difficulties uniquely related to complexity.

In this chapter we exploit some well-known methods for solving systems of linear differential equations for the purpose of solving systems of linear fractional differential equations. As in the classical case, the solution to the simple fractional differential equations provide the building blocks for more complex systems, such as the propagation of waves, the diffusion of information, and the relaxation of excitations, all in inhomogeneous fractal media. These extensions are discussed in due course.

The exponential matrix that provides the solution to a set of linear rate equations is replaced with a Mittag-Leffler matrix (MLM) that provides the solution to a set of linear fractional rate equations. Just as in the traditional case the arguments of the MLM, when expressed in a linear eigenfunction representation, are the linear eigenvalues. Aspects of the fractional harmonic oscillator are worked out in detail in Section 5.2.3 including the driven case. These examples demonstrate how the fractional case differs from the classical one and how in many ways it is the same, if viewed in the right way.

A discussion of Hamilton's equations generalized to the fractional case is postponed until Chapter 7.

Of course we do not limit the discussion in this chapter to formal or informal mathematical properties of dynamic equations. We also return to the generalized Langevin equation (GLE) and demonstrate that a fractional environment can model a kernel with an IPL decay of memory. Such an influence results in a fractional dissipation that still satisfies a fluctuation-dissipation relation. More importantly the assumed properties of the environmental fluctuations are shown to be determined experimentally using single-molecule fluorescence spectroscopy. The resulting dynamics of the autocorrelation function are shown to be a harmonic oscillator with fractional dissipation, a dynamic prototype that successfully models a number of phenomena, as we saw in the last chapter.

An application of the fractional Langevin equation (FLE) to a harmonically bound particle with fractional dissipation is used to explain the observation of chemical reactions on biologically relevant time scales. The random driving force is given by zero-centered FGn, whose autocorrelation function gives rise to the appropriate memory kernel in the FLE. The solution to the average of the FLE is given by a MLF that provides an excellent fit to experimental data.

Our interest is providing fractional calculus descriptions of complex phenomena is two fold. On the one hand, we want to understand the behavior of the phenomena and this is the science. On the other hand we also want to control the behavior of the phenomena and this is the engineering. Of course, some would argue that if we cannot control a process then we do not understand it. Consequently, we discuss how the traditional theory of filters needs to be modified to include allometric filters, those that are optimal for filtering time series generated by fractional dynamics. This opens the door to the study of fractional control processes that we can only briefly touch on.

The final methodology presented in this chapter puts together the fractional calculus and nonlinear dynamics. The analytic solution to the fractional logistic equation is obtained using an operator method and the results are shown to agree well with the numerical integration of the fractional nonlinear rate equation for values of the fractional index less than one.

5.1   What We Think We Know; Linear Systems

The formal solution to the initial value problem defined for the set of linear rate equations

$$\frac{d\mathbf{Q}(t)}{dt} = \mathbb{C}\mathbf{Q}(t), \tag{5.1}$$

subject to the initial condition $\mathbf{Q}(0)$ is

$$\mathbf{Q}(t) = e^{\mathbb{C}t}\mathbf{Q}(0). \tag{5.2}$$

Consider the dynamic observable $\mathbf{Q}(t)$ to be an $n-$component vector

$$\mathbf{Q}(t) = \begin{pmatrix} Q_1(t) \\ \cdot \\ \cdot \\ Q_n(t) \end{pmatrix} \tag{5.3}$$

and $\mathbb{C}$ is an $n \times n$ matrix of constant coupling coefficients. Appendix 5.8.1 contains a technique for explicitly constructing the exponential matrix given in the solution to the linear initial value problem. This technique is subsequently used to solve low-order sets of linear fractional differential equations.

Moler and Van Loan [34] observe that the exponential of a matrix, such as that in the solution of a set of linear rate equations, could be computed in many ways. The title of their paper on the subject provides an overview as to what they have to say: "Nineteen Dubious Ways to Compute the Exponential of a Matrix, Twenty-Five Years Later", which is the sequel to an earlier paper [27]. The adjective 'dubious' indicates that not any of the nineteen techniques they discuss is completely satisfactory in providing a practical way to explicitly calculate a matrix where n is no greater than a few hundred. The methods for calculating

$$e^{\mathbb{C}t} = 1 + \mathbb{C}t + \frac{\mathbb{C}^2 t^2}{2!} +, \tag{5.4}$$

are legion and they list the properties of various algorithms in their order of importance: generality, reliability, stability, accuracy, efficiency, storage requirements, ease of use and simplicity. They go on to say [34]:

> We would consider an algorithm completely satisfactory if it could be used as the basis for a general purpose subroutine which meets the standards of quality software now available for linear algebraic equations, matrix eigenvalues, and initial value problems for nonlinear ordinary differential equations. By these standards, none of the algorithms we know of are completely satisfactory, although some are much better than others.

However this is not a mathematics text so you may be wondering why I bring up these details regarding the practicality of evaluating a quantity that we believe we understand intuitively. Moreover, I have not even bothered to define terms like generality, reliability and so on, which I would be obliged to do if my purpose was to tell you how these issues might be resolved or at least circumvented, but that is not my intent. My sole ambition here is to sensitize you to the fact that there are a number of problems that arise in solving even linear rate equations that we also encounter in solving linear fractional rate equations. These are exactly the same problems of stability, reliability and so on and we point them out in the fractional context, because they are like old familiar faces spotted in an unfamiliar neighborhood and should be source of relief and not anxiety.

5.2   Fractional Linear System

Consider a linear dynamic system consisting of n elements that are interconnected. Here we are interested in fractional differential equations and write

$$\partial_t^\alpha \left[\mathbf{Q}(t) \right] = \mathbb{C}\mathbf{Q}(t) \tag{5.5}$$

where $\partial_t^\alpha \left[\cdot \right]$ is the Caputo or the MRL fractional derivative with $0 < \alpha \leq 1$ and t is a dimensionless parameter. The formal solution to this linear differential initial value problem is assumed to be the Mittag-Leffler matrix function (MLMF):

$$\mathbf{Q}(t) = \mathbf{E}_\alpha \left(\mathbb{C}t^\alpha \right) \mathbf{Q}(0), \tag{5.6}$$

which requires a definition in analogy with the exponential matrix used in the solution to a system of ordinary differential equations. A proof that Eq.(5.6) is the solution to the fractional rate equation Eq.(5.5) is presented in Appendix 5.8.2.

In the situation $\alpha = 1$ the Mittag-Leffler function reduces to an exponential as we have seen. In an analogous way the MLMF reduces to the matrix exponential

$$\lim_{\alpha \to 1} \mathbf{E}_\alpha \left(\mathbb{C}t^\alpha \right) = e^{\mathbb{C}t}. \tag{5.7}$$

This last equality suggests that we can generalize the method reviewed in Appendix 5.8.1 for the matrix exponential to obtain a closed form expression for the MLMF.

5.2.1   One-dimensional Case

We determined earlier that the solution to the simple fractional rate equation has a MLF solution, which was useful for describing viscoelastic effects in Section 4.3.2. Now consider a slightly more complicated fractional rate equation to describe the growth of a population to saturation. An equation in which a population $N(t)$ grows at a constant rate a, but eventually that constant birth rate is quenched by a linear death term

$$\partial_t^\alpha \left[N(t) \right] = a - bN(t). \tag{5.8}$$

It is worth pointing out, since we are discussing a measurable process, the population of a species, that the units of the variables are important. Consequently, introducing a fractional time derivative requires that we also introduce a time scale with units having the appropriate fractional power. If t in Eq.(5.8) has the units of [T], those of the parameters a and b must have units $[\text{T}]^{-\alpha}$ to retain the dimensional homogeneity discussed in Section 2.4. Moreover, the form of this equation is the same as Eq.(4.51) for the $z-$component of magnetization. Solutions for a variety of such linear problems will be found in the excellent text by Magin [28].

Laplace transform method of solution

The Laplace transform of Eq.(5.8) when $0 < \alpha < 1$, or when $1 < \alpha < 2$ and $\overset{\bullet}{N}(0) = 0$, yields

$$s^\alpha \widehat{N}(s) - s^{\alpha-1} N(0) = \frac{a}{s} - b\widehat{N}(s),$$

which after some algebra gives us

$$\widehat{N}(s) = \frac{s^{\alpha-1}}{s^\alpha + b} N(0) + \frac{a}{s} \frac{1}{s^\alpha + b}. \tag{5.9}$$

The inverse Laplace transform of the first term on the right in this equation gives the MLF, the second term is not so straight forward. However one can show that the inverse Laplace transform is a generalized MLF [37, 49]

$$N(t) = N(0) E_\alpha(-bt^\alpha) + at^\alpha E_{\alpha,\alpha+1}(-bt^\alpha). \tag{5.10}$$

Here the reason for requiring dimensional homogeneity in the fractional rate equation becomes apparent, since this is how the argument of the MLF, that being bt^α, remains dimensionless.

A more transparent form of the solution is obtained by rearranging terms in the series for the doubly indexed generalized MLF

$$E_{\alpha,\beta}(-bt^\alpha) = \sum_{k=0}^{\infty} \frac{(-bt^\alpha)^k}{\Gamma(k\alpha + \beta)}, \tag{5.11}$$

and shifting the index $j = k + 1$ we obtain

$$E_{\alpha,\alpha+1}(-bt^\alpha) = -\frac{1}{bt^\alpha} \left[E_\alpha(-bt^\alpha) - 1 \right].$$

When this last equation is inserted into Eq.(5.10) we have

$$N(t) = N(0) E_\alpha(-bt^\alpha) + \frac{a}{b} \left[1 - E_\alpha(-bt^\alpha) \right], \tag{5.12}$$

where the influence of the initial state clearly diminishes slowly in time, but eventually the equilibrium state is reached

$$N_{eq} = \frac{a}{b}. \tag{5.13}$$

The question is whether this equation describes the growth to saturation of a real population.

However, before we discuss the utility of the solution to Eq.(5.8) let us use an operator method to solve it. This method will later prove to be useful when the dynamics are not linear.

Operator method of solution

The solution to Eq.(5.8) using the operator formalism is deceptively simple; it is the MLF

$$N(t) = E_\alpha(a\mathcal{V}t^\alpha)N \tag{5.14}$$

in terms of the operators:

$$\mathcal{V} \equiv \mathcal{X} + \mathcal{Y}; \tag{5.15}$$

$$\mathcal{X} \equiv \frac{\partial}{\partial N}; \tag{5.16}$$

$$\mathcal{Y} \equiv -\frac{b}{a}N\frac{\partial}{\partial N}. \tag{5.17}$$

The formal solution Eq.(5.14) to the equation can also be expressed as

$$N(t) = \sum_{j=0}^{\infty} \frac{(at^\alpha)^j}{\Gamma(j\alpha+1)} (\mathcal{X} + \mathcal{Y})^j N. \tag{5.18}$$

However, the expansion in Eq.(5.18) requires a little thought to evaluate because the two operators satisfy the commutator relation

$$[\mathcal{X}, \mathcal{Y}] = \mathcal{X}\mathcal{Y} - \mathcal{Y}\mathcal{X} = \mathcal{Y} \tag{5.19}$$

Rather than writing out the general expression for the summand we calculate the terms in Eq.(5.18) one by one

$$\begin{aligned}
(\mathcal{X} + \mathcal{Y})^j N &= (\mathcal{X} + \mathcal{Y})^{j-1}\left(1 - \frac{b}{a}N\right) \\
&= (\mathcal{X} + \mathcal{Y})^{j-2}\left(-\frac{b}{a} + \left(\frac{b}{a}\right)^2 N\right) \\
&= (\mathcal{X} + \mathcal{Y})^{j-3}\left(\left(\frac{b}{a}\right)^2 - \frac{b}{a}\left(\frac{b}{a}\right)^2 N\right) \\
&= \left(-\frac{b}{a}\right)^j N + \left(-\frac{b}{a}\right)^{j-1}(\delta_{j,0} - 1),
\end{aligned}$$

which when inserted into Eq.(5.18) yields

$$N(t) = \sum_{j=0}^{\infty} \frac{(at^\alpha)^j}{\Gamma(j\alpha+1)}\left(-\frac{b}{a}\right)^j N + \sum_{j=1}^{\infty} \frac{(at^\alpha)^j}{\Gamma(j\alpha+1)}\left(-\frac{b}{a}\right)^{j-1}. \tag{5.20}$$

We have taken N to be the initial condition, so that relabeling the second sum we can write this expression as

$$N(t) = E_\alpha(-bt^\alpha)N(0) + at^\alpha E_{\alpha,\alpha+1}(-bt^\alpha)$$

in agreement with Eq.(5.10), obtained using the method of Laplace transforms.

What the solution means

First of all we note that we have not restricted the index for the fractional derivative in the fractional rate equation, except in the specification of the initial conditions. Consequently we can consider a range of values $0 < \alpha < 2$. In Figure 5.1 the growth to saturation starting from an arbitrarily small initial population is seen for multiple values of α. For $\alpha \leq 1$ the growth is monotonic from the initial value to saturation, without the inflection point typically seen in sigmoidal curves. For $\alpha > 1$ and the initial time derivative of the population set to zero, there is an overshoot of the equilibrium level followed by a relaxation to that level. A periodic relaxation of the population to saturation with increasing amplitude and frequency with increasing α is clearly observed. Note that at $\alpha = 2$ the solution becomes analogous to that of a driven harmonic oscillator.

This effect of overshoot in the growth of population is interesting, but is it observed? Searching on the internet I found the laboratory growth of a culture of Daphnia depicted in Figure 5.2. The traditional explanation is that the abundant food supply allows the population to overshoot the carrying capacity of ecological network. After the overshoot the population settles down to its steady-state value. The problem with this explanation is that it presupposes the existence of a logistic equation for population growth. I have not attempted to fit the overshoot and oscillation of the laboratory data to the fractional growth model, but it is clear that this may be an exemplar of the fractional growth depicted in Figure 5.1.

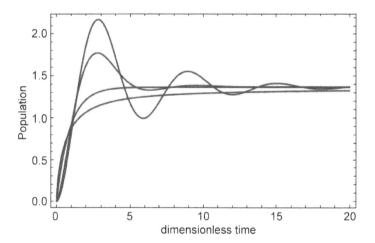

Figure 5.1 Eq.(5.12) with $N(0)$ very small, $a = 1.5, b = 1.1$ and from the top curve to the bottom $\alpha = 1.75, 1.5, 1.0, 0.75$. For $\alpha > 1$ the population overshoots the equilibrium value but relaxes back to it asymptotically.

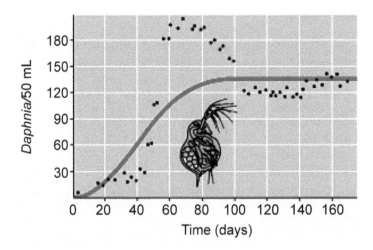

Figure 5.2 The population of Daphnia grown in the laboratory compared with the traditional sigmoidal growth curve. We thought that fitting the curve using Eq.(5.12) would be ill advised without an accompanying theoretical discussion. http://bio1152.nicerweb.com/Locked/media/ch53/growth.html

It very well be worth the time of an ambitious student to locate data sets for population growth that violate the traditional sigmoidal curve. An analysis of those data to determine whether the complexity suggested here by the fractional derivative can account for the discrepancy would be very interesting indeed. Understanding why data do not conform to traditional wisdom is the start of new wisdom.

The phenomena of overshoot and oscillatory relaxation to an equilibrium state has been observed in all manner of physical phenomena, including signal processing, control theory and electronics, but to see it in an ecological system is quite a different matter. The simple growth models that are usually used to describe the difference between birth and death processes, or the density dependence of the birth and death rates that are typically introduced do not incorporate the memory necessary to explain the overshoot effect. I have not attempted to fit the data with Eq.(5.12) since my intent is to provide motivation for this new way of thinking and not to explain any particular data set in detail.

5.2.2 2×2 Case

Consider the simplest coupled system consisting of two elements as done in detail by Kimeu [23] for a different definition of the fractional derivative. The fractional rate equation is of the form Eq.(5.5) with

$$\mathbf{Q}(t) = \begin{pmatrix} Q_1(t) \\ Q_2(t) \end{pmatrix}; \tag{5.21}$$

$$\mathbb{C} = \begin{pmatrix} a & b \\ c & d \end{pmatrix}, \tag{5.22}$$

with the solution given by Eq.(5.6). Now we explicitly construct the MLMF in the two variable case.

Consider the two eigenvalues λ_1 and λ_2 characterizing the 2×2 coupling matrix given by the characteristic polynomial for Eq.(5.22):

$$\lambda_1 = \frac{1}{2} \left[a + d - \sqrt{a^2 - 2ad + 4bc + d^2} \right], \tag{5.23}$$

$$\lambda_2 = \frac{1}{2} \left[a + d + \sqrt{a^2 - 2ad + 4bc + d^2} \right]. \tag{5.24}$$

The two eigenvalues are not necessarily distinct and depend on the values of the coupling elements.

From Appendix 5.8.2 we write the MLMF in terms of the components of an auxiliary vector as

$$\mathbf{E}_\alpha \left(\mathbb{C} t^\alpha \right) = q_1(t) \mathbf{M}_0 + q_2(t) \mathbf{M}_1, \tag{5.25}$$

with the 2×2 matrices defined as

$$\mathbf{M}_0 = \mathbf{I} = \begin{pmatrix} 1 & 0 \\ 0 & 1 \end{pmatrix} \text{ and } \mathbf{M}_1 = \mathbb{C} - \lambda_1 \mathbf{I} = \begin{pmatrix} a - \lambda_1 & b \\ c & d - \lambda_1 \end{pmatrix}.$$

The vector $\mathbf{q}(t)$ satisfies the auxiliary system of fractional differential equations

$$\partial_t^\alpha [\mathbf{q}(t)] = \begin{pmatrix} \lambda_1 & 0 \\ 1 & \lambda_2 \end{pmatrix} \begin{pmatrix} q_1(t) \\ q_2(t) \end{pmatrix} \tag{5.26}$$

subject to the initial condition

$$\mathbf{q}(0) = \begin{pmatrix} 1 \\ 0 \end{pmatrix}. \tag{5.27}$$

In this way we have defined the 2×2 fractional initial value problem.

General solution

Now consider the first component of the fractional differential equation of the auxiliary vector

$$\partial_t^\alpha [q_1(t)] = \lambda_1 q_1(t) \tag{5.28}$$

with the initial value $q_1(0) = 1$. The Laplace transform of Eq.(5.28), recalling that we are using Caputo or MRL fractional derivatives here, yields

$$s^\alpha \widehat{q}_1(s) - s^{\alpha-1} = \lambda_1 \widehat{q}_1(s)$$

which after some algebra yields

$$\widehat{q}_1(s) = \frac{s^{\alpha-1}}{s^\alpha - \lambda_1}. \tag{5.29}$$

The inverse Laplace transform of this equation is, of course, the MLF

$$q_1(t) = E_\alpha\left(\lambda_1 t^\alpha\right), \tag{5.30}$$

where again the eigenvalues have dimension $[T]^{-\alpha}$.

The second component of the fractional differential equation of the auxiliary vector

$$\partial_t^\alpha\left[q_2(t)\right] = q_1(t) + \lambda_2 q_2(t),$$

and inserting the solution we just obtained for the first component yields

$$\partial_t^\alpha\left[q_2(t)\right] = E_\alpha\left(\lambda_1 t^\alpha\right) + \lambda_2 q_2(t). \tag{5.31}$$

The Laplace transform of Eq.(5.31) yields

$$s^\alpha\widehat{q}_2(s) = \frac{s^{\alpha-1}}{s^\alpha - \lambda_1} + \lambda_2\widehat{q}_2(s),$$

where we have used $q_2(0) = 0$. The equation of which to take the Laplace inverse is

$$\widehat{q}_2(s) = \frac{s^{\alpha-1}}{(s^\alpha - \lambda_1)(s^\alpha - \lambda_2)},$$

which factors to

$$\widehat{q}_2(s) = \frac{s^{\alpha-1}}{\lambda_1 - \lambda_2}\left[\frac{1}{s^\alpha - \lambda_1} - \frac{1}{s^\alpha - \lambda_2}\right], \tag{5.32}$$

provided that the eigenvalues are not degenerate. The inverse Laplace transform of Eq.(5.32) yields for the second component of the auxiliary vector

$$q_2(t) = \frac{1}{\lambda_1 - \lambda_2}\left[E_\alpha(\lambda_1 t^\alpha) - E_\alpha(\lambda_2 t^\alpha)\right] \tag{5.33}$$

Combining the two components into the MLMF gives

$$\mathbf{E}_\alpha\left(\mathbb{C}t^\alpha\right) = \begin{pmatrix} E_\alpha\left(\lambda_1 t^\alpha\right) + (a - \lambda_1)\Delta_\alpha\left(t\right) & b\Delta_\alpha\left(t\right) \\ c\Delta_\alpha\left(t\right) & E_\alpha\left(\lambda_1 t^\alpha\right) + (d - \lambda_1)\Delta_\alpha\left(t\right) \end{pmatrix}, \tag{5.34}$$

and we have introduced the notation for the difference between MLFs

$$\Delta_\alpha\left(t\right) = \frac{1}{\lambda_1 - \lambda_2}\left[E_\alpha(\lambda_1 t^\alpha) - E_\alpha(\lambda_2 t^\alpha)\right]. \tag{5.35}$$

Thus, we can write the solution to the 2×2 initial value problem defined by the fractional differential equation Eq.(5.21) as

$$\mathbf{Q}(t) = \begin{pmatrix} E_\alpha\left(\lambda_1 t^\alpha\right) + (a - \lambda_1)\Delta_\alpha\left(t\right) & b\Delta_\alpha\left(t\right) \\ c\Delta_\alpha\left(t\right) & E_\alpha\left(\lambda_1 t^\alpha\right) + (d - \lambda_1)\Delta_\alpha\left(t\right) \end{pmatrix}\begin{pmatrix} Q_1(0) \\ Q_2(0) \end{pmatrix}.$$

The insight gained from this general solution is of limited value, so let us consider a simplification.

Equal coupling strengths

Consider the case $a = d = -\varepsilon$ and $b = c$, such that the eigenvalues are $\lambda_1 = -\varepsilon - b$ and $\lambda_2 = -\varepsilon + b$. The solution then reduces to

$$\mathbf{Q}(t) = \begin{pmatrix} C_{11}(t) & C_{12}(t) \\ C_{21}(t) & C_{22}(t) \end{pmatrix} \begin{pmatrix} Q_1(0) \\ Q_2(0) \end{pmatrix}. \tag{5.36}$$

where the coupling matrix elements are

$$C_{11}(t) = C_{22}(t) = \frac{1}{2} \{E_\alpha \left(-(\varepsilon + b) t^\alpha\right) + E_\alpha \left(-(\varepsilon - b) t^\alpha\right)\}, \tag{5.37}$$

$$C_{12}(t) = C_{21}(t) = -\frac{1}{2} \{E_\alpha \left(-(\varepsilon + b) t^\alpha\right) - E_\alpha \left(-(\varepsilon - b) t^\alpha\right)\}. \tag{5.38}$$

If the initial condition is chosen at random and the two components chosen independently, then the components of the coupling matrix can be determined from the average over an ensemble of realizations of initial conditions

$$\langle \mathbf{Q}(t)\mathbf{Q}^T(0) \rangle = \begin{pmatrix} \langle Q_1(t)Q_1(0) \rangle & \langle Q_1(t)Q_2(0) \rangle \\ \langle Q_2(t)Q_1(0) \rangle & \langle Q_2(t)Q_2(0) \rangle \end{pmatrix}. \tag{5.39}$$

Note that $Q_1(t)$ depends on both initial conditions as does $Q_2(t)$.

The diagonal matrix elements of Eq.(5.39) are the components of the autocorrelation function

$$C_{11}(t) \equiv \frac{\langle Q_1(t)Q_1(0) \rangle}{\langle Q_1(0)^2 \rangle}, \tag{5.40}$$

$$C_{22}(t) \equiv \frac{\langle Q_2(t)Q_2(0) \rangle}{\langle Q_2(0)^2 \rangle}. \tag{5.41}$$

Whereas the off-diagonal matrix elements of Eq.(5.39) are the components of the cross-correlation function

$$C_{12}(t) \equiv \frac{\langle Q_1(t)Q_2(0) \rangle}{\langle Q_2(0)^2 \rangle}, \tag{5.42}$$

$$C_{21}(t) \equiv \frac{\langle Q_2(t)Q_1(0) \rangle}{\langle Q_1(0)^2 \rangle}. \tag{5.43}$$

Consequently, the autocorrelation are positive for $\varepsilon > b$ and asymptotically decay as the IPL $t^{-\alpha}$, as do the cross-correlations.

Let us now consider the global variable given by the average over the two components

$$\xi(t) = \frac{1}{2} \sum_{j=1}^{2} Q_j(t), \tag{5.44}$$

which in terms of the solutions given in Eq.(5.36) is

$$\xi(t) = \frac{1}{2} [C_{11}(t) + C_{21}(t)] Q_1(0) + \frac{1}{2} [C_{12}(t) + C_{22}(t)] Q_2(0), \tag{5.45}$$

or inserting the correlation functions from Eqs.(5.37) and (5.38) yields

$$\xi(t) = E_\alpha\left(-(\varepsilon - b)t^\alpha\right)\left[\frac{Q_1(0) + Q_2(0)}{2}\right] = E_\alpha\left(-(\varepsilon - b)t^\alpha\right)\xi(0). \quad (5.46)$$

In this way we find the autocorrelation function for the global variable obtained by averaging over an ensemble distribution of initial conditions

$$\frac{\langle\xi(t)\,\xi(0)\rangle}{\langle\xi(0)^2\rangle} = E_\alpha\left(-(\varepsilon - b)t^\alpha\right), \quad (5.47)$$

which is not so different from that for the individual elements. The asymptotic autocorrelation function decays as an IPL in time $t^{-\alpha}$.

The integer derivative of a harmonic function is a harmonic function. This concept generalizes here to the fractional derivative of a MLF is another MLF. However the usual fractional derivative of a harmonic function is not exactly a harmonic function. Kaslik and Sivasundaram [21] prove that for the Caputo, Riemann-Liouville and Grunwald-Letnikov definitions of a fractional derivative there are no exact periodic solutions in a wide class of fractional-order dynamic systems. They emphasize that even in the case of a simple fractional-order neural network, where a limit cycle is apparently obtained in the numerical simulation, that the solution cannot be an exact periodic solution of the system of equations.

5.2.3 Harmonic Oscillator

The workhorse of theoretical physics has been the linear harmonic oscillator, since Newton first predicted the speed of sound in air over three hundred years ago. It is therefore reasonable to expect the fractional linear harmonic oscillator to play a similarly useful role in fractional kinetics. Here we introduce the fractional linear harmonic oscillator for discussion and consider its justification subsequently.

Scaling the time

The initial value problem for a fractional harmonic oscillator in one dimension is given by the two variable fractional differential equation

$$\theta^{-\alpha}\partial_t^\alpha\begin{pmatrix} P(t) \\ Q(t) \end{pmatrix} = \begin{pmatrix} 0 & -\omega_0^2 \\ 1 & 0 \end{pmatrix}\begin{pmatrix} P(t) \\ Q(t) \end{pmatrix}, \quad (5.48)$$

and we have introduced the parameter θ with the dimensions of $[T]^{1/\alpha - 1}$ in order to retain the standard physical definition of mass and frequency in the fractional harmonic oscillator equation. Introducing this parameter is equivalent to replacing the time variable with $\tau = \theta t$. In terms of the elements in the coupling matrix used in the last section we have $a = 0, b = -\omega_0^2, c = 1$

and $d = 0$, so that the eigenvalues of the system are the complex conjugate pair $+i\omega_0$ and $-i\omega_0$. The first component of the auxiliary vector given by Eq.(5.30) in terms of the complex MLF is

$$q_1(\tau) = E_\alpha \left(i\omega_0\tau^\alpha\right), \tag{5.49}$$

and the second component given by Eq.(5.33) is in terms of the MLF with complex arguments

$$q_2(\tau) = \frac{1}{2i\omega_0} \left[E_\alpha \left(i\omega_0\tau^\alpha\right) - E_\alpha \left(-i\omega_0\tau^\alpha\right)\right], \tag{5.50}$$

and $\omega_0\tau^\alpha = \omega_0\theta^\alpha t^\alpha$ is dimensionless. Inserting the two components from Eqs.(5.49) and (5.50) into Eq.(5.36) yields the solution to the initial value problem for the fractional harmonic oscillator in terms of generalized trigonometric functions

$$P(\tau) = P(0)\cos_\alpha\left(\omega_0\tau^\alpha\right) - \omega_0 Q(0)\sin_\alpha\left(\omega_0\tau^\alpha\right), \tag{5.51}$$

$$Q(\tau) = \frac{P(0)}{\omega_0}\sin_\alpha\left(\omega_0\tau^\alpha\right) + Q(0)\cos_\alpha\left(\omega_0\tau^\alpha\right). \tag{5.52}$$

We have introduced the notation for generalized trigonometric functions [49]

$$\sin_\alpha\left(\omega_0\tau^\alpha\right) = \frac{1}{2i} \left[E_\alpha \left(i\omega_0\tau^\alpha\right) - E_\alpha \left(-i\omega_0\tau^\alpha\right)\right], \tag{5.53}$$

and

$$\cos_\alpha\left(\omega_0\tau^\alpha\right) = \frac{1}{2} \left[E_\alpha \left(i\omega_0\tau^\alpha\right) + E_\alpha \left(-i\omega_0\tau^\alpha\right)\right]. \tag{5.54}$$

These functions reduce to the familiar sine and cosine when $\alpha = 1$, where the complex MLF becomes the complex exponential. Note that we also have the generalized Euler relations for the complex MLF

$$E_\alpha \left(\pm i\omega_0\tau^\alpha\right) = \cos_\alpha\left(\omega_0\tau^\alpha\right) \pm i\sin_\alpha\left(\omega_0\tau^\alpha\right). \tag{5.55}$$

It should be getting clear why we have made such a big deal out of the MLF. In fractional differential equations it plays the same fundamental role that the exponential plays in ordinary differential equations.

The solution for the oscillator displacement Eq.(5.52) can be used in conjunction with Eq.(5.163) given in Appendix 5.8.2 to obtain a "second-order" or $2\alpha-$ fractional derivative expression for the harmonic oscillator. Using

$$\partial_\tau^\alpha \left[E_\alpha \left(\pm i\omega_0\tau^\alpha\right)\right] = \pm i\omega_0 E_\alpha \left(\pm i\omega_0\tau^\alpha\right),$$

it is straight forward to obtain the fractional derivatives of the generalized trigonometric functions

$$\partial_\tau^\alpha \left[\sin_\alpha\left(\omega_0\tau^\alpha\right)\right] = \omega_0 \cos_\alpha\left(\omega_0\tau^\alpha\right), \tag{5.56}$$

and

$$\partial_\tau^\alpha \left[\cos_\alpha (\omega_0 \tau^\alpha)\right] = -\omega_0 \sin_\alpha (\omega_0 \tau^\alpha). \tag{5.57}$$

These relations along with the Caputo or MRL fractional derivative of the solution to the fractional harmonic oscillator equation enables us to write for the relation between the variables

$$\partial_\tau^\alpha \left[Q(\tau)\right] = P(\tau), \tag{5.58}$$

where the oscillator effective mass has been set to unity and the momentum is the fractional derivative of the oscillator displacement. Of course, this equation is part of the fractional harmonic oscillator given by Eq.(5.48), but consistency is reassuring. Equation (5.58) suggests that fractional kinetics might have a basis in a generalization of Hamiltonian dynamics, as we subsequently discuss. A second Caputo derivative of the displacement, along with the fractional derivative of the momentum, yields the $2\alpha-$fractional equation of motion for the harmonic oscillator

$$\partial_\tau^{2\alpha} \left[Q(\tau)\right] = -\omega_0^2 Q(\tau). \tag{5.59}$$

The same $2\alpha-$ fractional dynamic expression is obtained for the oscillator momentum in this case where the mass has been set to unity.

The solution to the fractional harmonic oscillator has been obtained by a number of investigators [43, 46]. In Figure 5.3 is depicted the time series for the fractional oscillator displacement. It is clear from the trajectory for the three values of the fractional derivative shown that the oscillator orbit decays to the origin in time suggesting that the fractional derivative induces a kind of 'dissipation' into the dynamics. Turchetti *et al.* [46] use a Caputo derivative and point out that as the trajectory approaches the asymptotic $(t \to \infty)$ stable equilibrium point it decays algebraically and not exponentially as would a physical dissipative system. Therefore the decay of the trajectory is due to a mechanism other than linear dissipation.

MLF with complex arguments

A curious point arises when solving the $2\alpha-$ fractional equation of motion for the fractional harmonic oscillator such as done by Stanislavsky [43]. As in the differential calculus the $2\alpha-$ equation Eq.(5.59) has two independent solutions for $1 \leq 2\alpha \leq 2$, which can be obtained from the relations

$$\partial_\tau^\alpha \left[E_{2\alpha} \left(-\omega_0^2 \tau^{2\alpha}\right)\right] = -\omega_0 \tau^\alpha E_{2\alpha,\alpha+1} \left(-\omega_0^2 \tau^{2\alpha}\right), \tag{5.60}$$

$$\partial_\tau^\alpha \left[\tau^\alpha E_{2\alpha,\alpha+1} \left(-\omega_0^2 \tau^{2\alpha}\right)\right] = \omega_0 E_{2\alpha} \left(-\omega_0^2 \tau^{2\alpha}\right), \tag{5.61}$$

where the two parameter MLF is defined in Eq.(5.11). Equations (5.60) and (5.61) taken together allows us to express the solution of the $2\alpha-$ equation as

$$Q(\tau) = Q(0)E_{2\alpha} \left(-\omega_0^2 \tau^{2\alpha}\right) + P(0)\tau^\alpha E_{2\alpha,\alpha+1} \left(-\omega_0^2 \tau^{2\alpha}\right), \tag{5.62}$$

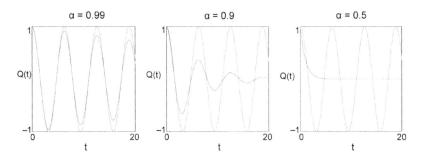

Figure 5.3 Time evolution of the displacement $Q(t)$ of the fractional harmonic oscillator in the interval $0 < t \leq 20$ for $\alpha = 0.99, 0.9, 0.5$ compared with $\alpha = 1$ for the ordinary time derivative. The solution is given by Eq.(5.52) with $\omega_0 \theta^\alpha = 1$ and is shown for four periods of oscillation (adapted from [46]).

and applying ∂_τ^α twice to $Q(\tau)$ in Eq.(5.62) yields Eq.(5.59). This solution looks very different from the simple form given by Eq.(5.52). So which one is correct?

To answer this question consider the series expansion for the generalized sine function

$$\sin_\alpha (\omega_0 \tau^\alpha) = \sum_{k=0}^\infty \frac{(\omega_0 \tau^\alpha)^k}{\Gamma (k\alpha + 1)} \frac{i^k - (-i)^k}{2i}. \tag{5.63}$$

In this series expansion we have

$$\frac{i^k - (-i)^k}{2i} = \sin (k\pi/2) = \begin{cases} 1 & \text{for } k \text{ odd} \\ 0 & \text{for } k \text{ even} \end{cases},$$

allowing us to change the summation index to include only odd terms $k = 2m + 1$ and obtain

$$\sin_\alpha (\omega_0 \tau^\alpha) = \sum_{m=0}^\infty \frac{\left(-\omega_0^2 \tau^{2\alpha}\right)^m \omega t^\alpha}{\Gamma ((2m + 1)\alpha + 1)} = \omega \tau^\alpha E_{2\alpha, \alpha+1} \left(-\omega^2 \tau^{2\alpha}\right). \tag{5.64}$$

The same procedure can also be applied to the generalized cosine function

$$\cos_\alpha (\omega_0 \tau^\alpha) = \sum_{k=0}^\infty \frac{(\omega_0 \tau^\alpha)^k}{\Gamma (k\alpha + 1)} \frac{i^k + (-i)^k}{2}. \tag{5.65}$$

In this series expansion we have

$$\frac{i^k + (-i)^k}{2} = \cos (k\pi/2) = \begin{cases} 0 & \text{for } k \text{ odd} \\ 1 & \text{for } k \text{ even} \end{cases},$$

allowing us to change the summation index to include only even terms $k = 2m$ and obtain

$$\cos_\alpha (\omega_0 \tau^\alpha) = \sum_{m=0}^\infty \frac{\left(-\omega_0^2 \tau^{2\alpha}\right)^m}{\Gamma (2m\alpha + 1)} = E_{2\alpha} \left(-\omega_0^2 \tau^{2\alpha}\right). \tag{5.66}$$

Consequently, inserting Eqs.(5.66) and (5.64) into the $2\alpha-$ solution Eq.(5.62) yields the expression in terms of the generalized trigonometric functions. Therefore the two solutions are identical. One expressed in terms of complex conjugate MLFs and the other in terms of two-parameter MLFs.

5.2.3.1 Asymptotic MLF

The asymptotic behavior of the complex MLF has the form

$$E_\alpha(i\omega_0\tau^\alpha) = \frac{1}{\alpha}\exp\left[\omega_0^{1/\alpha}\tau e^{\frac{i\pi}{2\alpha}}\right] + \frac{i}{\omega_0}\frac{1-\alpha}{\Gamma(2-\alpha)}\frac{1}{\tau^\alpha} + O\left(t^{-2\alpha}\right), \qquad (5.67)$$

where the first term is the short-time behavior and the second is the lowest-order term in the long-time expansion of the MLF. Using $\alpha = 1 - \epsilon$ for $\epsilon << 1$ we can approximate the first term using $cos[\pi/2\alpha] \approx -\pi\epsilon/2$ so the magnitude of the first term is proportional to $exp\left[-\epsilon\pi\omega_0^{1/\alpha}\tau/2\right]$ and is therefore dominant over the second term for $1 << t \le \epsilon^{-1}$. Over this early time $|E_\alpha(i\omega_0\tau^\alpha)|$ follows an exponential decay, that is,

$$|E_\alpha(i\omega_0\tau^\alpha)| \propto exp[-\gamma t] \qquad 1 << t \le \epsilon^{-1}, \qquad (5.68)$$

with an effective decay rate of

$$\gamma \equiv \pi\epsilon\omega_0^{1/\alpha}\theta/2 > 0. \qquad (5.69)$$

Note that $\omega_0^{1/\alpha}\theta$ has the units of inverse time and the parameter θ determines the perceived frequency of the oscillator. This type of asymptotic analysis of the properties of the solutions to linear fractional differential equations is given by Gorenflo and Mainardi [16].

The phase space portraits in Figure 5.4 allow for additional insight into the fractional harmonic oscillator dynamics. The early time behavior is evident with the definition of what constitutes early time becoming more and more contracted with decreasing values of the fractional derivative index as indicated in Eq.(5.68). However the early interpretation made by a number of investigators that the fractional harmonic oscillator is equivalent to a damped oscillator due to the spiraling trajectory into the equilibrium fixed point were premature. The interpretation was mistaken in part because the MLF only decays exponentially at early times. The long-time decay is IPL

$$|E_\alpha(i\omega_0\tau^\alpha)| \propto \frac{1}{\tau^\alpha} \qquad \tau > \frac{1}{\epsilon^2}, \qquad (5.70)$$

and this change in behavior has no friction-based dissipative analog. An explanation of this relaxation can be given in terms of the interference of the realizations of an ensemble of solutions to the harmonic oscillator equation. This latter interpretation makes sense when the fractional derivative is generated by a subordination process as described in Chapter 6. We apply this reasoning in a subsequent subsection to the Hamiltonian method for generating equations of motion.

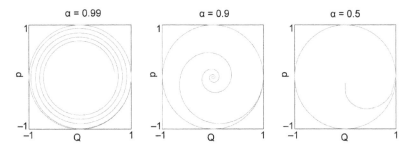

Figure 5.4 The (q, p) phase portrait for the fractional linear oscillator for the same parameter values used in Figure 5.3 with the initial condition $Q(0) = 1$ and $P(0) = 0$.

Fractional normal modes

As a final comment on the one-dimensional fractional oscillator let us consider the complex fractional normal mode amplitude defined by

$$a_\alpha(\tau) = a_\alpha(0) E_\alpha(-i\omega_0 \tau^\alpha), \tag{5.71}$$

and its complex conjugate. The Caputo or MRL fractional derivative of the fractional normal mode yields the equation of motion

$$\partial_\tau^\alpha [a_\alpha(\tau)] = a_\alpha(0)\partial_\tau^\alpha [E_\alpha(-i\omega_0 \tau^\alpha)] = -i\omega_0 a_\alpha(\tau), \tag{5.72}$$

along with its complex conjugate

$$\partial_\tau^\alpha [a_\alpha^*(\tau)] = a_\alpha^*(0)\partial_\tau^\alpha [E_\alpha(i\omega_0 \tau^\alpha)] = i\omega_0 a_\alpha^*(\tau). \tag{5.73}$$

These are the equations of motion for a fractional oscillator and are the same as those determined using the harmonic oscillator equation.

We can express the displacement of the fractional oscillator as

$$Q_\alpha(\tau) = \frac{1}{2\sqrt{\omega_0}} [a_\alpha(\tau) + a_\alpha^*(\tau)], \tag{5.74}$$

along with its momentum

$$P_\alpha(\tau) = \frac{\sqrt{\omega_0}}{2i} [a_\alpha(\tau) - a_\alpha^*(\tau)], \tag{5.75}$$

to obtain after some algebra

$$a_\alpha(\tau) = \sqrt{\omega_0} \left[Q_\alpha(\tau) - \frac{i}{\omega_0} P_\alpha(\tau) \right]. \tag{5.76}$$

Inserting the solutions to the fractional harmonic oscillator equation into Eq.(5.76) collapses this expression to Eq.(5.71) where we started.

Consequently, $a_\alpha(\tau)$ and $a_\alpha^*(\tau)$ are the linear eigenmodes of the fractional linear harmonic oscillator.

In Chapter 2 we established how a system of linear harmonic oscillators could be used to model the environment coupled to a system of interest in terms of a GLE. The dynamics of the environment (heat bath) were shown to be determined by its spectral properties, that is, whether the memory kernel was delta correlated in time or whether it could reach far back into the past to influence the present-time system's dynamics. The fluctuation-dissipation relation of the second kind relates this memory behavior to the autocorrelation of the fluctuating force driving the system dynamics. This interpretation was based on the existence of an underlying Hamiltonian that coupled the system to the environment. We now ask whether that same strategy can be adopted when the system is described by fractional normal modes?

Consider a Hamiltonian consisting of the fractional displacement Q_α and momentum P_α

$$H_\alpha = \frac{1}{2}\left(P_\alpha^2 + \omega_0^2 Q_\alpha^2\right), \tag{5.77}$$

and defined such the equations of motion are

$$\partial_\tau^\alpha [Q_\alpha] = \frac{\partial H_\alpha}{\partial P_\alpha} = P_\alpha, \tag{5.78}$$

and

$$\partial_\tau^\alpha [P_\alpha] = -\frac{\partial H_\alpha}{\partial Q_\alpha} = -\omega_0^2 Q_\alpha. \tag{5.79}$$

These equations of motion are equivalent to the fractional harmonic oscillator dynamics discussed previously. Proving the existence of this generalization to Hamilton's equations was done by Stanislavsky [43] and requires introducing a number of concepts from probability theory. We will do this subsequently, however if you must see the proof now then jump ahead to Section 7.1.

It is also possible to express the fractional Hamiltonian in terms of the fractional normal modes

$$H_\alpha = \omega_0 a_\alpha^* a_\alpha, \tag{5.80}$$

such that the equations of motion are

$$\partial_\tau^\alpha [a_\alpha] = -i\frac{\partial H_\alpha}{\partial a_\alpha^*} = -i\omega_0 a_\alpha, \tag{5.81}$$

and its complex conjugate. The last few sections have been intended to whet the reader's appetite for the formal utility of the fractional calculus.

Driven oscillator

Recall the observation we made that Kaslik and Sivasundaram [21] proved the non-existence of exact periodic solutions in a class of fractional-order dynamical systems. Another researcher [13] has shown that it is possible to obtain exact periodic solutions to impulsively driven fractional-order dynamic systems by choosing the correct impulses at the right moments of time. As an example, Duan [13] solved the equation for a fractional-order harmonic oscillator driven by a periodic function having a RL-fractional derivative with an infinite lower bound to obtain the Weyl fractional derivative [49]:

$$-\infty D_t^\beta [Q(t)] + \omega_0^\beta Q(t) = b \cos \Omega t. \tag{5.82}$$

Note that the intrinsic frequency is here raised to the power of the fractional derivative. This was done to retain the proper dimension for the intrinsic frequency. Given this choice of fractional derivative the solution does not depend on initial conditions and directly yields the steady-state condition for the fractional oscillator.

Using Fourier transforms it is a simple matter to write

$$\mathcal{FT}\left[-\infty D_t^\beta [Q(t)] ; \omega\right] = (-i\omega)^\beta \widetilde{Q}(\omega), \tag{5.83}$$

$$\mathcal{FT}\left[e^{\pm i\Omega t}; \omega\right] = \delta(\Omega \pm \omega), \tag{5.84}$$

so that the Fourier transform of the solution $\widetilde{Q}(\omega)$ to the driven harmonic oscillator equation is

$$\widetilde{Q}(\omega) = \frac{b}{2} \frac{\delta(\Omega + \omega) + \delta(\Omega - \omega)}{\omega_0^\beta + (-i\omega)^\beta}.$$

The inverse Fourier transform of this expression yields

$$Q(t) = A \cos(\Omega t - \phi), \tag{5.85}$$

with the amplitude given by

$$A = \frac{b}{\sqrt{\omega_0^{2\beta} + \Omega^{2\beta} + 2\Omega^\beta \omega_0^\beta \cos[\beta\pi/2]}}, \tag{5.86}$$

and the phase by

$$\phi = \tan^{-1}\left(\frac{\Omega^\beta \sin[\beta\pi/2]}{\omega_0^\beta + \Omega^\beta \sin[\beta\pi/2]}\right). \tag{5.87}$$

Note that Eq.(5.85) is the form of the equation that would be obtained asymptotically if there were damping in the original dynamics. It would be the asymptotic or steady-state solution after all transients had decayed away and whose dynamics depend only on the driving frequency and not the intrinsic frequency of the oscillator. What is different from the driven dissipative oscillator solution is the dependence of the amplitude and phase on the order of the fractional derivative, as well as, on the intrinsic frequency. However, in the case $\beta = 2$, the solution recaptures properties of the familiar driven oscillator including a 'resonance' when the intrinsic and driver frequencies match. These details and many others are discussed by Duan [13].

5.2.4   Open to a Fractional Environment

In Chapter 2 we used an infinite set of linear harmonic oscillators to model a system of interest coupled to the environment (heat bath). The resulting dynamics were described by a GLE given by Eq.(2.79). We noted that the memory kernel in the GLE reduces to a delta function in time for a broad flat spectrum of environmental oscillators, which would model a rather featureless random environment. In this case the environment is well represented by a single temperature and the Einstein relation, the ratio of the strength of the fluctuations to the dissipation parameter, provides a measure of that temperature. This is the zero memory situation in which the environment was first identified with a heat bath. Let us now turn our attention to the case where the environment has some structure.

Fractional Langevin Equation (FLE)

When the environment is no longer simple, as we found in our discussion of mesoscopic fluctuations in Brownian motion in Section 4.3.4, the broad flat discrete spectrum of environmental oscillators is no longer a sufficient description. The memory-induced effects of correlated environmental dynamics becomes important. When memory is present the spectrum of the environmental oscillators has color, which is to say, the discrete sum for the memory kernel given by Eq.(2.80) can be generalized to a continuum of frequencies:

$$K(t) = \int_0^\infty d\omega g(\omega) \Gamma(\omega) \cos \omega t. \tag{5.88}$$

This extension requires that the ratio quantity $\Gamma_\nu^2 (m_\nu \omega_\nu^2)^{-1} \to \Gamma(\omega)$ in the sum given in Eq.(2.80) is a sufficiently smooth function of the index ν as the number of environmental modes becomes infinite, see Lindenberg and West [25] for a complete discussion. Thus, $\Gamma(\omega)$ is the distribution of coupling strengths of the system of interest to the environmental modes and $g(\omega)$ is the density of environmental states. Consequently, in the general situation we have an infinite number of environmental oscillators and a continuous spectrum.

As pointed out by Tarasov [44] and two decades earlier by Lindenberg and West [25] there are a variety of environments that give rise to any given form of the product $g(\omega)\Gamma(\omega)$; there is no unique combination of distributions of coupling strengths and densities of states. For example when this product has the form of a Cauchy distribution

$$g(\omega)\Gamma(\omega) = \frac{\gamma/\tau_c}{1 + \omega^2 \tau_c^2},$$

the memory kernel is the exponential [25]

$$K(t) = \frac{\gamma}{\tau_c} e^{-t/\tau_c}, \tag{5.89}$$

with relaxation time τ_c. As the length of the memory contracts $\tau_c \to 0$ the memory kernel becomes a delta function in time as we discussed earlier.

It is more interesting for our present purposes to consider a IPL spectral density for which:

$$g(\omega)\Gamma(\omega) = A\omega^{\eta-1} \; ; \; 0 < \eta < 1. \tag{5.90}$$

What is of interest here is that this choice can be shown using a Tauberian theorem to yield an IPL memory kernel $K(t) \sim 1/t^\eta$, which has a remarkable influence on the GLE. We explicitly introduce the effect of this complex environment into the GLE by determining the cosine transform of the IPL

$$\int_0^\infty dt t^{-\eta} \cos \omega t = \frac{\pi}{2\Gamma(\eta)\cos \pi\eta/2}\omega^{\eta-1},$$

such that the memory kernel can be expressed as

$$K(t) = \frac{A}{2\Gamma(1-\eta)\cos[\pi(1-\eta)/2]}\frac{1}{t^\eta}. \tag{5.91}$$

This expression for the memory kernel when inserted into the GLE Eq.(2.79) yields

$$\frac{dP}{dt} + U'_m(Q) + \frac{A}{2\Gamma(1-\eta)\cos[\pi(1-\eta)/2]}\int_0^t \frac{1}{(t-\tau)^\eta}P(\tau)\,d\tau = f(t),$$

which using the definition of the Caputo fractional derivative enables us to write the FLE for a unit mass particle:

$$\frac{d^2Q}{dt^2} + U'_m(Q) + \Gamma\partial_t^\eta[Q] = f(t), \tag{5.92}$$

with the effective dissipation rate

$$\Gamma \equiv \frac{A}{2\cos[\pi(1-\eta)/2]}. \tag{5.93}$$

Note that this is precisely the form given for the extension of Brownian motion to the mesoscale discussed in Section 4.3.4, when the fluctuations in the ambient fluid environment have a spectrum with the fractional-index $\eta = 1/2$. It is worth mentioning that an extensive discussion of the solutions of the GLE with the inertial term also having a fractional derivative in addition to the memory term is given by Sandev *et al.* [40].

It is probably warranted to note that the GLE can be extended to the quantum domain following the analysis of Lindenberg and West [25]. The fully coupled oscillator system closely parallels the classical case except of course the fact that the variables are replaced by operators and we have operator fluctuations rather than fluctuations of functions. However since this is a somewhat more specialized application of these ideas we refer the reader to page 334 of [25] and urge him/her to develop the expression for the quantum memory function using different spectra for the environment. One thing that becomes apparent is that fluctuations and dissipation can have quite distinct time scales, contrary to the almost universal assumption that their characteristic time scales are the same.

5.3   Applications of FLE

A number of experiments have been done that have been well modeled by the fractional dissipative harmonic oscillator. Among them is the single-molecule fluorescence spectroscopy that allows for the observation of chemical reactions on biologically relevant time scales. Single electron transfer is used to probe changes in single protein molecular structure on scales of the order of Angströms [53]. Here we sketch how the linear equations of the last section have been used to explain the observed variation of fluorescein-tyrosine distance over time [32].

5.3.1   Fractional Dissipative Harmonic Oscillator

The equation of motion for the average displacement of a harmonically bound particle in a complex environment is given from Eq.(5.92):

$$\frac{d^2 \langle Q(t) \rangle}{dt^2} + \Gamma \partial_t^\eta \left[\langle Q(t) \rangle \right] + \omega_0^2 \langle Q(t) \rangle = 0 \tag{5.94}$$

where $0 < \eta < 1$ and ω_0 is the intrinsic frequency of the oscillator. Also implicit in Eq.(5.94) is the assumption that the fluctuations are zero-centered fractional Gaussian noise (FGn) satisfying the fluctuation-dissipation relation [9]:

$$\langle f(t) \rangle = 0 , \quad \langle f(t)f(t') \rangle = \frac{D}{|t - t'|^\eta}, \tag{5.95}$$

which for notational convenience in physical systems we choose the strength of the fluctuations D to satisfy the Einstein relation

$$\frac{D}{\Gamma} = k_B T.$$

In the case $\eta = 1$ Eq.(5.94) is replaced by the standard evolution for a linearly damped harmonic oscillator.

Burov and Barkai [6] give an extensive analysis of the FLE for a harmonically bound particle. Their motivation was, in part, to explain the experimental results on protein dynamics that used a single-molecule electron transfer to probe protein conformational dynamics [53]. The ensemble averaged measurements could not determine if the multiple exponential kinetics were the result of static heterogeneity or dynamic fluctuations. This problem was resolved by determining the correlation properties of the flavin-tyrosine distance over time and the autocorrelation function of this distance was found to be given be a stretched exponential [53].

The experimentally measured autocorrelation function is obtained by taking the product of displacements $Q(t)$ and $Q(0)$ and averaging over an ensemble of realization of the initial state $Q(0)$ to obtain

$$C_{QQ}(t) = \frac{\langle Q(t)Q(0)\rangle}{\langle Q^2(0)\rangle}. \tag{5.96}$$

The equation for the autocorrelation function for the fractional dissipative linear harmonic oscillator is obtained by inserting the potential $U(Q) = \frac{1}{2}\omega_0^2 Q^2$ into Eq.(5.92), multiplying the resulting expression with the initial condition $Q(0)$ and taking the average over an ensemble of initial states yields

$$\frac{d^2 C_{QQ}(t)}{dt^2} + \Gamma \partial_t^\eta \left[C_{QQ}(t) \right] + \omega_0^2 C_{QQ}(t) = 0. \tag{5.97}$$

This equation is identical to the dynamic equation for the average displacement given by Eq.(5.94). Thus, we obtain the dynamics of the physical observable for the single protein dynamics.

In the normal case there is over-damped and under-damped motion depending on the ratio ω_0/Γ. In the under-damped case $\omega_0 \gg \Gamma$ the average displacement and the autocorrelation function oscillate, taking on both positive and negative values. In the over-damped case $\omega_0 \ll \Gamma$ both measures of the dynamics are monotonically decaying without crossing the zero axis. As pointed out by Burov and Barkai [7] in the case $\eta = 1$ there exists a critical frequency $\omega_c = \Gamma/2$ which separates the two types of motion. They analyzed the fractional dissipative oscillator and found a similar kind of separation of the motion along with a number of transitions in the system dynamics indicative of critical values in the fractional-order exponent.

The solution to the fractional dissipative oscillator equation for the autocorrelation function is obtained by taking the Laplace transform of Eq.(5.97) to obtain

$$\widehat{C}_{QQ}(s) = \frac{s + \Gamma s^{\eta-1}}{s^2 + \Gamma s^\eta + \omega_0^2}, \tag{5.98}$$

that must be inverted to obtain the time-dependent autocorrelation function. To carry out the Laplace inversion requires that there be a polynomial in the Laplace variable in the denominator of Eq.(5.98). Since $0 < \eta < 1$ a function

$R(s)$ is introduced such that

$$\widehat{C}_{QQ}(s) = \frac{\left(s + \Gamma s^{\eta-1}\right) R(s)}{P(s)},$$ (5.99)

where $P(s)$ is a polynomial in s. The solution to the inverse Laplace transform of Eq.(5.99) is determined by an analysis of the poles, that is, the zeros of the denominator

$$P(s) = \left(s^2 + \Gamma s^{\eta} + \omega_0^2\right) R(s) = 0.$$

Burov and Barkai [7] provide an extensive analysis of the solutions from a determination of the complementary polynomial $R(s)$. Their analysis requires the assumption $\eta = \frac{p}{q}$ where $q > p > 0$ are integers and their ratio is irreducible, using techniques given for solving linear fractional differential equation with rational exponents by Miller and Ross [31].

5.3.2 Experimental Realization

A more intuitive use of the FLE was given by Min *et al.* [32], who experimentally determined the properties of the fluctuations; properties that were assumed in the above analysis. A complete mathematical treatment that includes this case is given by Sandev *et al.* [40]. The memory kernel $K(t)$ was determined by experiment to be IPL with index $\eta = 0.51 \pm 0.07$ over a broad range of time scales from 10^{-3} sec to 10 sec. The FLE for the autocorrelation function in the experiments of Min *et al.* [32] was for the distance between a fluorescein-tyrosine pair within a single protein complex, as directly monitored in real time by a photo-induced electron transfer. Assuming the dynamics are described by the over-damped FLE they use

$$\Gamma \int_0^t dt' K(t - t') \frac{dC_{QQ}(t')}{dt'} + \omega_0^2 C_{QQ}(t) = 0,$$ (5.100)

with an unknown memory kernel that is to be determined from experimental data.

The Laplace transform of Eq.(5.100) yields after some algebra

$$\widehat{K}(s) = \frac{\omega_0^2}{\Gamma} \frac{\widehat{C}_{QQ}(s)}{C_{QQ}(0) - s\widehat{C}_{QQ}(s)}.$$ (5.101)

Min *et al.* [32] determine the Laplace transform of the memory kernel empirically by taking the Laplace transform of the autocorrelation data depicted by the circles in Figure 5.5, and inserting that function numerically into Eq.(5.101) along with the empirical constant $C_{QQ}(0) = 0.22$ Å. The resulting solution for the Laplace transform of the memory kernel in depicted in Figure 5.6 so that $\widehat{K}(s) \propto s^{-0.49}$. The inverse Laplace transform of $\widehat{K}(s)$ yields the memory kernel $K(t) \propto t^{-0.51}$.

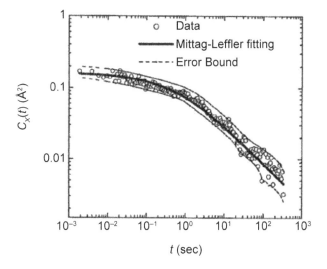

Figure 5.5   Autocorrelation function of distance fluctuations $C_{QQ}(t)$ (open circles, average of 13 molecules under the same experimental condition), determined with high time resolution(from [32] with permission).

They go on to draw a number of conclusions from these and other data they obtained concerning the statistical properties of the fluctuations. Since the data for the displacement indicate that $Q(t)$ is a stationary, Gaussian process and the fact that the FLE is linear, requires the random fluctuations to share those properties. Moreover, the fluctuation-dissipation relation indicates that the IPL decay of the memory kernel implies time scaling invariance of the autocorrelation of the fluctuations $\langle f(t)f(t')\rangle$. They conclude that the only mathematical process satisfying these conditions is FGn [29], consistent with the theoretical assumption made in Section 5.2.4.

The random force driving the fractional dissipative oscillator is FGn and can be expressed as, using the notation of Min *et al.* [32]:

$$f(t) = \sqrt{2D}\frac{dB_H(t)}{dt},\tag{5.102}$$

resulting in the kernel

$$K(t-t') = 2D\left\langle\frac{dB_H(t)}{dt}\frac{dB_H(t')}{dt'}\right\rangle$$

$$= 2DH(2H-1)\left|t-t'\right|^{2H-2}.\tag{5.103}$$

Comparing the IPL index in Eq.(5.103) to that in Eq.(5.91) the data enables us to write

$$\eta = 2 - 2H = 0.51 \pm 0.07,$$

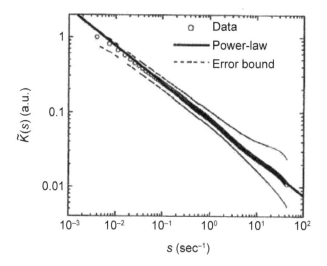

Figure 5.6 Normalized $\widehat{K}(s)$ calculated from $C_{QQ}(t)$ in Figure 5.5 using Eq.(5.101) (open circles). The full line is the fit of u^θ with $\theta = -0.49$. Its inverse Laplace transform yields $K(t) \propto t^{-0.51}$. The dashed lines are error bounds (from [32] with permission).

and the empirical Hurst exponent is $H = 0.75 \pm 0.03$.

The autocorrelation function can be constructed from Eq.(5.101) with the Laplace transform of the kernel from Eq.(5.103) inserted to be

$$\widehat{C}_{QQ}(s) = \frac{s^{1-2H}}{s^{2-2H} + \gamma_0} C_{QQ}(0),$$

which when inverse Laplace transformed yields the autocorrelation function in terms of a MLF

$$C_{QQ}(t) = C_{QQ}(0) E_\eta \left(- [\gamma_0 t]^\eta\right), \tag{5.104}$$

where the characteristic time scale for the system is given by the rate

$$\gamma_0 = \left[\frac{\omega_0^2 \Gamma(2H+1)}{\Gamma D}\right]^{1/\eta}, \tag{5.105}$$

and the exponent is $\eta = 2 - 2H$. Using the empirical value of $H = 0.75$ the MLF simplifies to

$$C_{QQ}(t) = C_{QQ}(0) e^{\gamma_0 t} \operatorname{erf} c\left(\sqrt{\gamma_0 t}\right) \tag{5.106}$$

where erfc is the complimentary error function. Figure 5.5 depicts the fitting of Eq.(5.106) to the data using $H = 0.75$, $\Gamma D/\omega_0^2 = 0.7 s^{1/2}$ and $1/\gamma_0 = 0.9 s$ and is seen to be excellent.

Min *et al.* [32] have consequently characterized equilibrium fluctuations occurring over a broad spectrum of time scales on which protein reactions take place. They conjecture that the IPL memory kernel in proteins might be general. If this is the case then it would be natural to take the next step and conjecture that the dynamics with long-time memory entails dynamic disorder and fractional kinetics for biochemical reactions.

5.4 Control of Complexity

It is not a coincidence that our understanding of the complex technological society in which we live is mirrored by the modern view of how the human body operates. It is readily accepted that a motion detector dims lights when no one is in a room, that a voice can turn a radio or DVR on and off, as well as activate a phone without touching a button, and that a thermostat maintains a constant temperature in a building regardless of season. These and countless other devices are the elegantly simple control processes modern engineering has successfully introduced into western society and correspondingly, if unintentionally, shaped our view of the world.

The control processes that nature has converged on, through the process of adaptation and natural selection, appear to lack the simplicity of human engineering, but they are no less elegant. In medicine such control was identified and labeled homeostasis over a century ago. Many consider homeostasis to be the guiding principle of medicine, whereby every human body has multiple automatic inhibition mechanisms that suppress disquieting influences of the environment, even though it is not always evident how a particular suppressing response is related to a specific antagonism. Part of what masks the connection between the stimulus and the response is the multiplicity of scales involved in the complex dynamics of physiological systems. These scales are evident in the scaling of time series and in allometry relations so abundant in living systems [51].

One strategy for modeling complexity has been the application of the ideas from network science. In Chapter 6 we briefly touch on some aspect of networks, but for the moment we are interested in control. In this regard we mention the work of Liu *et al.* [26], who note that the ultimate proof of understanding complex systems lay in our ability to control them. They present a method of control of self-organized networks based on the structure of network. A distinctly different approach is provided by Cornelius *et al.* [10] who use the interactive nonlinear dynamic nature of complex networks, inherent in real phenomena, to direct a network to a target state. The idea is to control a network through application of a perturbation that bumps the network's dynamics into the basin of attraction of the target state, after which the internal dynamics, and without further external encouragement drives the network to the target state. Another approach is based on Haken's theory of *Synergetics* as well as the fractional calculus approach in modern control theory is reviewed by Lazarevoc [24]. The latter technique exploits the importance of criticality and self-organization in the process of control.

The differential calculus enabled the systematic study of the dynamics of mechanical systems, extending the description of phenomena from static geometrical structure to the geometry of dynamical attractors. Stochastic differential equations pushed the boundaries back further, by considering discontinuous and intermittent processes that the smooth behavior of discrete differential functions cannot describe. The frontiers of empirical descriptions were moved again through the application of the fractional calculus to incorporate into the evolutionary description the memory of past behavior of system dynamics in a way only vaguely anticipated by the integer-order differential equations. The approach to controllability discussed herein emphasizes these latter aspects of complex dynamic networks and the influence these aspects have on the response to a control signal, over that of the network's response to changes in its topological structure. The separate approaches to control are not completely unrelated, but for that level of detail we refer the reader back to the literature.

The first artificial control mechanism of the modern era was the centrifugal flyball governor constructed by J. Watt in 1788, for regulating the speed of the rotary steam engine. This control mechanism heralded the onset of the Industrial Revolution and set society on the path we still travel today. The first mathematical description of Watt's governor was created by the Scottish physicist J.C. Maxwell eighty year later, when he linearized the differential equations describing the dynamics and determined their characteristic values. The solutions to the linear differential equations (control) are stable when the eigenvalues have negative real parts and in this way the language for the control of dynamical systems was initiated.

In physics the Langevin equation may be interpreted as describing a natural, if somewhat erratic physical, control process [3, 38]. The motion of a Brownian particle is determined by the fluctuations and dissipation generated by the environment and the linear Langevin equation drives the Brownian particle to its equilibrium state. When the environment is not so simple, as in Section 5.3, the scales contributing to the fluctuation-dissipation process are coupled and in that analysis the fluctuations were represented as FGn. In this section we introduce some ideas to couple the fractional differential equations and scaling to the control of complex phenomena.

Innumerable physical, social and living phenomena have been classified as complex, in part, because their patterns have been described by scaling laws and scaling exponents. Scafetta and West [41] argue that a large set of these observations emerge, because complex systems might often process much simpler stimuli, and the complexity of the response depends on both the properties of the input and the complexity of the processing mechanism. They show that allometric filters and relaxation processes, which are quite common in nature, might naturally produce fractal patterns from simple signals, such as random fluctuations. We review some of their arguments in the next couple of sections.

5.4.1 Allometric Filters

Section 5.2.4 focuses on how the complete dynamics of a large-scale system can be partitioned into the network of interest interacting with a complex environment, which we here refer to as the network and the external driver. We saw that it was possible to represent the dynamics by means of a memory kernel, which we do here in simplified form relating the input $Q_i(t)$ to the output $Q_o(t)$ of the network. In the engineering literature this is referred to as the transfer integral representation

$$Q_o(t) = \int_0^t K(t - t')Q_i(t')dt', \qquad (5.107)$$

and the transfer function $K(t)$ is determined by the internal dynamics of the network, as found earlier in the construction of the GLE.

Consider the first of two physical examples of allometric filters. Others are given in a more engineering context [28, 35], but are not discussed herein. The first is that of a low-pass filter constructed from an RC-circuit with a relaxation time given by $\tau = RC$, where R is the resistance in *ohms* and C is the capacitance in *farads*. The rate equation describing the dynamics of the circuit output subject to a given input is

$$\frac{dQ_o(t)}{dt} = -\frac{1}{\tau}Q_o(t) + \frac{c}{\tau}Q_i(t), \qquad (5.108)$$

where c is a constant. The Fourier transform of this equation in time yields the solution in Fourier space

$$\tilde{Q}_o(\omega) = \tilde{K}(\omega)c\tilde{Q}_i(\omega), \qquad (5.109)$$

whose inverse Fourier transform provides the transfer integral Eq.(5.107) with the transfer function being

$$K(t) = \mathcal{FT}^{-1}\left\{\tilde{K}(\omega); t\right\} = \mathcal{FT}^{-1}\left\{\frac{1}{1 + i\tau\omega}; t\right\}. \qquad (5.110)$$

To understand why this is called a low-pass filter we express the solution in terms of the corresponding spectra

$$P_o(\omega) = \frac{1}{1 + (\tau\omega)^2}P_i(\omega). \qquad (5.111)$$

The network dynamics do not modify the input signal as measured by the input spectrum $P_i(\omega)$ for $\tau\omega \ll 1$, whereas it drives the output signal as measured by the output spectrum $P_o(\omega)$ to zero for $\tau\omega \gg 1$, when $P_i(\omega)$ is finite as $\omega \to \infty$. In this way the filter allows the low frequency part of the signal to pass and blocks the high-frequency part.

The second example is that of a high-pass filter, obtained from a differently configured RC-circuit whose dynamics are described by

$$\frac{dQ_o(t)}{dt} = -\frac{1}{\tau}Q_o(t) + c\frac{dQ_i(t)}{dt}.$$

(5.112)

In this case the Fourier transform of the transfer kernel is

$$\widetilde{K}(\omega) = \frac{i\tau\omega}{1 + i\tau\omega},$$

(5.113)

thereby relating the input and output spectra by

$$P_o(\omega) = \frac{(\tau\omega)^2}{1 + (\tau\omega)^2}P_i(\omega).$$

(5.114)

In this case the network dynamics does not modify the input spectrum for $\tau\omega \gg 1$, whereas it drives the output spectrum to zero for $\tau\omega \ll 1$, when the input spectrum is finite as $\omega \to \infty$. In this way the filter allows the high-frequency of the signal to pass and blocks the low-frequency part.

These two filters, taken together, can selectively block or pass any band of frequencies important for the operation of a particular application. Moreover, they can be generalized by making the relaxation time dependent on the frequency $\tau(\omega)$.

5.4.2   Fractional Filter

The issue we address is the control of complexity through variability. Controlling variability is one of the goals of medicine to ensure the proper operation of the human body, in manufacturing to reduce product failure, and to produce other desirable outcomes in uncertain environments. We distinguish between the traditional negative feedback of homeostasis and allometric control mechanisms. The former is both local in time and instantaneous, whereas the latter is a relatively new concept that can take into account long-time memory. The long-time memory is manifest in correlations that are IPL in time, as well as, long-range in complex phenomena as manifest in IPL PDFs in the system variable. Allometric control can introduce fractal characteristics into otherwise featureless random time series to enhance the robustness of the control mechanisms.

Consider the response of a fractional system to a random time series input

$$\partial_t^{\alpha_o}[Q_o(t)] = Q_i(t),$$

(5.115)

which we can also write as

$$Q_o(t) - Q_o(0) = \frac{1}{\Gamma(\alpha)} \int_0^t \frac{Q_i(t')\,dt'}{(t-t')^{1-\alpha_o}}. \tag{5.116}$$

Assume that the input time series is Gaussian noise such that it satisfies the scaling relation in distribution

$$Q_i(\lambda t) = \lambda^{-\alpha_i} Q_i(t), \tag{5.117}$$

and $\alpha_i = 1/2$ for a simple Gauss process, but in general the scaling index covers the interval $0 < \alpha_i \leq 1$. This scaling of the input enables us to deduce the scaled version of the output from Eq.(5.116):

$$
\begin{aligned}
Q_o(\lambda t) - Q_o(0) &= \frac{1}{\Gamma(\alpha)} \int_0^{\lambda t} \frac{Q_i(t')\,dt'}{(\lambda t - t')^{1-\alpha_o}} \\
&= \lambda^{\alpha_o - \alpha_i} \left[Q_o(t) - Q_o(0) \right]. \tag{5.118}
\end{aligned}
$$

Using the trick of writing the scaling parameter as $\lambda = 1/t$ we can express the solution, after some rearrangement, in the form

$$Q_o(t) - Q_o(0) = t^{\alpha_o - \alpha_i} \left[Q_o(1) - Q_o(0) \right], \tag{5.119}$$

where the coefficient of the time factor is time-independent and the variance $\sigma_o^2(t) \equiv \left\langle \left[Q_o(t) - Q_o(0) \right]^2 \right\rangle$ becomes

$$\sigma_o^2(t) = \sigma_o^2(1)\, t^{2(\alpha_o - \alpha_i)}. \tag{5.120}$$

The bracket denotes an average over an ensemble of realizations of the random input. The time dependence of the variance of the output signal agrees with that obtained for anomalous diffusion, if we identify $H = \alpha_o - \alpha_i$, and H with the Hurst exponent. With the fractional derivative index $\alpha_o \leq 1$, we have $1/2 \geq H > 0$, if the stochastic driver is that of classical diffusion, that is, $\alpha_i = 1/2$. Consequently, the process described by Eq.(5.115), the dissipation-free FLE, is anti-persistent.

Anti-persistent behavior of time series was observed by Peng *et al.* [36] for the differences in time intervals between heart beats. They interpreted their time series, as have a number of subsequent investigators, in terms of random walks with $H < 1/2$. However, we can see from Eq.(5.120) that the FLE without dissipation is a phenomenologically equivalent description to the anti-persistent random walk. The scaling behavior alone cannot distinguish between these two models, particularly since the FLE model may well be a continuous version of a FRW. What is needed is the complete PDF and not just the time-dependence (scaling) of one or two moments.

5.4.3 Frequency-dependent Relaxation Time

When the input signal is monofractal noise with spectral exponent $2\alpha_i$, the output can present a bi-scaling behavior with two scaling spectral exponents $2\alpha_o$ and $2\alpha_i$. As Scafetta and West [42] point out this suggests that a bi-scaling signal might indeed be the results of a particular response of a system to a simpler mono-scaling stimulus. By using the low-pass allometric filter mechanism, if the power spectrum of the input is

$$P_i(\omega) = \frac{1}{\omega^{2\alpha_i}}, \tag{5.121}$$

by choosing the frequency-dependent relaxation time to be

$$\tau(\omega) = \tau_0 \omega^{(\alpha_o - \alpha_i - 2)/2}, \tag{5.122}$$

we obtain the power spectrum of the output

$$P_o(\omega) = \frac{\omega^{-\alpha_i}}{1 + \tau_0^2 \omega^{\alpha_o - \alpha_i}} \begin{cases} \omega^{-\alpha_i} & \text{for } \tau_0^2 \omega^{\alpha_o - \alpha_i} \ll 1 \\ \omega^{-\alpha_o} & \text{for } \tau_0^2 \omega^{\alpha_o - \alpha_i} \gg 1 \end{cases}, \tag{5.123}$$

The transition between low-frequency and high-frequency regimes depends on the value of τ_0. In the case $\alpha_o > \alpha_i$ ($\alpha_i > \alpha_o$) the allometric low-pass filter modifies the high- (low-) frequency component of the input. Thus, the scaling exponent referring to high frequencies (small time scale) is always larger than the scaling exponent referring to low frequencies (large time scales). Because the input signal might be white noise, $\alpha_i = 0$, this modeling suggests that the processing mechanism might easily transform input random fluctuations into IPL correlated output.

Under the same conditions, by using the allometric high-pass filter mechanism, we choose the frequency-dependent relaxation time to be

$$\tau(\omega) = \tau_0 \omega^{(\alpha_i - \alpha_o - 2)/2}, \tag{5.124}$$

and obtain the output power spectrum

$$P_o(\omega) = \frac{\omega^{-\alpha_o}}{1 + \tau_0^2 \omega^{\alpha_i - \alpha_o}} \begin{cases} \omega^{-\alpha_i} & \text{for } \tau_0^2 \omega^{\alpha_i - \alpha_o} \gg 1 \\ \omega^{-\alpha_o} & \text{for } \tau_0^2 \omega^{\alpha_i - \alpha_o} \ll 1 \end{cases}. \tag{5.125}$$

The transition between low and high frequency regimes depends on the value of τ_0. In the case $\alpha_o > \alpha_i$ ($\alpha_i > \alpha_o$) the allometric high-pass filter modifies the high- (low-) frequency component of the input signal.

5.4.4 Scaling and Optimal Control

The simplest dynamical network is given by Eq.(5.107) in terms of a convolution over the external driver, or in terms of Laplace transforms of the input and output signals, as well as, the transfer function. If the external driver is a stochastic process then its control can be characterized in terms of a cost function. The cost function is defined in terms of error, that being the difference between the two time series

$$\varepsilon(t) = Q_o(t) - \int_0^t K(t-t')Q_i(t')\,dt', \tag{5.126}$$

where the transfer function is to be selected to provide optimal control of the dynamics. This optimal control is typically defined by a minimum error condition. The stationary autocorrelation function for the output signal is

$$C_o(\tau) = \langle Q_o(t+\tau)Q_o(t)\rangle, \tag{5.127}$$

and the brackets again denote an average over an ensemble of realizations of the fluctuations. The optimal transfer function is obtained as the solution to an integral equation of the first kind; an equation obtained by minimizing the mean-squared error [5]:

$$C_o(t) - \int_{-\infty}^{\infty} K(t-t')C_i(t')\,dt' = 0, \tag{5.128}$$

and $C_i(t)$ is the autocorrelation function of the input signal.

The solution to Eq.(5.128) yields a transfer function in terms of the Fourier transform of the autocorrelation function for the input and output. In the causal case the filter is online and used in real time, so one has the restriction $t \geq 0$ and Eq.(5.128) becomes a Wiener-Hopf equation. Solving this latter equation is a formidable, but straight forward, task and will not be discussed further. In the non-causal case, the filter is offline and the full data stream is analyzed, so the time is unrestricted $-\infty \leq t \leq \infty$. In the latter case the solution to the equation is simpler and is given by the ratio of the spectrum of the input to the spectrum of the output

$$\tilde{K}(\omega) = \frac{P_o(\omega)}{P_i(\omega)}. \tag{5.129}$$

It is now clear that if the statistical process being considered has long-term memory, then both spectra in Eq.(5.129) are IPL. Suppose the spectral indices for input and output are the same as those used in the last section, then the optimal transfer function is

$$K(t-t') = \mathcal{FT}^{-1}\left\{\frac{P_o(\omega)}{P_i(\omega)}; t-t'\right\} \propto |t-t'|^{2(\alpha_i-\alpha_o)-1}. \tag{5.130}$$

This form of the transfer function strongly suggests the fractional calculus as the proper way to model stochastic processes with long-time memory, in which the memory kernel of the former is the transfer function of the latter.

5.4.5 Fractional Control

Classical control theory is a mature discipline and we do not propose to review it here. However, we must bear in mind that its starting point is dramatically different from that discussed for dynamic systems, that is, control theory starts with data and without knowing the dynamics for the system we endeavor to control it. We therefore proceed by implication and consider the simplest feedback system expresses in Laplace space:

$$\widehat{Q}_o\left(s\right) = \frac{\widehat{K}\left(s\right)}{1 + \widehat{K}\left(s\right)} \widehat{Q}_i\left(s\right), \tag{5.131}$$

further discussion of the formulation of this equation can be found elsewhere [50]. Eq.(5.131) describes a unity feedback system, with a conventional PID-controller, meaning that the transfer function has an integer-integrator and integer-differentiator controller. It is important to realize that the structure of the model, which is to say, the functional form of the transfer function is postulated to conform to the input data. For example in chemical engineering a first-order integral and first-order derivative controller is very common, so the typical transfer function has the Laplace form

$$\widehat{K}\left(s\right) = c_1 + c_2 s + c_3 s^{-1}. \tag{5.132}$$

Note that the Laplace transform of the transfer function given here is a combination of the low-pass and high-pass filters given earlier in Fourier space. The coefficients of the model are determined by fitting them to real world data.

If the phenomenon being investigated is fractal in nature, which is to say, it has dynamics described by a fractional-differential equation, then attempting to control it with an integer-order feedback, such as given by Eq.(5.132) leads to extremely slow convergence, if not divergence, of the system output [37]. On the other hand, a fractional-order feedback, with the indices appropriately chosen, lead to rapid convergence of the output to the desired signal. Thus, one might anticipate that dynamic complex phenomena with scaling properties, since they can be described by fractional dynamics would have fractional-differential control systems. We have referred to such feed back in the past as allometric control [48].

We can transcribe the fractional calculus into classical control theory by replacing the transfer function given by Eq.(5.132) with

$$\widehat{K}\left(s\right) = c_1 + c_2 s^{\gamma}, \tag{5.133}$$

corresponding to a fractional-order controller, called the PD$^{\gamma}$-controller. Podlubny [37] demonstrates that to obtain the best results the mathematical model of the original system and that of the controller must be compatible. Consequently, an integer-order system ought to have an integer-order controller for best results and correspondingly, a fractional-order system ought to have fractional-order controller for best results. Balacandran and Kokila [2] address the controllability of both linear and nonlinear fractional dynamic systems and point out that Chen *et al.* [8] proposed robust controllability for interval fractional-order linear time invariant systems, whereas Adams and Hartley [1] studied finite time controllability for fractional systems. There is an ever increasing literature on fractional-order control including those generated by variation calculation for fractional optimal control [15, 17] to which we invite the reader to contribute.

5.5 Fractional Logistic Equation

Over forty years ago Robert May [30] introduced the general scientific community to what was then the new concept of chaos, whereby a discrete deterministic nonlinear dynamical systems generated apparently random fluctuations. In the intervening years the fields of biology, ecology and sociology, as well as the physical sciences, have reshaped themselves to accommodate this new view of dynamical modeling of complex phenomena. This strategy for understanding complex dynamics has as its basis the specification of an initial state from which the system is predicted to evolve to a given final state. Chaos, of course, is a sensitive dependence on those initial conditions, such that a small change diverges to a totally unexpected final state. This remarkable property of nonlinear systems has occupied the attention of a significant number of scientists in essentially every scientific discipline.

In Section 2.3.2 the continuous logistic equation was solved analytically using the Carleman embedding technique. In Section 5.2.1 we generalized a linear growth equation to fractional form and found that in addition to the saturation properties of the logistic equation, there were a number of new properties that emerged for the fractional-order index greater than one. Here we join these analyses to blend the fractional calculus with nonlinear dynamics and incorporate some of the complexity discussed by May through the use of fractional derivatives. Many creative approximation techniques have been developed to solve fractional nonlinear rate equations [4, 11, 12, 14, 20, 33], but we do not present them here. Instead we provide a limited discussion of the first analytic solution for the fractional form of the logistic equation.

Consider the fractional logistic equation (FLogE)

$$\partial_t^\alpha [Q(t)] = k^\alpha Q(t) [1 - Q(t)], \qquad (5.134)$$

where $Q(t)$ is the fraction of the total population that can be supported, k^α is the growth rate and we use a Caputo or MRL fractional derivative for the dynamics. This equation can be reexpressed as a linear equation in the dimensionless time $\tau = kt$ using the operators introduced in Section 2.3.3

$$\partial_\tau^\alpha [Q] = \mathcal{O}Q, \qquad (5.135)$$

with the operator defined as

$$\mathcal{O} \equiv (Q - Q^2) \frac{\partial}{\partial Q}. \tag{5.136}$$

The formal solution to the linear operator equation is

$$Q(\tau) = E_\alpha (\mathcal{O}_0 \tau^\alpha) Q_0, \tag{5.137}$$

where the operator $\mathcal{O}_0$ acts on the initial state $Q_0 = Q(0)$. The explicit solution is

$$Q(\tau) = \sum_{l=0}^{\infty} \frac{\tau^{l\alpha}}{\Gamma(l\alpha + 1)} \mathcal{O}_0^l Q_0, \tag{5.138}$$

where the difficulty is giving the operator $\mathcal{O}_0^l$ an explicit form that facilitates constructing an analytic time-dependent solution.

We introduce a new technique for expressing the operator equation in terms of analytic functions by implementing a transformation of the variable $Q_0 \to A$ such that

$$\frac{\partial}{\partial A} = \frac{dQ_0}{dA} \frac{\partial}{\partial Q_0}, \tag{5.139}$$

is obtained from the nonlinear dynamics. The derivative in Eq.(5.139) is identified with the operator $\mathcal{O}_0$ to yield

$$\frac{dQ_0}{dA} = Q_0 (1 - Q_0), \tag{5.140}$$

which can be solved to obtain

$$e^A = \frac{Q_0}{1 - Q_0} \Rightarrow Q_0 = \frac{1}{1 + e^{-A}}.$$

Consequently, we can write for the operator term in the series solution Eq.(5.138) as

$$\mathcal{O}_0^l Q_0 = \left(\frac{\partial}{\partial A} \right)^l \left(\frac{1}{1 + e^{-A}} \right),$$

and since $e^{-A} < 1 \Rightarrow Q_0 > 1/2$ we can expand the term on the right-hand side in a Taylor series to obtain

$$\mathcal{O}_0^l Q_0 = \left(\frac{\partial}{\partial A} \right)^l \sum_{n=0}^{\infty} (-e^{-A})^n = \sum_{n=0}^{\infty} (-1)^n (-n)^l e^{-An}.$$

This last expression can be inserted into the series form of the solution to the FLogE yielding

$$Q(\tau) = \sum_{l=0}^{\infty} \frac{\tau^{\alpha l}}{\Gamma(l\alpha + 1)} \sum_{n=0}^{\infty} (-1)^n (-n)^l e^{-An},$$

which after some algebra reduces to

$$Q(t) = \sum_{n=0}^{\infty} (-1)^n e^{-An} E_\alpha \left(-nk^\alpha t^\alpha\right).$$

(5.141)

An even further reduction is obtained in terms of the initial condition

$$Q(t) = \sum_{n=0}^{\infty} \left(\frac{Q_0 - 1}{Q_0}\right)^n E_\alpha \left(-nk^\alpha t^\alpha\right).$$

(5.142)

This solution has also been obtained using the matrix method discussed in Section 2.3.2 suitably extended to fractional form [52]. It might be criticized that this approach requires the use of the Leibniz rule for derivatives; a rule that is violated by the fractional derivatives. However, in the next section we show that using Jumarie's definition of the MRL fractional derivative the Leibniz rule is satisfied.

An independent verification of the form of the solution to the FLogE is the known analytic result for the logistic equation. The limiting case $\alpha = 1$ is obtained from Eq.(5.142) using the property of MLFs

$$\lim_{\alpha \to 1} E_\alpha \left(-nk^\alpha t^\alpha\right) = e^{-nkt},$$

when inserted into Eq.(5.142) results in the explicitly summable series

$$Q(t) = \sum_{n=0}^{\infty} \left(\frac{Q_0 - 1}{Q_0}\right)^n e^{-nkt} = \frac{Q_0}{Q_0 + (1 - Q_0) e^{-kt}},$$

(5.143)

which is the exact solution to the logistic equation. The asymptotic properties $Q_0 = Q(0)$ and $Q(\infty) = 1$ are shared by the solution to the FLogE given by Eq.(5.142), although the rate of approach to these asymptotic values are quite different in the two cases; being exponential in time for the logistic equation, but inverse power law in time for the MLF.

The FLogE is also integrated numerically and these solutions are compared with the analytic solution for four values of the fractional-order parameter α depicted in Figure 5.7. It is clear that there is excellent agreement between the analytic solution and the numerical integration of the FLogE at early times for all values of the fractional order parameter. At late times there is a modest deviation from the analytic solution for $\alpha \leq 0.5$, which we attribute to the fact that we have used the same integration parameters in all the calculations, rather than optimizing the numerics. This is compelling evidence that we have obtained the exact solution to the initial value problem for the FLogE.

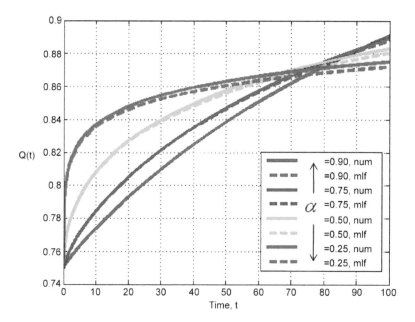

Figure 5.7 The solid curves denote the numerical integration of Eq.(5.134) and the dashed curves the analytic solution given by Eq.(5.142) for the four values of the fractional index shown.

5.6 Fractional Leibniz Rule

One of the surprising properties of the various definitions of fractional derivatives including Riemann-Liouville, Caputo, Marchaud, Weyl, Reisz and others [22, 39] is their failure to satisfy the Leibniz rule [37, 49]:

$$D_t^\alpha\left[G(t)F(t)\right] \;=\; \sum_{k=0}^{\infty} D_t^{\alpha-k}\left[G(t)\right] D_t^k\left[F(t)\right] \tag{5.144}$$

$$\neq\quad D_t^\alpha[G(t)]F(t) + G(t)D_t^\alpha\left[F(t)\right], \quad 0 < \alpha < 1 \tag{5.145}$$

with the rule applying when the inequality in Eq.(5.145) becomes an equality. In pointing this out Tarasov [45] went so far as to present a proof that if the Leibniz rule is not violated there can be no fractional derivative. He maintained that this violation of the rule is one of the characteristic properties of fractional derivatives. The proof he constructed established that if the Leibniz rule is assumed to be true, a consequence would be $\alpha = 1$ and therefore the rule is inconsistent with fractional derivatives.

On the other hand, Weberszpil [47] presented a counter argument to show that if a function is non-differentiable then in a coarse-grained medium its fractional derivative is compatible with the Leibniz rule. He agreed that Eqs.(5.144) and (5.145) are appropriate for the fractional derivatives identified, but maintained that the MRL fractional derivative introduced by Jumarie [18, 19] satisfies the fractional Leibniz rule:

$$\mathcal{D}_t^\alpha [G(t)F(t)] = \mathcal{D}_t^\alpha [G(t)] F(t) + G(t)\mathcal{D}_t^\alpha [F(t)], \quad 0 < \alpha < 1. \quad (5.146)$$

Following the logic of Tarasov, Weberszpil also assumed that a given function satisfied the Leibniz rule and was defined in a Hölder space. He pointed out that a Hölder space and nowhere differentiable functions are related, giving a Weierstrass function as an example.

The definition of a fractional derivative given as the limit of a fractional difference in Section 4.2.1 relies on a Taylor series expansion of an exponential operator and is therefore only valid when the function $Q(t)$ is differentiable. When the function is non-differentiable it is necessary to generalize the Taylor series expansion. A pedestrian way to achieve such a generalization is by identifying the continuous forward shift operator expansion as

$$(B_\tau - 1)^\alpha Q(t) \equiv \sum_{k=0}^\infty (-1)^k \binom{\alpha}{k} Q(t + (\alpha - k)\tau), \quad (5.147)$$

and the definition of MRL fractional derivative is

$$\mathcal{D}_t^\alpha [Q(t)] = \lim_{\tau \to 0^+} \frac{(B_\tau - 1)^\alpha}{\tau^\alpha} Q(t). \quad (5.148)$$

Jumarie [18] showed that the shift operator has the form given in terms of the MLF with an operator argument

$$Q(t + \tau) = E_\alpha (\tau^\alpha \mathcal{D}_t^\alpha) Q(t), \quad (5.149)$$

which replaces the exponential operator in Eq.(4.19). Thus, the limit for the fractional derivative given by Eq.(4.20) is replaced with

$$\mathcal{D}_t^\alpha [Q(t)] = \lim_{\tau \to 0^+} \frac{Q(t + \tau) - Q(t)}{\tau^\alpha} = \lim_{\tau \to 0^+} \frac{E_\alpha (\tau^\alpha \mathcal{D}_t^\alpha) - 1}{\tau^\alpha} Q(t), \quad (5.150)$$

where again the exponential operator is replaced by the ML operator.

The generalization of the Taylor expansion to non-differentiable functions is therefore given by

$$Q(t + \tau) = \sum_{k=0}^\infty \frac{\tau^{k\alpha}}{\Gamma(k\alpha + 1)} \mathcal{D}_t^{k\alpha} [Q(t)], \quad (5.151)$$

which is an explicit expansion of Eq.(5.149) for a non-differentiable $Q(t)$.

5.7   After Thoughts

This chapter has given some indication of how linear equations of motion may be generalized to their fractional counterparts and solved. This enables the folding in of certain kinds of complexity into, as well as going beyond, the linear mathematical models with which we are familiar. We have discussed three sources of complexity: statistical fluctuations, nonlinearity and fractional dynamics. However we have seen that the nonlinearity may be incorporated into an infinite-order linear description. Thus, it might be possible to generally represent complexity by infinite-order sets of fractional differential equations. Of course, the expectation that obtaining an infinite-order linear representation of a complex nonlinear problem entails its solution is naive. If that were the case the Hilbert space of quantum mechanics would have made the mathematical description of quantum phenomena straight forward, and that is certainly not the case. But having such a representation does make a class of previously intractable problems somewhat more manageable.

The general solution of a fractional linear system of N elements was expressed through the introduction of an $N \times N$ Mittag-Leffler matrix (MLM) as a generalization of the traditional exponential matrix solution to a linear system of rate equations. In the one-dimensional case a linear fractional equation for the growth of a population was shown to overshoot the saturation level and to decay with oscillations back to saturation. This modeling of the overshoot phenomenon in a biological context seems to be original and awaits a bright student for its development.

The fractional two-dimensional case was used to introduce a general method for solving coupled fractional rate equations using linear eigenvalue techniques. This approach was then applied to the physical example of a fractional harmonic oscillator. A number of authors have studied the fractional harmonic oscillator and some of their results were reviewed in Section 5.2.3, including the properties of MLF's with complex arguments. These fractional oscillators naturally led to the concept of fractional normal modes, which in turn provide a fractional environment with which to model FLEs. The IPL memory kernel of the FLE was shown to provide the rich dynamics observed in the three-scale Brownian motion.

The idea of generalizing control theory to incorporate the memory entailed by the fractional calculus into the controller is perhaps the most significant. A complex phenomenon characterized by a fractal time series can be described by a fractal function. Such a function was shown earlier to have divergent integer-order derivatives. Consequently, traditional control theory, involving as it does integer-order differential and integral operators, cannot be used to determine how feedback is accomplished, but it can be generalized to fractional-order operators as shown. Therefore, it seems reasonable that one strategy for modeling the dynamics and control of such complex phenomena is through the application of the fractional calculus. The fractional calculus has been used to model the interdependence, organization, and concinnity of complex phenomena ranging from the vestibulo-oculomotor system to the electrical impedance of biological tissue to the biomechanical behavior of physiologic organs, see, for example, Magin [28] for an excellent review of such applications.

5.8 Appendix Chapter 5

5.8.1 Putzer Algorithm

In this appendix the proof of the Putzer Algorithm for constructing analytic expressions for matrix exponentials is reproduced for completeness. This method uses only eigenvalues and components in the solution to an n-component linear system. It is found to be particularly useful because of its generalization to a n-component fractional differential linear system provided in the text. Consider the set of linear equations

$$\dot{\mathbf{Q}}(t) = \mathbb{C}\mathbf{Q}(t) \tag{5.152}$$

where $\mathbf{Q}(t)$ is the n-component vector

$$\mathbf{Q}(t) = \begin{pmatrix} Q_1(t) \\ \cdot \\ \cdot \\ \cdot \\ Q_n(t) \end{pmatrix} \tag{5.153}$$

and the dot over the function denotes its derivative with respect to time. The elements of the $n \times n$ matrix $\mathbb{C}$ that couples the vector components together are time independent. The solution to Eq.(5.152) is, or course, formally given by

$$\mathbf{Q}(t) = e^{\mathbb{C}t}\mathbf{Q}(0) \tag{5.154}$$

where $\mathbf{Q}(0)$ is the initial configuration of the vector. We are interested in constructing a representation for the matrix exponential.

We assume the linear system can be diagonalized to produce a set of eigenvalues $\lambda_1, \lambda_2, \cdots, \lambda_n$ of the matrix $\mathbb{C}$ that are not necessarily distinct from one another. These eigenvalues are used to construct the sequence of matrices

$$\mathbf{M}_k = \prod_{j=1}^{k} (\mathbb{C} - \boldsymbol{\lambda}_j \mathbf{I}) = (\mathbb{C} - \boldsymbol{\lambda}_k \mathbf{I}) \mathbf{M}_{k-1} \tag{5.155}$$

and setting $\mathbf{M}_0 = \mathbf{I}$ enables the following formal expression for the matrix exponential:

$$e^{\mathbb{C}t} = \sum_{k=0}^{n-1} q_{k+1}(t)\mathbf{M}_k \tag{5.156}$$

where the components of the vector

$$\mathbf{q}(t) = \begin{pmatrix} q_1(t) \\ \cdot \\ \cdot \\ \cdot \\ q_n(t) \end{pmatrix} \tag{5.157}$$

are taken to be the solution to the rate equations

$$\dot{\mathbf{q}}(t) = \begin{pmatrix} \lambda_1 & 0 & 0 & \cdot & \cdot & 0 \\ 1 & \lambda_2 & 0 & \cdot & & 0 \\ 0 & 1 & \lambda_3 & & & 0 \\ \cdot & 0 & 1 & \cdot & \cdot & \\ \cdot & & & \cdot & \cdot & \\ 0 & & & 0 & 1 & \lambda_n \end{pmatrix} \begin{pmatrix} q_1(t) \\ \cdot \\ \cdot \\ \cdot \\ \cdot \\ q_n(t) \end{pmatrix}, \tag{5.158}$$

subject to the initial condition

$$\mathbf{q}(0) = \begin{pmatrix} 1 \\ 0 \\ 0 \\ \cdot \\ 0 \end{pmatrix}. \tag{5.159}$$

The validity of Eq.(5.156) can be proven in a straightforward manner using the Cayley-Hamilton Theorem. Consider the matrix sum given by the right hand side of Eq.(5.156)

$$\mathbf{G}(t) = \sum_{k=0}^{n-1} q_{k+1}(t)\mathbf{M}_k \tag{5.160}$$

from which we can evaluate the time derivative $\dot{\mathbf{G}}(t)$ using the definition

$$\dot{q}_1(t) = \lambda_1 q_1(t), \tag{5.161}$$

$$\dot{q}_j(t) = q_{j-1}(t) + \lambda_j q_j(t) \quad, \quad j > 1, \tag{5.162}$$

to yield

$$\dot{\mathbf{G}}(t) - \mathbb{C}\mathbf{G}(t) = \sum_{k=0}^{n-1} \dot{q}_{k+1}(t)\mathbf{M}_k - \mathbb{C}\sum_{k=0}^{n-1} q_{k+1}(t)\mathbf{M}_k.$$

Inserting Eq.(5.161) into this expression and using Eq.(5.155) leads to

$$\dot{\mathbf{G}}(t) - \mathbb{C}\mathbf{G}(t) = \lambda_1 q_1(t) + \sum_{k=1}^{n-1} \left[\lambda_{k+1} q_{k+1}(t) + q_k(t)\right]\mathbf{M}_k$$

$$- \sum_{k=0}^{n-1} q_{k+1}(t)\left[\mathbf{M}_{k+1} + \lambda_{k+1}\mathbf{M}_k\right],$$

which by relabeling terms in the second sum simplifies to

$$\dot{\mathbf{G}}(t) - \mathbb{C}\mathbf{G}(t) \;=\; \sum_{k=1}^{n-1} q_k(t)\mathbf{M}_k - \sum_{k=0}^{n-1} q_{k+1}(t)\mathbf{M}_{k+1}$$

$$=\; -q_n(t)\mathbf{M}_n.$$

The proof is complete with the recognition that from the Cayley-Hamilton Theorem $\mathbf{M}_n = 0$, since this defines the characteristic polynomial for the linear system and therefore the matrix sum rate equation

$$\dot{\mathbf{G}}(t) = \mathbb{C}\mathbf{G}(t),$$

is equivalent to the initial value problem given by Eq.(5.152).

5.8.2 Proof of MLMF

A quick proof of the validity of the fact that the MLMF satisfies the system of fractional differential equations is readily obtained. Consider the fractional derivative of the series representation of the MLMF

$$\partial_t^\alpha \left[\mathbf{E}_\alpha \left(\mathbb{C}t^\alpha \right) \right] = \partial_t^\alpha \left[\sum_{k=0}^\infty \frac{\mathbb{C}^k t^{k\alpha}}{\Gamma\left(k\alpha+1\right)} \right] = \sum_{k=0}^\infty \frac{\mathbb{C}^k}{\Gamma\left(k\alpha+1\right)} \partial_t^\alpha \left[t^{k\alpha} \right].$$

The fractional derivative of the monomial in time can be determined using the Laplace transform of the fractional derivative

$$\mathcal{LT}\left\{ \partial_t^\alpha \left[t^{k\alpha} \right]; s \right\} = s^\alpha \frac{\Gamma\left[k\alpha+1\right]}{s^{k\alpha+1}} - s^{\alpha-1}\delta_{k,0} = \Gamma\left[k\alpha+1\right] s^{1-(k-1)\alpha} - s^{\alpha-1}\delta_{k,0}$$

and its inverse

$$\mathcal{LT}^{-1}\left\{ \Gamma\left[k\alpha+1\right] s^{1-(k-1)\alpha} - s^{\alpha-1}\delta_{k,0}; t \right\} = \frac{\Gamma\left[k\alpha+1\right]}{\Gamma\left[(k-1)\alpha+1\right]} t^{(k-1)\alpha} - \frac{t^{-\alpha}}{\Gamma\left(1-\alpha\right)}.$$

Inserting this value for the derivative into the summation, canceling the $k = 0$ terms, and setting $k = m + 1$, yields

$$\partial_t^\alpha \left[\mathbf{E}_\alpha \left(\mathbb{C}t^\alpha \right) \right] \;=\; \sum_{k=0}^\infty \frac{\mathbb{C}^k}{\Gamma\left(k\alpha+1\right)} \frac{\Gamma\left[k\alpha+1\right]}{\Gamma\left[(k-1)\alpha+1\right]} t^{(k-1)\alpha} - \frac{t^{-\alpha}}{\Gamma\left(1-\alpha\right)}$$

$$=\; \mathbb{C}\sum_{m=0}^\infty \frac{\mathbb{C}^m}{\Gamma\left(m\alpha+1\right)} t^{m\alpha} = \mathbb{C}\mathbf{E}_\alpha\left(\mathbb{C}t^\alpha\right), \tag{5.163}$$

which when multiplied on the right by $\mathbf{Q}(0)$ is the original linear system of fractional rate equations given by Eq.(5.5) with the solution given by the MLMF Eq.(5.6).

References

[1] Adams, J.L. and T.T. Hartley. 2008. *J. Comp. and Non. Dyn.* **3**, 021402-1.

[2] Balachandran, K. and J. Kokila. 2012. *Int. J. Appl. Math. Comput. Sci.* **22**, 523–533.

[3] Bechhoefer, J. 2005. *Rev. Mod. Phys.* **77**, 783.

[4] Bhalekar, S. and V. Daftardar-Gejji. 2012. *Int. J. Diff. Eq.* **2012,** doi:10.1155.

[5] Brown, R.G. 1983. *Introduction to Random Signal Analysis and Kalman Filtering,* Wiley & Sons, New York.

[6] Burov, S. and E. Barkai. 2008. *Phys. Rev. Lett.* **100**, 070601.

[7] Burov, S. and E. Barkai. 2008. *Phys. Rev. E* **78**, 031112.

[8] Chen, Y.W., H.S. Ahn and D. Xue. 2006. *Signal Processing* **86**, 2794.

[9] Coffey, W.T., Yu.P. Kalmykov and J.T. Waldron. 2004. *The Langevin Equation,* World Scientific, NJ.

[10] Cornelius, S.P., W.L. Kath and A.E. Motter. 2013. *Nature Comm.* **4**, 1.

[11] Das, S., P.K. Gupta and K. Vishal. 2010. *Applications & Appl. Math.* **5**, 605.

[12] Diethelm, K., N.J. Ford, A.D. Freed and Yu. Luchko. 2005. *Compt. Meth. Appl. Mech. Engin.* **194**, 743.

[13] Duan, J.-S. 2013. *Adv. Math. Phys.* **2013**, ID 869484.

[14] El-sayed, A.M.A., A.E.M. El-Mesiry and H.A.A. El-Saka. 2007. *Appl Math. Lett.* **20**, 817.

[15] Frederuco, G.S.F. and D.F.M. Torres. 2008. *Nonlinear Dynamics* **53**, 215.

[16] Gorenflo, R. and F. Mainardi. 1997. In *Fractals and Fractional Calculus in Continuum Mechanics*, (Eds.) A. Carpinteri and F. Mainardi, Springer-Verlag, Vienna and New York.

[17] Jarad, F., T. Abdeljawad and D. Baleanu. 2012. *Abs. & Appl. Analysis* **2012**, ID 890396.

[18] Jumarie, G. 2006. *Comp. and Math. with Appl.* **51**, 1367.

[19] Jumarie, G. 2013. *Cent. Eur. J. Phys.*, 1.

[20] Khader, M.M. and M.M. Babatin. 2013. *Math. Prob. Eng.* **2013**, ID 391901.

[21] Kaslik, E. and S. Sivasundaram. 2012. *Nonl. Anal.: Real World App.* **13**, 1489.

[22] Kilbas, A.A., H.M. Srivastava and J.J. Trujillo. 2006. *Theory and Applications of Fractional Differential Equations*, Elsevier, Amsterdam.

[23] Kimeu, J.M. 2009. *Masters Theses & Specialized Projects.* http://digitalcommmons.wku.edu/theses/115/

[24] Lazarevoc, M.P. 2014. *Int. J. Nonlinear Mech.* (http://dx.doi.org/10.1016/j.ijnonlinmec.2014.11.011).

[25] Lindenberg, K. and B.J. West. 1990. *The Nonequilibrium Statistical Mechanics of Open and Closed Systems*, VCH Publishers, NY.

[26] Liu, W., R. Yan, W. Jing, H. Gong and Pl. Liang. 2011. *Protein & Cell* **2**, 764.

[27] Loan, C.F. Van. 1977. *SIAM J. Numer. Anal.***14**, 971.

[28] Magin, R.L. 2006. *Fractional Calculus in Bioengineering*, Begell House Inc., NY.

[29] Mandelbrot, B.B. and J.W. Van Ness. 1968. *SIAM Review* **10**, 422.

[30] May, R.M. 1976. *Nature* **261**, 459.

[31] Miller, K.S. and B. Ross. 1993. *An Introduction to the Fractional Calculus and Fractional Differential Equations*, John Wiley & Sons, NY.

[32] Min, W., G. Luo, B.J. Cherayil, S.C. Kou and X.S. Xie. 2005. *Phys. Rev. Lett.* **94**, 198302.

[33] Mohamed, M.S. 2014. *Appl. Compt. Math.* **3**, 27.

[34] Moler, C. and C. Van Loan. 2003. *SIAM Review* **45**, 3.

[35] Ortigueira, M.D. 2008. *IEEE Cir. Sys. Mag.* , 3rd quarter, 19.

[36] Peng, C.K., J. Mistus, J.M. Hausdorff, S. Havlin, H.E. Stanley and A.L. Goldberger. 1993. *Phys. Rev. Let.* **70**, 1343.

[37] Podlubny, I. 1999. *Fractional Differential Equations, Mathematics in Science and Engineering Vol.* **198**, Academic Press, San Diego.

[38] Reynolds, D.E. 2003. arXiv:cond:mat/0309116v1.

[39] Samko, S.G., A.A. Kilbas and O.I. Marichev. 1993. *Fractional Integrals and Derivatives Theory and Applications*, Gordon and Breach, NY.

[40] Sandev, T., R. Metzler and Z. Tomovski. 2014. *J. Math. Phys.* **55**, 023301-1.

[41] Scafetta, N. and B.J. West. 2006. *Geophys. Res. Lett.* **33**, 17.

[42] Scafetta, N. and B.J. West. 2007. *Eur. Phys. Lett.* **79**, 30003-p1.

[43] Stanislavsky, A.A. 2011. *Eur. Phys. J. B* **49**, 93.

[44] Tarasov, V.E. 2012. *Cent. Eur. J. Phys.* **10**, 382.

[45] Tarasov, V.E. 2013. *Comm. Nonl. Sci. & Num. Sim.* **18**, 2945.

[46] Turchetti, G., D. Usero and L. Vazquez, Hamitonian systems with fractional time derivatives, unpublished

[47] Weberszpil, J. 2014. arXiv:1405.4581v1.

[48] West, B.J. and L. Griffin. 1998. *Fractals* **6**, 101.

[49] West, B.J., M. Bologna and P. Grigolini. 2003. *Physics of Fractal Operators*, Springer, Berlin.

[50] West, B.J. 2009. In *Progress in Motor Control*, D. Sternad (Ed.), Springer, NY.

[51] West, D. and B.J. West. 2013. *Phys. of Life* **10**, 210.

[52] West, B.J. 2015. *Physica A* **429**, 103.

[53] Yang, H., G. Luo, P Karnchanaphanurach, T. Louie, I. Rech, S. Cova, L. Xun and X.S. Xie. 2003. *Science* **302**, 262.

CHAPTER 6

Fractional Cooperation

Another aspect of complexity is revealed through cooperative behavior observed in complex dynamic networks. The flocking of birds [22], the schooling of fish [44], the swarming of insects [120], the epidemic spreading of diseases [7], the spatiotemporal activity of the brain [9, 23, 33], the flow of highway traffic [5], and the cascades of load shedding on power grids [21], these and many more complex phenomena demonstrate collective behavior. This behavior in a societal context was brilliantly articulated in the 1852 classic book *Memoirs of Extraordinary Popular Delusions and the Madness of Crowds* by Charles Mackay [52]:

> In reading the history of nations, we find that, like individuals, they have their whims and their peculiarities, their season of excitement and recklessness, when they care not what they do. We find that whole communities suddenly fix their minds upon one object, and go mad in its pursuit: that millions of people become simultaneously impressed with one delusion, and run after it, till their attention is caught by some new folly more captivating than the first. We see one nation suddenly seized from its highest to its lowest members, with a fierce desire of military glory: another as suddenly become crazed upon a religious scruple: and neither of them recovering its senses until it has shed rivers of blood and sowed a harvest of groans and tears, to be reaped by its posterity. At an early age in the annals of Europe its population lost their wits about the sepulchre of Jesus, and crowded in frenzied multitudes to the Holy Land; another age went mad for fear of the devil......Men, it has well been said, go mad in herds, while they recover their senses slowly, and one by one...

Setting aside the eloquent imagery for the moment, a physical scientist might see all these phenomena as being reminiscent of the physics of particle dynamics near a critical point, whereas a social scientist would observe a tipping point; each sees a dynamic system undergoing a phase transition. Explaining such violent social transitions in terms of phase transitions was, in fact, done by two physicists, Callen and Shapero in 1974 [19]. They put together the concepts of social imitation and critical behavior a generation before Gladwell popularized the concept of the tipping point [34]. In a more modern setting critical behavior suggests itself, in part, because this is one area of complexity where the mathematical theory, for example the scaling predictions of renormalization group theory, and experimental data dovetail. The phenomenon of a phase transition, although complex, is fairly well understood, in that both experiment and theory can follow the transition of the short-range particle interactions to long-range interactions as a control parameter of the system is adjusted to the critical point. However, in some phenomena criticality is the normal operating point and the control parameter is the lever the scientist has available to reestablish normality when the systems loses the functionality for which it was designed. We provide a number of applications of the fractional calculus and then return to how it may appear in a social context.

One aspect of critical phenomena that is often overlooked outside the physical sciences is the influence of fluctuations. At the critical point that influence is amplified, not suppressed, and the smallest fluctuation may be amplified to the largest effect. Therefore the non-Gaussian behavior of statistical fluctuations become very important. It has been known for a long time that a physiologic phenomena with demonstrated non-Gaussian statistics is the variability of the beat to beat interval of the human heart. The empirical heart rate variability (HRV) has been shown to have Lévy stable statistics; the latter being the solution to a fractional equation of motion for the PDF.

As humans our senses respond to a number of sensor excitations: seeing, hearing, touching and smelling. The stimuli for two of these four sensor modes propagate as waves. Our eyes are stimulated by light waves, with different frequencies experienced as the spectrum of colors. Our ears respond to sound waves, which at sufficiently low frequencies is not heard, but is experienced as the tactile vibrations of a rock band. The latter is touch and not hearing. Simple linear waves were discussed earlier, and in this chapter we examine some of the ways wave equations become fractional, in either how they propagate, or the way in which they dissipate energy.

Physical complexity can be seen in the apparently stationary patterns of water flowing around a rock in a stream. In a strong current the water can rise in a hump and swirl in vortices in the lee of the rock. On close inspection it can be seen that intermittently generated bubbles in the flow follow erratic paths, but with the large-scale structure persisting for perhaps hours on end. This flow is said to be turbulent and its general description remains a mystery, although most who study such things would say the fluid motion is turbulent and is ultimately described by the Navier-Stokes equations for fluid velocity. In this chapter we address a property of the simplest kind of turbulent flow, that being the stationary statistical nature of homogeneous isotropic turbulence. We show that a fractional equation for the fluid density captures the statistical properties observed in wind gusts over the ocean. The discussion is extended to incorporate turbulent plasmas as well.

The human brain is another area of phenomenological complexity, where physical scientists collaborating with physicians are making inroads into understanding how the various physiologic networks cooperate. Progress has been made through the use of machines to carry out magnetic resonance imaging (MRI) to probe complex, porous and heterogeneous living matter, looking for such things as tumors in the brain. Herein we step beyond the myriad successes made in this area over the past thirty years and examine the extension of theory into the domain where the differential field equations for the magnetization of living matter become fractional in space and time, in order to explicitly model one of the many forms of complexity observed in living tissue.

Of course it is not only physical mechanisms that lead to fractional descriptions of dynamics. It turns out that an erratic flight path or random flight, was thought to provide the most efficient strategy for a predator to explore a spatial area foraging for prey, in the absence of information to constrain the path. But this turned out not to be the case as we subsequently explain. The assumed Rayleigh random flight pattern is replaced by a fractional one, a Lévy flight, in which extraordinary long flight paths appear in the search pattern. The introduction of Lévy flights and Lévy walks in the exploration of phenomena from turbulence to the formation of search strategies, has lead to some controversy that we briefly outline. The observational evidence seems to support the idea that the ordinary diffusion equation ought to be replaced by a fractional diffusion equation to account for the variability in both the predator's search strategy and in the dispersal of prey.

It is also found that the same arguments for how predators forage for prey may also apply to how searches for a stored memory in the brain are made. The experimental evidence strongly suggest that the fractional search hypothesis (FSH) may be one of those general principles for which we are always on the alert. It is also possible that since a search describes how the information from one network is retrieved by a second network, that the FSH is a corollary of the PCM introduced earlier.

These examples are intended to show that the macroscopic fluctuations observed in complex networks display emergent properties of spatial and/or temporal scale-invariance, manifest in IPLs of connectivity and waiting-times. These IPLs cannot be inferred from the equations describing the nonlinear dynamics of the individual elements of the network and are often obtained through large-scale numerical calculations. Despite the advances made by renormalization group and self-organized criticality theories, that have shown how scale-free phenomena emerge at critical points, the issue of determining how the emergent properties influence the microscopic dynamics at criticality is only partly resolved [112]. Put another way, we may understand how criticality changes the large-scale properties of a network, going from the microscopic to the macroscopic, but it is not so clear how that criticality modifies the behavior of the individual, that is going down the scale from the macroscopic to the microscopic.

The utility of the fractional calculus is further demonstrated by capturing the dynamics of the individual elements within a complex network from the information quantifying that network's global behavior. The phase transitions of complex networks suggest the wisdom of using a generic model from the Ising universality class to characterize the network dynamics. West *et al.* [112] demonstrated, using a subordination argument, that the individual's trajectory response to the collective motion of the network is described by a linear fractional differential equation. The solution to these linear fractional equations appears to retain the full influence of the nonlinear network dynamics on the individual, when compared with the numerical simulation. The measurable behavior change in the individual remains undetected by that person and the implications of this mathematical model await the ambitious student for their complete exploration.

6.1 HRV and Lévy Statistics

Peng *et al.* [65] found that the successive increments in the cardiac beat-to-beat intervals of healthy subjects display scale-invariant, long-range anti-correlations. They also found that the histogram for the heartbeat interval increments is well described by Lévy stable PDF as shown in Figure 6.1. For a group of subjects with severe heart disease, they find that the distribution remains unchanged, but the long-range correlations vanish. Therefore, the different scaling behaviors in health and disease are related to the underlying cardiac dynamics.

The steady-state PDF of interbeat intervals is fit to a Lévy stable form:

$$P_{ss}(q) = L_\beta(q) \equiv \mathcal{FT}^{-1}\left\{\exp[-D_\beta |k|^\beta]; q\right\} \tag{6.1}$$

The steady-state form of the Lévy PDF was obtained by West and Seshadri [104] as the asymptotic solution to a fractional equation for the PDF corresponding to a linear dissipative process driven by Lévy noise. The series representation for the Lévy PDF with Lévy index β is given in a number of places, see for example [64]:

$$P_{ss}(q) = \sum_{n=1}^{\infty} \frac{(-1)^{n+1}\,\Gamma\left[\beta n + 1\right] \sin\left[\beta n\pi/2\right]}{\pi\Gamma\left(n\right)} \frac{1}{|q|}\left(\frac{D_\beta}{|q|^\beta}\right)^n \quad ; \quad -\infty < q < \infty,$$

(6.2)

whose lowest order term is the IPL in "space":

$$\lim_{|q|\to\infty} P_{ss}(q) = \frac{\Gamma\left[1 + \beta\right] \sin\left[\beta\pi/2\right]}{\pi} \frac{D_\beta}{|q|^{\beta+1}}.$$

(6.3)

It bears mentioning that space, as used here, is the change in time intervals between successive heart beats.

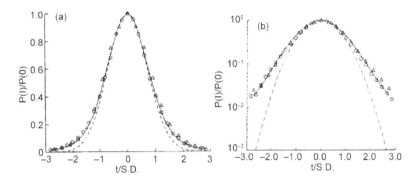

Figure 6.1 The HRV increment data is fit with a Lévy distribution (solid curve) and compared with a Gauss distribution (dashed curve) with the same mean and variance. Healthy (circles) and diseased (triangles) individuals are depicted with the data normalized to the standard deviation and the probability density to $P(0)$. The same data are plotted in (a) and (b) only the vertical axis has been changed to a logarithm in the latter to emphasize the separation between the Gauss and empirical distributions (Adapted from [65] with permission).

The beat-to-beat time series are denoted $B(n)$ for the beat number n. Peng *et al.* [65] explain that the resulting HRV time series are non-stationary as a consequence of the competing neuroautonomic inputs. Parasympathetic stimulation decreases the firing rate of pacemaker cells in the heart's sinus node; sympathetic stimulation has the opposite effect. The competition between these two branches of the involuntary nervous system is the postulated mechanism for much of the erratic variability recorded in healthy subjects [35, 36].

Peng *et al.* [65], in order to remove the non-stationarity in the time series, introduced the difference in the beat interval $I(n) = B(n + 1) - B(n)$, the interbeat increments, which they heuristically determined to be stationary. The second moment of the interbeat increment time series data scale with time as an IPL and the spectrum scales as a power law in frequency

$$S(f) \propto f^\mu \tag{6.4}$$

where $\mu = 1 - 2H$ and the mean-square level of the interbeat fluctuations increases as n^{2H}, as depicted in Figure 6.2. Here $H = 0.5$ corresponds to Brownian motion, so that $\mu = 0$ indicates the absence of correlations in the time series $I(n)$ ("white noise"). They observed that for a diseased data set that μ is approximately zero in the low-frequency regime confirming that the $I(n)$ are not correlated over long times. On the other hand, they also observed that for the healthy data set μ is slightly less than 1 indicating a long-time correlation in the interbeat interval differences. The anti-correlated property of $I(n)$ are consistent with a nonlinear feedback system that "kicks" the heart rate away from extremes. This tendency operates on a wide range of time scales not on a beat-to-beat basis.

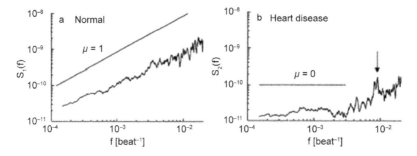

Figure 6.2 The power spectrum for the interbeat interval increments sequences over a 24 hour period. (a) Data from healthy adult. The best-fit line for the low-frequency region has a slope 0.93. (b) Data from a patient with severe heart failure. The best-fit line of the low-frequency region has slope 0.14 (Adapted from [65] with permission).

The spectrum reflects the correlation of the heart beat intervals. In normal health individuals shown in Figure 6.2a the heart beats are determined to be anti-correlated. Correspondingly, this correlation vanishes in patients with heart disease (dilated cardiomyopathy) as shown in Figure 6.2b. This loss of correlation in diseased individuals does not influence the statistics of the heartbeat increments. Both healthy and diseased individuals are described by a Lévy stable PDF as depicted in Figure 6.1 and it is not possible to distinguish between the healthy and diseased individuals statistically.

More recent studies conclude that although the statistics of HRV have heavy tails they need not be strictly Lévy [38] as we discuss in Section 7.6. The physiologic mechanism producing the effect of keeping the heart rate away from extremes, the greatest interbeat intervals, is modeled in Chapter 7 by modifying the fractional phase space equation (FPSE) to obtain a solution that turns out to be an exponentially truncated Lévy PDF.

6.2   Fractional Wave Equations

Wave motion is arguably one of the most familiar forms of organized behavior observed in the physical domain; aside from phase transitions, of course. Waves are seen as ripples emanating from your foot as you step into the bath, or heard in the sound of music from the CD as you lower yourself into the hot water. The water surface moves up and down as the ripples travel, but it is the transverse wave not the water that propagates across the surface of the bath. The water molecules undergo circular motion about their average position and do not travel with the wave. The same is true of the wind generated waves, observed from satellites, on the surface of the earth's oceans and lakes. The acoustic waves, carrying music in air, is another form of wave motion, where the air molecules oscillate around their original location as the longitudinal sound wave passes. It is the organized periodic motion in space and time of the molecules of the supporting medium that constitute wave motion. Recall Newton's calculation of the speed of sound in air.

Of course we are interested in constructing a fractional generalization of this linear wave motion. Let us begin at the point where Lagrange attempted to find the flaw in Newton's argument for obtaining the value for the speed of sound in air. He constructed the discrete equations for a string of linearly coupled oscillators whose displacements are determined by Eq.(2.1), which we rewrite in slightly different form:

$$\frac{m}{\Delta}\frac{d^2\xi(n,t)}{dt^2} = (k\Delta)\frac{[\xi(n+1,t) - 2\xi(n,t) + \xi(n-1,t)]}{\Delta^2} \qquad (6.5)$$

In the continuum limit the space differential vanishes $\Delta \to 0$, the continuous space variable becomes $q = n\Delta$, and the limit yields the constant parameters; $m/\Delta \to \varrho$ the linear mass density, $k\Delta \to \mathcal{Y}$ Young's modulus, and $\xi(n,t) \to \xi(q,t)$ the wave field amplitude in space and time. This limit results in Eq.(6.5) becoming the linear wave equation

$$\frac{1}{c^2}\frac{\partial^2\xi(q,t)}{\partial t^2} = \frac{\partial^2\xi(q,t)}{\partial q^2} \qquad (6.6)$$

where the phase speed of the wave is defined

$$c = \frac{\mathcal{Y}}{\varrho}. \qquad (6.7)$$

An alternative derivation could include a linear dissipation mechanism, so that the wave equation would be replaced with the telegrapher's equation

$$\frac{1}{c^2}\frac{\partial^2\xi(q,t)}{\partial t^2} + \frac{1}{D}\frac{\partial\xi(q,t)}{\partial t} = \frac{\partial^2\xi(q,t)}{\partial q^2} \qquad (6.8)$$

with D being appropriately defined. It is known that at early times an initial impulse propagates as a wave, while at later times it propagates as a diffusion packet. This phenomenon was observed in the early days of telegraphy. Signal diffusion reduced the data rate in long cables such as the early Atlantic cable, requiring the installation of booster stations. Subsequent applications of the telegrapher's equation, when the "wave amplitude" is interpreted as a PDF [64], has been to the propagation of impulses in nerves and to exciton transport in photosynthetic units.

Note that the telegrapher's equation interpolates between the hyperbolic wave equation for $D \to \infty$ with c finite and the parabolic diffusion equation for $c \to \infty$ with D finite. Consider a process by which to introduce the square root of the diffusion equation in one dimension:

$$\left[\partial_t - D\partial_q^2\right]\xi(q,t) = \mathcal{O}_+\mathcal{O}_-\xi(q,t) = 0, \tag{6.9}$$

where the operators are defined in terms of the fractional time derivative and the integer space derivative

$$\mathcal{O}_\pm \equiv \partial_t^{1/2} \pm \sqrt{D}\partial_q. \tag{6.10}$$

Eq.(6.9) may be factored to obtain the two component equations

$$\begin{aligned}
\mathcal{O}_+\xi_+(q,t) &= 0, \\
\mathcal{O}_-\xi_-(q,t) &= 0
\end{aligned}$$

and solved subject to the appropriate boundary conditions. The original diffusion equation may now be expressed in the form of a fractional Dirac equation as

$$\left[\mathbf{A}\partial_t^{1/2} + \mathbf{B}\sqrt{D}\partial_q\right]\mathbf{\Psi}(q,t) = 0 \tag{6.11}$$

where $\mathbf{A}$ and $\mathbf{B}$ are the 2×2 Dirac matrices [66]:

$$\mathbf{A} = \begin{bmatrix} 0 & 1 \\ 1 & 0 \end{bmatrix} \text{ and } \mathbf{B} = \begin{bmatrix} 0 & 1 \\ -1 & 0 \end{bmatrix} \tag{6.12}$$

and $\mathbf{\Psi}(q,t)$ the two-component vector

$$\mathbf{\Psi}(q,t) = \begin{bmatrix} \xi_+(q,t) \\ \xi_-(q,t) \end{bmatrix}. \tag{6.13}$$

The components of the vector are scalar solutions to the standard diffusion equation. Pierantozzi and Vazquez [66] point out that these scalar solutions can be interpreted as PDFs with structure associated with internal degrees of freedom of the system. Thus, Eq.(6.11) is the square root of the classical diffusion equation.

A version of this technique, implementing Laplace transforms, was applied to fluid mechanics by Kulish and Lage [48]. They used the form of the fractional equation that is first order in space and $1/2$-order RL fractional derivative in time to obtain closed-form analytic solutions for the flux and scalar response at a fluid-solid interface. The results were validated considering the known solutions for the first and second Stokes problems.

6.2.1 Dirac's Method Fractionalized

The fractional generalization of Eq.(6.11) can be written [66]

$$[\mathbf{A}\partial_t^\alpha + \mathbf{B}\lambda\partial_q] \boldsymbol{\Psi} (q,t) = 0, \tag{6.14}$$

with $0 < \alpha \le 1$ and with the definitions of the Dirac matrices given above this is the square root of the fractional diffusion-wave equation (FDWE):

$$\partial_t^{2\alpha} [\xi (q,t)] = \lambda^2 \partial_q^2 [\xi (q,t)]. \tag{6.15}$$

Here the parameter λ^2 can be interpreted as the diffusion coefficient as $\alpha \to 1/2$ and as the phase speed of the wave as $\alpha \to 1$. Mainardi [56, 57] presented the solution to the FDWE in the case $0 < \alpha \le 1$. In the region $0 < \alpha \le 1/2$ the equation is considered a fractional diffusion equation and in the domain $1/2 < \alpha \le 1$ it is a fractional wave equation. Somewhat earlier Mainardi [55] had explained that a fractional wave equation is entailed by the propagation of mechanical diffusive waves in viscoelastic media exhibiting power-law creep. Whenever there is a hereditary mechanism of an IPL type in diffusive or wave phenomena the appearance of fractional derivatives is anticipated.

Mainardi [56, 57] used Laplace transforms to solve Eq.(6.15) in two separate cases. The first is the *Cauchy problem,* which is an initial value problem when the data are assigned at $t = 0^+$ on the space axis $-\infty < q < \infty$. The second is the *Signaling problem,* which is an initial boundary-value problem when the data are assigned both at $t = 0^+$ on the semi-infinite space axis $q > 0$ (initial data) and at $q = 0^+$ on the semi-infinite time axis $t > 0$ (boundary data). For tutorial purposes we restrict attention to the Cauchy problem and consider:

$$\begin{aligned} \xi (q,0^+) &= g(q), & -\infty < q < \infty, \\ \xi (\pm\infty, t) &= 0 & t > 0. \end{aligned} \tag{6.16}$$

However we note that for $1/2 < \alpha \le 1$ the value of the time derivative of the field variable must be added to the constraints

$$\partial_t [\xi (q,t)]_{t=0^+} = 0, \tag{6.17}$$

since in this parameter region the solution to Eq.(6.15) requires two linearly independent solutions.

The Fourier transform of Eq.(6.15) yields

$$\partial_t^{2\alpha} \left[\tilde{\xi} (k,t) \right] = -\lambda^2 k^2 \tilde{\xi} (k,t),$$

whose solution can be expressed in terms of the inverse Fourier transform of the MLF

$$\xi(q,t) = \mathcal{FT}^{-1}\left\{E_\alpha\left(-\lambda^2 k^2 t^{2\alpha}\right)\tilde{g}(k);q\right\}. \tag{6.18}$$

However, as we learned in Section 5.2 in the study of the fractional harmonic oscillator that the MLF can be written

$$E_\alpha\left(-\lambda^2 k^2 t^{2\alpha}\right) = \frac{1}{2}\left[E_\alpha\left(i\lambda k t^\alpha\right) + E_\alpha\left(-i\lambda k t^\alpha\right)\right], \tag{6.19}$$

which when inserted into Eq.(6.18) yields for $g(q) = \delta(q)$:

$$\xi(q,t) = \frac{1}{2t^\alpha}W\left(-\frac{|q|}{\lambda t^\alpha};-\alpha,1-\alpha\right) = \frac{1}{2}\left[\xi_+(q,t) + \xi_-(q,t)\right], \tag{6.20}$$

which is a generalization of the D'Alembert formula for the classical wave equation. In this solution we have

$$\xi_+(q,t) = \begin{cases} 0, & q < 0 \\ \frac{1}{t^\alpha}W\left(-\frac{q}{\lambda t^\alpha};-\alpha,1-\alpha\right), & q \geq 0 \end{cases} \tag{6.21}$$

and

$$\xi_-(x,t) = \begin{cases} \frac{1}{t^\alpha}W\left(\frac{q}{\lambda t^\alpha};-\alpha,1-\alpha\right), & q \leq 0 \\ 0, & q > 0 \end{cases} \tag{6.22}$$

where $W(z;\alpha,\beta)$ is the Wright function [56]:

$$W(z;\alpha,\beta) = \sum_{n=0}^{\infty} \frac{z^n}{\Gamma(n+1)\Gamma(n\alpha+\beta)}. \tag{6.23}$$

Consequently, we conclude that the solution to the FDWE for the Cauchy problem is a linear combination of the solutions to the fractional Dirac-type equations, since Eq.(6.14) can be expressed as the equation pair [66]

$$\partial_t^\alpha[\xi_+] = -\lambda\partial_q[\xi_+], \tag{6.24}$$
$$\partial_t^\alpha[\xi_-] = \lambda\partial_q[\xi_-], \tag{6.25}$$

and whose solutions for the Cauchy problem are given by Eqs.(6.21) and (6.22), respectively. The $\alpha = 1/2$ case was explicitly calculated by Kulish and Lage [48]. As pointed out by Mainardi [56] and later by Pierantozzi and Vazquez [66] the fractional analysis provides a way of interpolating between the hyperbolic operator of the wave equation and the parabolic operator of the diffusion equation.

6.2.2 Additional Phenomenology

The FDWE given by Eq.(6.15) describes a broad class of physical processes that can be solved using the above method. The general physical process involves the fractional continuity equation in space such that

$$\partial_t^\alpha \left[\xi \left(\mathbf{r}, t \right) \right] + \nabla \cdot \mathbf{F} \left(\xi \left(\mathbf{r}, t \right) \right) = 0 \tag{6.26}$$

where $\xi \left(\mathbf{r}, t \right)$ is the physical field in three-dimensional space $\mathbf{r}$ and $\mathbf{F} \left(\xi \left(\mathbf{r}, t \right) \right)$ is the corresponding vector flux. The following phenomenological laws have been listed by Vazquez [98] as candidates for generalization:

Fourier's Law:

$$\mathbf{Q} = -\kappa \nabla T(\mathbf{r}, t) \tag{6.27}$$

where $\mathbf{Q}$ is the vector heat flux, κ is the thermal conductivity of the medium and $T(\mathbf{r}, t)$ is the temperature field in space and time.

Fick's Law:

$$\mathbf{F} = -D \nabla C(\mathbf{r}, t) \tag{6.28}$$

where $\mathbf{F}$ is the flux of the solute, D is the diffusion coefficient of the solute in the medium and $C(\mathbf{x}, t)$ is the concentration of the solute. In the anisotropic case, where the diffusion rate differs across direction, the diffusion coefficient can be replaced by a diffusion tensor.

Ohm's Law:

$$\mathbf{J} = -\sigma \nabla V(\mathbf{r}, t) \tag{6.29}$$

where $\mathbf{J}$ is the electrical current (flux of charge), σ is the electrical conductivity of the medium, and $V(\mathbf{r}, t)$ is the voltage. Here again the assumption of spatial homogeneity has been made.

Darcy's Law:

$$\mathbf{q} = -K \nabla h(\mathbf{r}, t) \tag{6.30}$$

where $\mathbf{q}$ is the flux of groundwater, K is the hydraulic conductivity, and $h(\mathbf{r}, t)$ is the potential related to the hydraulic head in three-dimensional space and time.

In each of these phenomenological laws the assumption has been made that the underlying process is differentiable. When this is not the case and the process is truly complex, the field variable is more than likely to be non-differentiable and consequently either the nabla operator or the fractional time derivatives are replaced with their fractional counterpart. This can be done systematically, as for example, in the recent study of anisotropic attenuated fractional wave equations [58]. For example, a fractional Fick's Law generates a fractional diffusion equation, whereas a fractional Fourier's Law produces a fractional heat equation. Langlands *et al.* [49] use a fractional Fick's Law to derive a fractional cable equation for the electrodiffusion of ions in nerve cells. They study two models: one with a D in Eq.(6.28) that is a monomial in time; the other has a D with a RL fractional derivative in time. Both models predict the same power-law diffusive spatial variance, each modeling a different aspect of memory generated under very different physical circumstances.

Let us look a little more closely at the physical processes that can give rise to fractional wave equations.

6.2.3   Fractal Scatterers

The remote sensing of the environment is usually accomplished through the decoding of information carried by waves that have been scattered by distant objects of interest and/or obscured by heterogeneous media. This decoding is routinely done passively as in the detection and processing of reflected sunlight or sound, but only a limited amount of information is obtainable from the sensing of such uncontrolled signals. A great deal more can be learned when the generation of the waves to be detected are controlled in both amplitude and frequency. The waves can be selected and tuned to probe particular properties of the medium of propagation, that is, they can be electromagnetic as in the use of radar, acoustic as in sonar, or even elastic as in the non-destructive testing of materials. The traditional approach has been to construct physical models, make predictions, and compare those predictions to experimental data. This has been a successful strategy since Lord Rayleigh first explained why the sky is blue in terms of the long-wavelength light scattered by air molecules [72].

The scattering of vector and scalar waves from rough surfaces or objects having complicated shapes is not well understood and the simple propagation of waves through inhomogeneous media is an area of active research. In particular when the inhomogeneities are fluctuations in the refractive index of the medium, the scattered wave field becomes stochastic. In the latter case the statistics of the transmitted wave contain information about the fluctuations in the traversed medium. As observed elsewhere [106], the distortion in the signal transmitted from satellites to Earth, the rapid variation in acoustic waves in the deep ocean, the interplanetary scintillation of quasar radio sources, and the twinkling of starlight, each depends on the irregularities in the intervening medium. If we are interested in the unperturbed wave, then the fluctuations imposed are considered to be noise and must be filtered out to obtain the signal. If, however, it is the medium that we wish to understand, then the fluctuations contain information that must be extracted in order to understand the causes of the irregularities in the medium.

Out of this vast field of study of wave propagation and scattering we select only a few topics for discussion in the present context. One that has particular relevance is the use of ultrasonic waves as probes into the structure of materials. Material properties, such as the distribution of grain sizes in polycrystalline materials, the degree of homogeneity, the existence of macroscopic cracks, inclusions, twin boundaries, and dislocations, all affect fracture micromechanics and fracture-control technology. The basis of the ultrasonic approach is the observation that the amplitudes of low-frequency (long wavelength) ultrasonic waves (of known amplitude and direction) are exponentially attenuated with distance, that is, the wave intensity decays as $exp[-\alpha(\omega)z]$, where z is the line-of-sight distance between the source and receiver and $\alpha(\omega)$ is the frequency-dependent attenuation factor.

Elastic wave attenuation in all manner of complex media including living tissue, polymers, rocks and rubber are characterized by attenuation factors that are power law in frequency

$$\alpha(\omega) \propto \omega^{\mu} \qquad 0 < \mu \leq 4. \tag{6.31}$$

More generally an absorption attenuation coefficient (from such effects as dislocation damping as well as magnetoelastic and thermoelastic hysteresis) and a scattering attenuation coefficient [from Rayleigh scattering $\mu = 4$ for wavelength $>>$ typical length of scatterer; and stochastic scattering $\mu = 2$ for wavelength $\leq$ typical length of scatterer]. The power-law index is restricted under various conditions; for example Holm and Sinkus [40] summarize that compressional wave attenuation in biological tissue commonly manifests exponents in the range $1 \leq \mu \leq 2$, whereas the shear waves in tissue are typically in the domain $0 \leq \mu \leq 1$. The non-integer value of the power-law index in the attenuation factor suggests that the volume of space occupied by the scatterers is also non-integer. West and Shlesinger [105] argue that the volume occupied by the scatterers is fractal and that μ is a direct measure of the dimensionality of that physical volume, that is, the power-law index is a measure of the non-integer fractal dimension. Wu [118] developed these ideas in the context of impedance inhomogeneity in the lithosphere of the Earth.

Holm and Sinkus [40] provide a wide range of attenuated wave equations of the general form

$$\nabla^2 \xi(\mathbf{r}, t) - \frac{1}{c^2}\frac{\partial^2 \xi(\mathbf{r}, t)}{\partial t^2} + \mathcal{O}\xi(\mathbf{r}, t) = 0 \tag{6.32}$$

where $\mathcal{O}$ is the loss operator and the equation is appropriate for both compressional and shear waves. We shall satisfy our discussion here with a single example drawn from their discussion of potential fractional loss terms. They discuss the loss operator

$$\mathcal{O}\xi(\mathbf{x}, t) \propto \partial_t^{\mu-1}\left[\nabla^2 \xi(\mathbf{r}, t)\right] \tag{6.33}$$

where $0 < \mu - 1 \leq 1$. This loss term gives a causal wave equation as pointed out by the scientist who introduced it [117]. Holm and Sinkus [40] point out that Eq.(6.33) is based on a constitutive equation which is a fractional stress-strain relation and was first used to derive a lossy wave equation by Caputo in 1967 [20].

6.3   Turbulence

Let us turn from the complexity of wave scattering to one of the great mysteries of classical physics, turbulent fluid flow. This physical example establishes a connection between intermittent statistics and the phase space equation for the PDF. We take up the phase space equation in the next chapter. Although not easily defined, turbulent fluid flow has the characteristics of being unpredictable, rapidly diffusive and to dissipate kinetic energy [89]. Almost a century ago G.I. Taylor [93] attempted to adapt the kinetic theory view point to predict the statistical properties of such fluid flow by taking velocity measurements at a point in space rather than following fluid parcels around in space. We review enough of this perspective to reveal some of the more obvious differences between the statistics associated with the diffusivity of homogeneous isotropic turbulence and the diffusion of a Brownian particle. Moreover, we indicate how the fractional calculus may provide a way to overcome some of the barriers to understanding this phenomenon encountered over the decades.

6.3.1   Richardson Dispersion

The experiments of Richardson [77], discussed in Section 3.3, gave rise to the growth equation for the variance of a passive scalar

$$\frac{d\left\langle |\mathbf{R}(t)|^2 \right\rangle}{dt} = \left\langle |\mathbf{R}(t)|^2 \right\rangle^{2/3}, \tag{6.34}$$

where $\mathbf{R}(t)$ the displacement vector of the passive scalar and from which the Lagrangian or Richardson dispersion

$$\left\langle |\mathbf{R}(t)|^2 \right\rangle \propto t^{3+}, \tag{6.35}$$

was determined. Here the exponent of time is slightly larger than three. As related by Holm [39], the original experiment on which this equation was based was the simultaneous release of ten thousand balloons at the *London Expo* on a windy day in 1925. Each balloon contained a note asking the finder to call and tell him the location and time when the balloon came to earth. The implications of the *Richardson Dispersion Law*, the solution to Eq.(6.34), are that the statistics of turbulent flow are non-Gaussian and intermittent.

An explanation of the statistical intermittence in turbulence was proposed by Shlesinger *et al.* [83] who showed that this law could be derived by assuming the statistical fluctuations were Lévy stable. A PDF is stable if it contains all the convolutions of the laws belonging to it, for example, the convolution of an arbitrary number of Lévy PDFs is another Lévy PDF. The scaling properties of a Lévy PDF and how those properties are related to the fractional partial differential equations are discussed subsequently.

Experiments, observations and data analysis have vindicated the assumption that homogeneous isotropic turbulent fluctuation have Lévy statistics. An exemplar of observational data in given in Figure 6.3 where the PDF for gusts of wind over the open ocean for short time intervals (4 sec) is depicted. The empirical wind field PDF shown in the figure is from the wind gusts off the coast of the German North Sea and is seen to have fluctuations that can be six orders of magnitude greater than a Gaussian PDF with the same mean and variance as indicated by the arrows. The statistics have an IPL tail, which can be modeled as an alpha–stable Lévy PDF. This explanation of intermittency has subsequently found wide application including the statistics of turbulent quantum fluids [4, 114], assessing the quality of wind turbine power [14] and in describing the intermittent fluctuations in the beat to beat intervals of the human heart [108], to name a few.

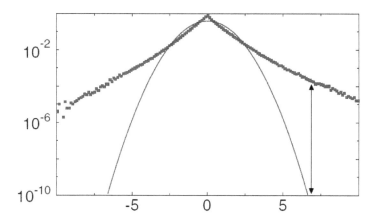

Figure 6.3 Measured probability density of changes of the wind speed over 4 sec; obtained from a wind measurement at the German North Sea coastline. The solid curve corresponds to a Gaussian distribution having the same mean and variance as the observational data. The velocity measurements are normalized to a standard deviation of 0.8 m/sec. (with permission from Boettcher *et al.* [16])

6.3.2 Probability Density Function

The mystery of turbulence is that hundreds of books have been written concerning the thousands of experiments that have been done, but there is no universally accepted set of equations that describe the full dynamic range of turbulent fluid flow. Therefore any discussion of turbulence begins with a set of simplifying assumptions, some more Draconian than others. Keeping this in mind the dynamics for the number density of a passive scalar (tracer) $\rho(\mathbf{r}, t)$ at a location $\mathbf{r}$ in a fluctuating velocity field $\mathbf{V}(t)$ at time t, such as Richardson's balloons or chimney smoke in a fluctuating wind field, can be expressed as [104]:

$$\frac{\partial \rho(\mathbf{r}, t)}{\partial t} + \mathbf{V}(t) \cdot \nabla \rho(\mathbf{r}, t) = f(t)\delta(\mathbf{r}). \tag{6.36}$$

The dramatic over-simplification made here is that the fluctuations in the velocity field are not spatially dependent, or alternatively that the velocity field is homogeneous on the spatial scale over which the dispersion of the passive scalar is being tracked. The number density can be written in terms of the phase space distribution function that keeps track of all the tracer particles

$$\rho(\mathbf{r}, t) = \sum_{n=1}^{N} \delta(\mathbf{r} - \mathbf{R}_n(t)), \tag{6.37}$$

where the location of the dynamic trajectory of the nth particle of the N passive scalar particles is given by $\mathbf{R}_n(t)$ and $\mathbf{r}$ is the corresponding phase space variable. In Eq.(6.36) $\mathbf{V}(t)$ is the fluctuating velocity field of the ambient fluid in which the passive scalar is embedded and is assumed to dominate over such influences as chemical reactions and molecular diffusion in the case of particles. Note that the source of the fluctuations in Brownian motion is thermal, the kinetic energy of the ambient fluid particles, whereas the fluctuations in turbulence are a consequence of the dynamics of the macroscopic flow field. The time-dependent source of the passive scalar given by $f(t)$ defines the origin of the coordinate system.

The solution to Eq.(6.36) is obtained by first taking the spatial Fourier transform, then solving the resulting linear rate equation and finally averaging the solution over the velocity fluctuations. We do not carry out these operations here, but in subsequent sections such details are shown in the solving of analogous problems. Boettcher *et al.* [16] did a statistical analysis of wind fields and determined them to be well described by an alpha−stable Lévy statistical process with an average Lévy index of $\beta = 1.5$. The averaging (denoted by a bracket) of the Fourier transformed solution to Eq.(6.36) is therefore carried out using the properties of alpha−stable Lévy statistics $dL(t)$ to obtain [26]:

$$\left\langle \exp\left[\int_0^t ig(t')dL(t')\right]\right\rangle = \exp\left[-K_\beta \int_0^t |g(t')|^\beta \, dt'\right], \tag{6.38}$$

from which West and Seshadri [104] determined the homogeneous solution to Eq.(6.36) in terms of the inverse three-dimensional Fourier transform $\mathcal{FT}^{-1}\{\cdot; \mathbf{r}\}$ to be

$$\langle \rho(\mathbf{r}, t)\rangle = \mathcal{FT}^{-1}\left\{\exp\left[-K_\beta |\mathbf{k}|^\beta t\right]\widetilde{\rho}(\mathbf{k}, 0); \mathbf{r}\right\}. \tag{6.39}$$

The average over the ensemble of trajectories when properly normalized determines the PDF

$$P(\mathbf{r}, t) \equiv \langle \rho(\mathbf{r}, t)\rangle, \tag{6.40}$$

of a tracer particle being in the interval $(\mathbf{r}, \mathbf{r} + d\mathbf{r})$ at time t. The time derivative of Eq.(6.39) yields the fractional diffusion equation that is isotropic in time

$$\frac{\partial P(\mathbf{r}, t)}{\partial t} = K_\beta \nabla^\beta P(\mathbf{r}, t), \tag{6.41}$$

which is solved subject to the source $f(t)$ located at $\delta(\mathbf{r})$. The fractional nabla operator $\nabla^\beta(\cdot)$ in this case is the three-dimensional Riesz-Feller fractional derivative [30] defined in terms of its Fourier transform:

$$\mathcal{FT}\left\{\nabla^\beta\left[G(\mathbf{r})\right]; \mathbf{k}\right\} \equiv -|\mathbf{k}|^\beta \, \tilde{G}(\mathbf{k}), \tag{6.42}$$

and K_β is a generalized diffusion coefficient; here K_β is the diffusivity of the turbulent fluid flow. See Appendix 7.9.2 for details on the Riesz-Feller fractional derivative. Note that if the velocity fluctuations are Gaussian then $\beta = 2$, in which case Eq.(6.41) reduces to the turbulent diffusion equation found in a number of standard texts, for example, Monin and Yaglom [63].

In general the self-similar scaling of the velocity in time is inextricably linked to the self-similar variability of the PDF of the passive scalar in space as manifest in the fractional spatial derivative. This is consistent with Richardson's observation that the turbulent velocity field is not differentiable. Introducing the mean energy dissipation rate ϵ the parameters in Eq.(6.41) can be replaced using $K_\beta |\mathbf{k}|^\beta = \epsilon^{1/3} |\mathbf{k}|^{2/3}$ to obtain for the Fourier representation of the turbulent diffusion equation [89]

$$\frac{\partial \widehat{P}(\mathbf{k}, t)}{\partial t} = -\epsilon^{1/3} |\mathbf{k}|^{2/3} \, \widehat{P}(\mathbf{k}, t). \tag{6.43}$$

The Richardson approximation allows us to introduce the turbulence diffusion coefficient $D(k)k^2 = \epsilon^{1/3} |\mathbf{k}|^{2/3}$ so that $D(k) = \epsilon^{1/3} |\mathbf{k}|^{-4/3}$, which is consistent with the Kolmogorov [46] eddy cascade model of the coupling between scales in homogeneous turbulence, which now has a fractional calculus basis.

The solution to the turbulent diffusion equation obtained for Eq.(6.43) using Eq.(6.39) is a Lévy PDF. Such statistics have been observed in laboratory experiments involving two-dimensional turbulent flow [87, 119], as predicted theoretically by Shlesinger *et al.* [83]. The dispersive particles (passive scalars) in the turbulent flow field were observed to resemble Lévy single particle trajectories, that is, they were interpreted by Xia *et al.* [119] to be small displacements of particles trapped within the forcing scale vortices for long times, only to be followed by long jumps. Solomon *et al.* [87] were able to experimentally determine that the particles are trapped within the vortices for trapping times having an IPL PDF with index 1.6 ± 0.3, and having some indication of an exponential cut-off of the tail at long times. We shall discuss the mechanisms resulting in such truncation of IPLs subsequently in the context of truncated Lévy distributions.

6.3.3   Plasma Turbulence

Turbulent transport does not only occur in neutral fluids, but in a variety of other physical problems including, as discussed by Sánchez et al. [78], in solar and atmospheric turbulence, combustion and many other areas where anomalous scalings are observed. They point out that many simulations of the radial transport of tracer particles in turbulent plasmas satisfy fractional transport equations:

$$\partial_t^\alpha \left[\rho(\mathbf{r}, t) \right] = K_{\alpha\beta} \nabla_r^\beta \left[\rho(\mathbf{r}, t) \right], \tag{6.44}$$

which, of course, is a generalization of Eq.(6.41). Here ρ is the mass density of the tracer particles, $K_{\alpha\beta}$ is the effective fractional diffusivity and $\nabla_r^\beta [\cdot]$ is the radially symmetric Riesz fractional derivative in space, whereas $\partial_t^\alpha [\cdot]$ is the Caputo fractional derivative in time. Sánchez *et al.* [78] provide the first formal derivation from a reasonable description of the microscopic dynamics, using a clever application of renormalization group ideas.

However, the arguments they [78] employ are specific to a physical environment, where the microscopic equation of motion for a single tracer particle is

$$\frac{d\mathbf{r}}{dt} = \mathbf{V}(\mathbf{r}, t), \tag{6.45}$$

and $\mathbf{V}(\mathbf{r}, t)$ is the incompressible turbulent velocity field. The tracer fluid density satisfies the continuity equation

$$\frac{\partial \rho(\mathbf{r}, t)}{\partial t} + \mathbf{V}(\mathbf{r}, t) \cdot \nabla \rho(\mathbf{r}, t) = 0, \tag{6.46}$$

which differs from Eq.(6.36) in that the velocity is here dependent on the spatial variable. Sánchez *et al.* [78] use the term "renormalization" for any transformation that generates a linear transport equation of the class defined by Eq.(6.44) from Eq.(6.46), with the diffusivity determined by the flow statistics. The general argument centers on the separation of variables into an average and a fluctuating part:

$$\rho(\mathbf{r}, t) = \langle \rho(\mathbf{r}, t) \rangle + \delta\rho(\mathbf{r}, t), \tag{6.47}$$
$$\mathbf{V}(\mathbf{r}, t) = \langle \mathbf{V}(\mathbf{r}, t) \rangle + \delta\mathbf{V}(\mathbf{r}, t). \tag{6.48}$$

Inserting these equations into the continuity equation they solve for the dynamics of the average and fluctuating mass density separately, using arguments that are fluid specific and not readily extendable to a more general setting. Consequently we refer the reader back to their excellent paper [78] to determine which arguments can and cannot be applied in the reader's area of expertise.

Magnetically confined fusion plasmas are known to deviate from the standard diffusion model as a consequence of a number of factors including nonlocality, nonlinearity, long-range correlations and non-Gaussian fluctuations. Del-Castillo-Negrete *et al.* [25] take these various contributions to complexity into account using fractional derivative operators. The strategy they adopt is to carry out the calculation to two distinct ways. First by direct numerical simulation of the coupled dynamics of the electrostatic potential and pressure of the plasma in a cylinder using the separation into average and fluctuating quantities such as given in Eqs.(6.47) and (6.48). The results of the numerical simulation are then treated as data to be explained using a simpler physical model. That simpler model is schematically of the form of Eq.(6.44) with a source term for the PDF for the radial displacement of the tracer particles. Their philosophy is very similar to the one adopted in the modeling the numerical results of the DMM network in Section 6.6.

The solution they obtain for the PDF is terms of a dimensionless radial variable x and dimensionless time t:

$$P(x,t) = \int_{-\infty}^{\infty} \mathcal{G}(x - x', t) P_0(x') dx', \tag{6.49}$$

where $P_0(x)$ is the initial PDF and the Green's function for the fractional diffusion equation [67] in terms of the similarity variable $\eta = x \left(K_{\alpha\beta}^{1/\alpha} t \right)^{-\alpha/\beta}$ is

$$\mathcal{G}(x,t) dx = \mathcal{K}(\eta) d\eta, \tag{6.50}$$

and

$$\mathcal{K}(\eta) = \operatorname{Re} \mathcal{FT} \left\{ E_\alpha \left(-z^\beta \right) ; \eta \right\}. \tag{6.51}$$

The Green's function captures the familiar results: for $\beta = 2, \alpha = 1$, $\mathcal{G}$ reduces to the Gaussian form; for $1 < \beta \leq 2, \alpha = 1$, $\mathcal{G}$ reduces to a symmetric Lévy PDF; for $\beta = 2, 0 < \alpha < 1$, $\mathcal{G}$ reduces to the solution of the subdiffusive fractional equation [59]. The analytic result is depicted by the solid curve in Figure 6.4 and the fit to the numerical results given by the triangles is quite good. What is clear is that the PDF does not have the parabolic form of a Gaussian PDF.

In the present context the nonlocality of the fractional derivatives is a consequence of non-Fickian effects that arise from avalanche-like events that induce large displacements of tracer particles. Just as in the hydrodynamic experiments of Solomon *et al.* [87] the tracer particles are either trapped in eddies for long times, or as pointed out by del-Castillo-Negrete *et al.* [25], they jump over several sets of eddies in a single flight. This is one source of anomalous diffusion in plasma turbulence.

6.4 Fractional Magnetization Equations

Another application of the fractional calculus appears at the nexus of physics and physiology in the measurement of healthy tissue and tumors. This is

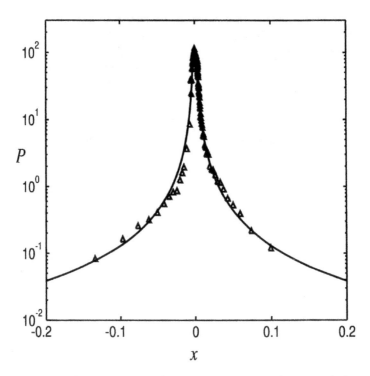

Figure 6.4 Non-Gaussian PDF of tracer particles in plasma turbulence. The triangles depict the numerical results from the turbulence model [25]. The solid line is their analytic solution using the fractional calculus model with $\alpha = 1/2, \beta = 3/4$ and $K_{1/2,3/4} = 0.09$. Adapted from [25]

where nuclear magnetic resonance (NMR) and magnetic resonance imaging (MRI) are used to regularly examine complex, porous and heterogeneous materials for both living and inanimate systems. The motivation to generalize the linear, dissipative and diffusive phenomenological Bloch equation description of magnetic precession and relaxation of nuclear spins was a new method that Bennett *et al.* [12, 13] used to describe diffusion in MRI implementing a 'stretched exponential' function $e^{-bt^{\alpha}}$ to fit experimental data. Rather than interpreting the stretched exponential as a way to improve curve fitting Magin *et al.* [53] saw it as a way to connect nanoscale fractal models of porous materials [43, 45, 116] and tissues with observable NMR relaxation and anomalous diffusion processes. But before looking into the fractional field equations, it might be useful for the reader to go back and review the phenomenological rate equations of Bloch, discussed in Section 4.3.3.

6.4.1 Fractional-Order Bloch-Torrey Equation

The validity of the phenomenological Bloch equations is well established in MRI studies for static magnetic fields up to 10 Tesla [60]. As Magin *et al* [53] point out, as MRI is applied, with increasing resolution in space and time, the description of spin dynamics needs to be modified to incorporate water diffusion heterogeneity. The diffusion mechanism becomes increasingly important at higher resolution and subsequently the Bloch equations introduced in Section 4.3.3 were extended [95] to include diffusion as captured in the Bloch-Torrey equation:

$$
\frac{\partial \mathbf{M}(\mathbf{r},t)}{\partial t} = \gamma \mathbf{M}(\mathbf{r},t) \times \mathbf{B}(\mathbf{r},t) - \frac{1}{T_2} [M_x(\mathbf{r},t)\mathbf{e}_x + M_y(\mathbf{r},t)\mathbf{e}_y]
$$
$$
- \frac{1}{T_1} [M_z(\mathbf{r},t) - M_0]\mathbf{e}_z + D \nabla^2 \mathbf{M}(\mathbf{r},t), \tag{6.52}
$$

for an isotropic medium with the same diffusion coefficient in each of the three directions $(\mathbf{e}_x, \mathbf{e}_y, \mathbf{e}_z)$. In the rotating frame of the magnetic field, neglecting relaxation and assuming $\mathbf{B}(\mathbf{r},t)$ to be only due to the time-varying magnetic field gradient $\mathbf{G}(t)$, hence $\mathbf{B}(\mathbf{r},t) = \mathbf{r} \cdot \mathbf{G}(t)\mathbf{e}_z$ reduces Eq.(6.52) to

$$
\frac{\partial M_\pm(\mathbf{r},t)}{\partial t} = \lambda(\mathbf{r},t) M_\pm(\mathbf{r},t) + D \nabla^2 M_\pm(\mathbf{r},t); \quad \lambda(\mathbf{r},t) \equiv -i\gamma \mathbf{r} \cdot \mathbf{G}(t). \tag{6.53}
$$

We shall not be concerned with solving this equation here, but point out that diffusion MRI, or diffusion tensor imaging, as it is known in the imaging of anisotropic materials, has been used successfully for over 30 years.

Magin *et al.* [53] used fractional-order differential operators to generalize the Bloch-Torrey equation in order to characterize neurodegenerative, malignant and ischemic diseases. The basis for their generalization was the CTRW, which we discuss in the next chapter, to provide the theoretical underpinnings for the introduction of fractional operators in both space and time to capture the complete spin dynamics that has been called "spin turbulence". The generalized Bloch-Torrey equation can be written schematically, to the same level of approximation as Eq.(6.53):

$$
\tau^{\alpha-1} \partial_t^\alpha [M_\pm(\mathbf{r},t)] = \lambda(\mathbf{r},t) M_\pm(\mathbf{r},t) + K_\beta \nabla^{2\beta} [M_\pm(\mathbf{r},t)], \tag{6.54}
$$

where we have used the Caputo derivative in time $\partial_t^\alpha [\cdot]$ and the Riesz-Feller nabla operator in three dimensional space $\nabla^{2\beta} [\cdot]$. Magin *et al.* [53] introduced the time parameter τ just as they did for the Bloch equation to preserve the proper time units and the fractional diffusion coefficient introduces the parameter μ, $K_\beta = \mu^{2(\beta-1)} D$ to preserve the proper units in space. We show how to integrate fractional diffusion equations of this kind in the next chapter.

It is worth pointing out here that the biophysical reasons for extending the original equations to fractional form are based on the complexity of the underlying tissue. Magin *et al.* [54] explain that the CTRW random walker (water molecules in proton MRI) in biological tissue (membranes, organelles and cells) is disrupted in both the length of the sojourn times between steps, as well as, in the stride length of those steps. This complexity leads to an equation for a particle undergoing anomalous diffusion and the solution to the fractional Bloch-Torrey equation is expressed in terms of MLFs. Here again we emphasize that it is the complexity of the tissue that is encoded in the fractional derivatives and this complexity provides new mechanisms for enhancing the contrast in diffusion-weighted MRI.

The solutions to the new Bloch-Torrey equation [53] were successfully compared with three kinds of experiments: (1) dextran polymer gel with many small interconnecting pores; (2) human articular cartilage plugs; (3) diffusion-weighted brain imaging on a healthy human volunteer. Here we show the quality of the solution for diffusion-weighted imaging experiments for $\alpha = 1$ and spatial heterogeneity by fitting the calculated decay of image intensity S:

$$S = S_0 \exp\left[-bD_m\right]; \; D_m^\beta = D\left(\frac{\Delta}{\mu^2}\right)^{1-\beta}, \tag{6.55}$$

for increasing b as defined in the caption of Figure 6.5.

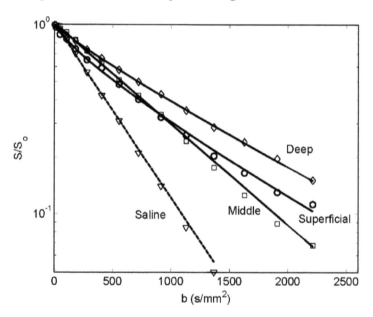

Figure 6.5 Normalized signal intensity plotted versus b, where $b = (\gamma G_z \delta)\Delta$, for selected regions of interest in the three different zones (superficial, middle and deep) of a human cartilage sample and in saline. The experimental data were fit to the fractional order stretched exponential model to determine D, β and μ for $\Delta = 25ms$ and $\delta = 1ms$. (adapted from [53] with permission)

The cartilage/bone sample consists of three distinct regions, the superficial, the middle and the deep, each of which, as explained by Magin *et al.* [53], has its own characteristic composition in terms of water content and structural organization, as manifest in the orientation of collagen fibers. The spatial fractional order parameter lies in the region $1 \geq \beta \geq 0.77$, where ordinary diffusion $\beta = 1$ is found for the saline solution. These results are found to be consistent with the overall structure of cartilage, which allows water to freely diffuse in the middle zone ($\beta = 1$), but restricts diffusion in the superficial ($\beta = 0.77$) and deep ($\beta = 0.9$) regions. The lower values of the fractional-order parameters in the latter two regions is indicative of the complexity of the tissue, with the lower β values indicating more complex or heterogeneous tissue.

They [53] emphasize that the utility of the fractional calculus lies in the encoding of information into the fractional operators about the molecular interactions of spin-labeled water that is embedded in the structure of polymers, membranes, and the extracellular matrix of cells and tissue. Finally, they point out that clinical applications of these techniques have been proposed for assessing the severity of stroke, cancer progression and spinal injury. In addition they emphasize that the fractional Bloch-Torrey equation will ultimately need to be justified through its utility in describing NMR phenomena.

Very recently studies into the development of numerical methods for solving the Bloch-Torrey equations in two and three spatial dimensions have been published [88, 121].

6.5   Fractional Search Hypothesis

The recognition that the Lévy PDF is a solution to a fractional differential equation [81] in 1982 was determined as part of a search initiated a few years earlier to model and understand anomalous transport using Lévy statistics [64]. Once introduced into the physicist's modeling tool kit the Lévy PDF was observed in all manner of physical [85], social [59] and biomedical phenomena [107], including intermittent search strategies [11]. A characteristic of FRWs resulting in a Lévy PDF is a scaled clustering of steps in space separated by large walker-free zones as suggested in Figure 6.6. We also encountered the Lévy PDF in the discussion of homogeneous isotropic turbulence and postponed a more detailed discussion of its mathematical properties until the next chapter.

It is useful to have in mind a concrete model of Lévy random movement as we discuss applications of the Lévy PDF to real swarm activity involving the dispersal of elements without communications. Consider a random walk in which the length of each step has a random direction and is drawn from an IPL of step lengths. If each step takes the same time duration, independently of the length of the step, the process is a Lévy flight. If however the velocity ties the step length to the duration of the time interval for the step, as was

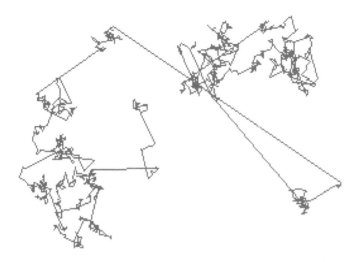

Figure 6.6 The Lévy flight differs from a RRW in that the mover can take arbitrarily long jumps in each time interval. The generalized central limit theorem can be used to show that a jump size q drawn from the IPL transition PDF $1/|q|^{\alpha}$ produces a random displacement PDF that converges on an alpha-stable Lévy PDF.

done in the description of homogeneous turbulence [83], the process is a Lévy walk.

6.5.1   Foraging

The first indication that Lévy random flights could be used as a foraging strategy due to the efficiency with which the flyer explores space was given by Shlesinger and Klafter [82]. This suggestion was subsequently applied to a natural system by Viswanathan *et al.* [99] for the foraging of a wandering albatross. It was assumed that an albatross flies until prey is spotted and then it lands to feed; the time intervals between landings was recorded and provided a measure of the distance between the locations of prey. A histogram of the flight durations was determined to be an IPL in time $t^{-\alpha}$, that being the tail of a Lévy PDF with $\alpha \approx 2$. They [99] interpreted this result to be a consequence of food being fractally distributed on the ocean surface for which a Lévy random flight was an efficient foraging strategy. Subsequently, a torrent of theoretical investigations were undertaken to prove that the Lévy foraging strategy was optimal under a variety of circumstances, see, for example, Lomholt *et al.* [51].

A decade after the analysis of Viswanathan *et al.* [99], Edwards *et al.* [28] showed that the original statistical analysis of the data was flawed and called into question the notion that an albatross actually uses a Lévy foraging strategy. They also questioned the inferences from the experimental findings in the search strategies of deer and bumblebees [100]. The controversy over the evidence of whether animals in ecosystems actually implement the optimal Lévy foraging strategy is well documented by Bénichou *et al.* [11], who offer alternative strategies to resolve the disagreement. However this is not the end of the saga.

Most recently Humphries *et al.* [41] have been able to show the success of the fractional foraging strategy in natural environments. They did this in two ways. One is by re-examining previously studied data sets using new, more robust, statistical techniques and the other is using a more recent GPS data track of wandering albatross. What they found is that the Lévy pattern did not always occur, but was in fact dependent on the environmental context, such as whether or not the prey were sparsely distributed in space, as had been anticipated by theory. When the prey are sparsely distributed the individual albatross are observed to adopt the fractional foraging strategy. On the other hand, when prey are plentiful the albatross adopts the generally less efficient RRW strategy. Consequently, when the foraging under these different strategies are combined into a single data ensemble, as had been done by previous investigators, the resulting distribution could and did look quite different from the Lévy random walk.

Evidence indicates that species in other ecological communities also apply Lévy walks and flights to foraging strategies. These others include the free-ranging spider money (*Ateles geoffroyt*) in the forest of the Yucatán Peninsula, México [71] and as pointed out by Viswanathan [101] in species ranging from dinoflagellates [8] to fish [84]. Using data on 55 individual fish including sharks, tuna, billfish and a sunfish they find that the search mode changes from Lévy flights in the open ocean where prey is scarce to RRWs where prey is plentiful. The evidence is not conclusive however, and so the switching between the two search strategies is still the subject of research. These efforts have culminated in a book on the physics of foraging [102].

Riascos and Mateos [75] take cognizance of the fact that the Lévy PDF observed in foraging is entailed by the Lévy navigation strategy and this strategy can give rise to a walker more efficiently covering a network than a walker using the traditional RRW. They go on to observe that this enhanced efficiency can have the effect of dynamically transforming a large-world network into a small-world network, by means of long-range steps. They subsequently develop this idea further [76], by showing analytically that a fractional random walk on a network, such as introduced in Section 4.1.2, can be used to model a Lévy random flight. Moreover, this is a discrete version of the fractional phase space equation for the PDF developed in Chapter 7. As they point out, their fractional analysis provides a framework to deal with a rich dynamics on complex networks that includes, among other things, Lévy flights.

6.5.2 Human Mobility

A consensus is emerging that fractional foraging is the application of an optimal search strategy in complex ecological systems. But this strategy is not only used by animals in search of prey, but it also appears that Lévy flights are genuinely intrinsic to human mobility [73], as well as to groups of individuals [18]. One measure of the mobility statistics of individuals that has captured the imagination of lay people, as well as scientists, is the circulation of bank notes in the United States. Brockmann *et al.* [17] use over one million displacements of a set of bank notes to determine that the geophysical displacement trajectories of human travel follow those of Lévy flights. A more detailed discussion of this process is presented in Chapter 7 where the idea of the evolution of the PDFs for fractional statistics is developed.

The fractional diffusion equation for the dynamics of human mobility are given in [17] by Eq.(7.37) based on the CTRW. Brockmann *et al.* [17] are able to solve the equation using the methods of fractional calculus, which we also do in Chapter 7 to obtain the scaled PDF given by Eq.(7.110). González *et al.* [37] emphasize that the dispersal of bank notes is a proxy for human mobility and comment that a Lévy flight patterns is consistent with observation.

However they [37] go on to critique the interpretation of this bank note exchange as being a reliable indicator of human mobility and conclude that it is not. Instead they use mobile phones as a more reliable proxy of human mobility, since a phone is carried by a single individual and does not involve hand-offs. They analyzed the statistical properties of the mobility patterns by measuring the displacement between consecutive calls for over 16 million displacements and found an exponentially truncated hyperbolic PDF:

$$P(r) \propto \frac{\exp\left[-r/\kappa\right]}{(r + r_0)^\beta}, \tag{6.56}$$

with the fitted parameter values $\beta = 1.75 \pm 0.15$, $r_0 = 1.5km$ and cutoff value $\kappa = 400km$ for one data set. They established that this distribution is the result of a convolution over two effects: 1) each individual follows a Lévy trajectory with jump size PDF given by Eq.(6.56) and 2) the differences between individuals produces a population-based heterogeneity. The resulting distribution is the PDF for a truncated Lévy flight. They go on to observe that the regularity displayed by individuals in the use of mobile phones, because they return to a few highly frequented location, does not apply to bank notes. Consequently, dollar bills diffuse, human do not.

This most recent vindication of the fractional search strategy in foraging suggests that the evolution of PDFs can be determined by fractional diffusion equations such as that given by Eq.(7.37). The identification of the search strategy and human mobility patterns necessary for the correct modeling of social networks with the fractional calculus further suggests that there may be a fundamental approach with which to replace the phenomenological random walk or random flight arguments and that is by fractional variational arguments [1, 42]. Such an approach would provide a generalization to Noether's theorem in mechanics to more generic complex phenomena.

6.5.3 Cognitive Searching

In a totally different context evidence is accumulating [69, 70, 74] that the human brain employs the same Lévy strategy in seeking to put a name to the face of the person coming towards you in a crowded room and who clearly recognizes you. Baronchelli and Radicchi [6] note that Lévy patterns are observed in memory retrieval and they argue that these patterns suggest that the brain regions involved in such activity are old in an evolutionary sense, that is, the regions of the brain engaged in memory searches do not involve the frontal cortex.

Radicchi *et al.* [70] obtain empirical evidence that players taking part in an online auction, a lowest unique pay-to-bid auction, explore the bidding space by means of a Lévy flight strategy arguing [69] that the exploration of the space is a mental search process. As is characteristic of Lévy flights the exploration of the bidding space occurs in bursts, that is, consecutive bids are close to one another in value, but intermittently there are a significantly larger bids made. The resulting PDF is IPL in the size of the bid. This is like an animal searching for scarce sources of food in space and like the foraging animal the strategy does not appear to be learned. They [70] emphasize that this strategy appears to be an intrinsic part of the mental search process in that the jump lengths in bid space follow the same IPL at any stage of the auction covering four orders of magnitude

6.6 The Network Effect

The regulatory dynamics of the brain [23], the cardiovascular system [91], and indeed most biological/sociological networks appear to be poised at criticality. The existence of phase transitions is so common, in part, because criticality is the most parsimonious way for a many-body system, with nonlinear interactions to exert self-control. Given the ever expanding collection of scientists that subscribe to this view it is worthwhile to have a clear picture of what is meant by a phase transition. Moreover, this view of complexity control is completely consistent with the generalizations of dynamic equations using the fractional calculus. Due to the torrent of papers published in the area of complex networks over the past decade in order to keep the discussion manageable we introduce the decision making model (DMM) as a concrete exemplar of a mathematical model possessing the properties of interest.

A network's influence on the dynamics of an individual within the network is determined in this section using a subordination argument. The network effect requires the development of some nomenclature to make it understandable and for this we need the mathematical concept of subordination. The idea of subordination that is introduced requires the existence of two different notions of time. One is the operational time τ, which is the internal time of a single individual. An individual is assumed to generate ordinary dynamics of a non-fractional system. The other notion of time is chronological time; the time as measured by the clock of an external observer. Note that the idea of operational time ought to be familiar. We encountered a version of it in the discussion of empirical physiologic time in which the time interval or frequency experienced by an organism is determined by the total body mass of that organism; this appears as an allometry relation [79, 110]. A similar distinction is made in psychology, where the subjective time experienced by an individual in the performance of tasks is separate and distinct from the objective time of the clock on the wall [109]. Given this empirical distinction between the reality of the individual and that of the collective, mathematicians recognized the need for a time that was intrinsic to a process, whose dynamics are regular, but that appears quite complicated to an observer measuring the process from outside. Consequently, a procedure was developed to transform intrinsically regular behavior to the experimentally observed complex behavior by relating operational time to chronological time.

6.6.1 Complexity and Criticality

The complex networks described by the DMM implements the echo response hypothesis (ERH), which assumes that the dynamic properties of a network of identical individuals is determined by individuals imperfectly copying the behavior of one another [112]. Formally an isolated individual is modeled as switching back and forth between two states with an exponential probability of making a transition in any interval of time with a rate g_0. The network structure has been configured as a two-dimensional lattice having an individual interacting with only its four nearest neighbors. A two-state master equation determines the probability that an individual switches states in a given time step.

In a DMM network the state of an individual $s_i(t)$ is described by a two-state master equation,

$$\frac{d\mathbf{p}^{(i)}(t)}{dt} = \mathbf{G}^{(i)}(t)\mathbf{p}^{(i)}(t), \qquad (6.57)$$

where $\mathbf{G}^{(i)}(t)$ is a 2×2 transition matrix:

$$\mathbf{G}^{(i)}(t) = \begin{bmatrix} -g_{+-}^{(i)}(t) & g_{-+}^{(i)}(t) \\ -g_{-+}^{(i)}(t) & g_{+-}^{(i)}(t) \end{bmatrix}, \qquad (6.58)$$

and the probability of being in one of two states $(+1, -1)$, is $\mathbf{p}^{(i)}(t) = (p_+^{(i)}, p_-^{(i)})$. Positioning N such individuals at the nodes of a lattice network yields a system of N coupled two-state master equations [96] which, contain time-dependent transition rates for each of the i individuals:

$$g_{+-}^{(i)}(t) = g_0 \exp\left[\frac{K}{N}\left(N_+^{(i)}(t) - N_-^{(i)}(t)\right)\right];$$

$$g_{-+}^{(i)}(t) = g_0 \exp\left[-\frac{K}{N}\left(N_+^{(i)}(t) - N_-^{(i)}(t)\right)\right], \tag{6.59}$$

where K is the strength of the interaction and is referred to as the control parameter. In the mean-field approximation, as the number of element of the network diverges

$$p_\pm^{(i)}(t) = \lim_{N \to \infty} \frac{N_\pm^{(i)}(t)}{N}, \tag{6.60}$$

where $N_\pm^{(i)}(t)$ is the number of elements in state ± 1 at time t for the neighbors of individual i. In this limiting case the transition rates become exponentially dependent on the state probabilities, resulting in a highly nonlinear master equation [112].

The existence of a phase transition in a complex network consisting of N individuals with a coupling parameter K is measured by the mean field amplitude,

$$\xi(K,t) = \frac{1}{N}\sum_{j=1}^{N} s_j(K,t), \tag{6.61}$$

where $s_j(K,t)$ is the value (± 1) of the element j, which is used to calculate the equilibrium global value

$$\xi_{eq} \equiv \langle|\xi(K,t)|\rangle. \tag{6.62}$$

The brackets denote an average over an ensemble of realizations of the trajectory of the global variable. When the coupling parameter $K > 0$, single elements of the network become more and more cooperative and for coupling values larger than a critical one, $K > K_c$, the interaction between individuals is sufficiently strong to give rise to a majority or consensus state, during which the majority of individuals adopt the same opinion at the same time. At the critical value $K = K_c$ there is a qualitative change in the dynamics. Thus, the global dynamics of the DMM is characterized by a phase transition with respect to the coupling parameter K, demonstrating that a network of identical imitating individuals is able to reach consensus, given sufficient influence of the imitation on their decisions [112].

The first observation to make, given the content of the previous section, is that a phenomena may have an IPL distribution in some observable that characterizes the process, and although criticality does entail an IPL in the global variable, an IPL in itself does not guarantee criticality. This needs to be stressed because of the growing influence network research is having on how complexity is interpreted in the social and life sciences. Complex networks are often characterized by their topology, as measured by the distribution in the number of links connecting its members. Those whose topology is IPL are said to be scale-free, and consequently they do not have a characteristic number of links. However as Beggs and Timme [10] explain in their brilliant Socratic dialogue on the nature of criticality in the brain, it is possible for non-scale-free networks to exhibit critical behavior and it is also possible for scale-free networks to not exhibit critical behavior. Consequently the connectivity and the criticality of a network are in fact very different properties.

Criticality is observed as a change from one form of matter (phase) to another form of that same matter, as a consequence of a parameter of the system being varied (control parameter), and this only occurs in systems that are sufficiently large. Critical behavior occurs in phenomena consisting of a large number of identical basic elements, whose interactions change due to changing the control parameter, resulting in a macroscopic change in the behavior of the systems (phase change), at the critical point. The existence of a critical point entails a dramatic change in the system dynamics as this point is approached, resulting in a change in the large-scale system behavior.

Traditionally the correlation function is considered to be an equilibrium property of a network [32], however, in order to study temporal fluctuations a distinction from the equilibrium correlation function must be made. For each instant of time we define the spatiotemporal correlation function:

$$C(r,t) = \langle [s_i(K,t) - \langle s_i(K,t) \rangle] [s_{i+r}(K,t) - \langle s_{i+r}(K,t) \rangle] \rangle .$$

The difference term $s_i(K,t) - \langle s_i(K,t) \rangle$ in the correlation function determines the fluctuation in the value of individual i from its average value, whereas $s_{i+r}(K,t) - \langle s_{i+r}(K,t) \rangle$ denotes a similar fluctuation in the opinion of an individual a distance r away. Therefore to make $C(r,t)$ large both the element i and the one at $i+r$ must fluctuate in a coordinated way, at the same time and in the same direction. If the two elements separated by a distance r are aligned, the dynamic correlation would be zero, but the static correlation would still be 1 as explained by Beggs and Timme [10]. Consequently, both fluctuations and cooperation are required to have a substantial dynamic correlation.

Figure 6.7 depicts lattice configurations in the neighborhood of the transition of the global variable $\xi(K,t)$ and their corresponding dynamic correlation functions. One state is indicated as white and the other as black in this figure. It is evident from panel (6.7a) that the majority of individuals in the network are in the white state, even though a not insignificant number hold the black opinion. These fluctuations away from a uniform state are due to the finite number of elements in the network. The spatial correlation function at this point in time is denoted by the lowest curve in the bottom panel of the figure. Note the rapid relaxation of $C(r,t)$ to vanishingly small values with increasing distance r.

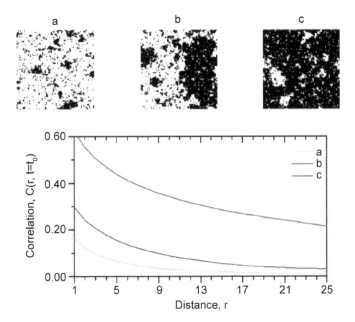

Figure 6.7 Configuration of the lattice (a–c) and corresponding correlation function (bottom panel). White areas correspond to the units in state "yes" and black to the units in the state "no". Lattice size is $N = 100 \times 100$, with the control parameter $K = 1.66$, and the transition rate is $g = 0.10$. (Adapted from [112] with permission.)

Similarly, in panel (6.7c) the vast majority of individuals are in the black state. The spatial correlation function for this configuration is given by the middle curve in the bottom panel. Here again the relaxation of $C(r,t)$ to zero is exponential. Panel (6.7a) is the dynamic state of the network just prior to transition and that of (6.7c) is the lattice configuration just after the transition has occurred. The transition is not just a switch however, it occurs with its own characteristic lattice configuration that persists for a short but finite time.

The slower than exponential relaxation of $C(r,t)$ with distance, given by the uppermost curve in the lower panel, corresponds to the lattice configuration depicted in panel (6.7b). In this configuration there are nearly an equal number of individuals in each of the two states, but they exist in dynamic clusters. A movie of the dynamics would reveal that these cluster migrate across the lattice, with some growing in size and others diminishing. The global variable is in the process of transitioning from the majority white state on the left to the majority black state on the right.

It should be emphasized that the difference in the spatial relaxation properties, manifest by the correlation functions shown here, occur very close together in time, but the transition is not infinitesimally fast. In the short time to transition from the white to the black majority state the network passes from a short-range relaxation process, to one that is long-range, and back to a short-range process. Note that in both of the majority states the process is short range, however during the brief period of transition or instability, the process becomes critical and the internal dynamics induce long-range correlations in the fluctuations.

This property of extending the correlation length is characteristic of systems undergoing a phase transition [32] at a critical point. However, for this to be critical behavior requires certain scaling conditions to be met. These are met for variations in the control parameter through the critical value [112], but have not yet been confirmed for the brief time during the transition between majority states. It is important to note that this extended correlation implies the emergence of dynamical coupling between individuals that are not nearest neighbors and therefore not directly linked in the master equation, even though it has not yet been verified that this is a critical state.

6.6.2 Subordination

Svenkeson *et al.* [92] point out that in operational time, an individual's behavior can appear deterministic, but to an experimenter observing the individual, their temporal behavior can appear to erratically grow in time, then abruptly to freeze in different states for extended time intervals. Here we use the notion of a complex dynamic network, consisting of a large number of individuals. Each individual in isolation fluctuates between two states, according to a simple rate equation. In the case of isolation, the operational and behavioral times coincide. The average behavior of the individual in this uncoupled case is determined by a Poisson distribution. When the network elements are allowed to interact the behavior of the individual changes and the two times become distinct. The dynamics of the network influences the dynamics and the statistics of the individual. Due to the random nature of the evolution of the individual in chronological time, the subordination process involves an ensemble average over many individuals, each evolving according to her own internal clock. The resulting ensemble average over a large number of individuals results in an average over a collection of independent random trajectories.

In the operational time frame the typical temporal behavior of an individual is regular and evolves exactly according to the ticks of that individual's clock. Therefore it is assumed that the opinion of individual i in operational time τ is well defined and given by $Q(i, \tau)$, which is the exponential solution to the linear rate equation Eq.(4.37). On the other hand, introducing the discrete time interval $\Delta\tau$ after n ticks of the individual's clock results in

$$Q_n^{(i)} = (1 - \lambda_0 \Delta\tau)^n Q_0^{(i)}. \tag{6.63}$$

In this discrete form the operational time τ is replaced by the discrete value n such that $Q_n^{(i)} \equiv Q^{(i)}(\tau = n\Delta\tau)$. However when the individual is part of a dynamic network the exponential no longer describes her behavior. Adopting the subordination interpretation the discrete index n is an individual's operational time that is stochastically connected to the chronological time t, in which the global behavior of the network is observed. So what is the behavior of the individual in chronological time? The answer to that question depends on the network's dynamics.

The dynamic network we adopt in this discussion is that generated by the DMM, which is a member of the Ising universality class. Being a member of the Ising universality class means that the solution to the DMM equations of motion has the same scaling behavior as does the Ising model for phase transitions in non-equilibrium statistical physics [110]. In the DMM an isolated individual randomly switches between two states according to the discrete operational time. The subordination argument, which is presented below, is applied to the individual to quantify the influence of the network on the individual's dynamics and the discrete index becomes tied to the occurrence of fluctuating events within the network [?]. The properties of these network fluctuations become strongly amplified as the control parameter approaches its critical value and the network dynamics undergoes a phase transition.

The solution to the discrete equation Eq.(6.63) is an exponential, in the limit $\lambda_0\Delta\tau << 1$, such that $n \to \infty$ and $\Delta\tau \to 0$, but in such a way that the product $n\Delta\tau$ becomes the continuous time τ. However, when the individual is influenced by the network the limit of the discrete solution is no longer exponential. Adopting the subordination interpretation, we define the discrete index n as an operational time that is stochastically connected to the chronological time t in which the global behavior is observed. We assume that the chronological time lies in the interval $(n-1)\Delta\tau \le t \le n\Delta\tau$, and consequently, the equation for the average dynamics of the individual is given by [68]:

$$\langle Q(t) \rangle = \sum_{n=0}^{\infty} \int_0^t G_n(t, t') Q_n dt', \qquad (6.64)$$

where we have suppressed the individual index i and understand we are discussing the response of a typical individual to the network dynamics.

The physical meaning of Eq.(6.64) is determined by considering each tick of the internal clock n as measured in experimental time, to be an event induced by the network. Since the observation is made in experimental time the time intervals between events define a set of independent identically distributed random variables. The integral in Eq.(6.64) is then built up according to renewal theory [103]. After the n-th event, the individual changes from state Q_{n-1} to Q_n, where she remains until the next event is generated. The sum over n takes into account the possibility that any number of events could have occurred prior to an observation at experimental time t, and $G_n\left(t, t'\right) dt'$ is the probability that the last event occurs in the time interval $\left(t', t' + dt'\right)$.

6.6.3　Network Survival

We assume that the waiting times between consecutive events in Eq.(6.64) are identically distributed independent random variables, so that the kernel is defined:

$$G_n\left(t, t'\right) = \Psi\left(t - t'\right) \psi_n\left(t'\right), \tag{6.65}$$

The probability that no event has occurred in a time t since the last event is given by the survival probability $\Psi(t)$. Individual events occur statistically with a waiting-time PDF $\psi(t)$ and taking advantage of their renewal nature, the waiting-time PDF for the n-th event in a sequence is connected to the previous event by a convolution integral

$$\psi_n(t) = \int_0^t dt' \psi_{n-1}(t') \psi(t - t'), \tag{6.66}$$

and $\psi_0(t) = \psi(t)$. The waiting-time PDF is related to the survival probability through:

$$\psi(t) = -\frac{d}{dt}\Psi(t). \tag{6.67}$$

The intermittent statistics of network events were shown by direct numerical integration of the DMM two-state master equation [96] to have an IPL waiting-time PDF as depicted in Figure 6.8. Here, we select the probability of no event occurring up to time t to be

$$\Psi\left(t\right) = \left(\frac{T}{T+t}\right)^\alpha, \tag{6.68}$$

which asymptotically is IPL. Consequently, the waiting-time PDF is renewal and also IPL. As mentioned, the rationale for choosing Eq.(6.68) for the survival probability is the temporal complexity observed in numerical calculation of the DMM for networks with up to 10^4 individuals. The elements in the DMM consist of two-state oscillators with the interaction between elements determined by a N-dimensional two-state master equation. Figure 6.8 illustrates the consensus survival probability $\Psi\left(t\right)$ corresponding to the critical value of the control parameter for two kinds of networks. The DMM lattice network that we use for the subordination argument is given by the solid black line. Although emerging from a simple spatial network, that is a network composed of a lattice on which the individuals are restricted to nearest neighbor interactions, and there is no structural complexity, the survival probability is scale-free with $\alpha = \mu - 1 \approx 0.55$ over more than four decades of time.

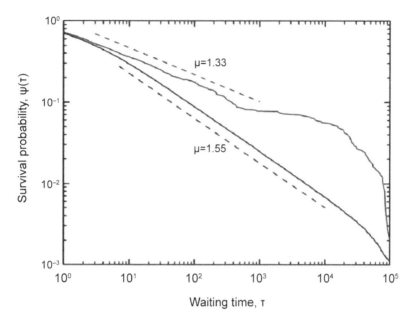

Figure 6.8 The survival probability versus waiting time on doubly logarithmic graph paper. Both curves are for a DMM network of 10^4 individuals, the dark curve is for a two-dimensional lattice with nearest neighbor interactions and the grey curve is for an *ad hoc* random network. Both DMM networks are calculated with the coupling parameter at its critical value. The dashed lines guide the eye to corresponding scaling exponents $\mu = 1.55$ and $\mu = 1.33$, respectively (from [112] with permission)

To find an analytical expression for the behavior of the individual in experimental time, it is convenient to study the Laplace transform of Equation (6.64):

$$\left\langle \widehat{Q}(s) \right\rangle = \widehat{\Psi}(s) \sum_{n=0}^{\infty} Q_n \left[\widehat{\psi}(s) \right]^n, \tag{6.69}$$

and the relation $\widehat{\psi}_n(s) = \left[\widehat{\psi}(s) \right]^n$ was used and results from the convolution structure of Eq. (6.66). With the operational discrete time solution Eq. (6.63) this last expression can be written as:

$$\left\langle \widehat{Q}(s) \right\rangle = \widehat{\Psi}(s) \sum_{n=0}^{\infty} (1 - \lambda_0 \Delta\tau)^n Q_0 \left[\widehat{\psi}(s) \right]^n. \tag{6.70}$$

Performing the sum and noting the relationship given by Eq. (6.67), we find:

$$\left\langle \widehat{Q}(s) \right\rangle = \frac{1 - \widehat{\psi}(s)}{s} \frac{1}{1 - (1 - \lambda_0 \Delta\tau)\widehat{\psi}(s)} Q_0, \tag{6.71}$$

This last equation can be expressed in the form:

$$\left\langle \widehat{Q}(s) \right\rangle = \frac{1}{s + \lambda_0 \Delta\tau \widehat{\Phi}(s)} Q_0, \tag{6.72}$$

with the newly defined function

$$\widehat{\Phi}(s) = \frac{s\widehat{\psi}(s)}{1 - \widehat{\psi}(s)}, \tag{6.73}$$

which is the Montroll–Weiss memory kernel obtained in CTRW [62], discussed in Chapter 7. In the asymptotic limit $s \to 0$, when the survival probability is given by Eq.(6.68), Eq. (6.72) simplifies to

$$\left\langle \widehat{Q}(s) \right\rangle = \frac{1}{s + \lambda s^{1-\alpha}} Q_0, \tag{6.74}$$

with the parameter value:

$$\lambda = \frac{\lambda_0 \Delta\tau}{T\Gamma(2 - \alpha)}.$$

It is evident from previous discussions that Eq.(6.74) has an inverse Laplace transform solution that is the MLF:

$$\langle Q(t) \rangle = Q_0 E_\alpha \left(-\lambda t^\alpha \right). \tag{6.75}$$

Consequently, the subordination process results in the ordinary rate equation through the inverse Laplace transform of Eq. (6.74), being replaced with the fractional rate equation [86, 92]:

$$D_t^\alpha \left[\langle Q(t) \rangle \right] - \frac{t^{-\alpha}}{\Gamma(1 - \alpha)} Q_0 = -\lambda \langle Q(t) \rangle, \tag{6.76}$$

where we have introduced the RL fractional operator defined in Section 4.2. Of course, we could just as well have defined the dynamic equation Eq.(6.76) in terms of the Caputo or MRL fractional derivative

$$\partial_t^\alpha \left[\langle Q(t) \rangle \right] = -\lambda \langle Q(t) \rangle, \tag{6.77}$$

in which the initial condition has been absorbed into the definition of the fractional derivative. Note that these fractional rate equations reduce to the ordinary rate equation in the limit $\alpha = 1$. Consequently, the solution to Eq. (6.76) reduces to the exponential in this limit, as does the solution to Eq.(6.77).

The subordination argument more generally yields a FLE in chronological time [112] in terms of the RL fractional derivative:

$$D_t^\alpha \left[Q(t)\right] - \frac{t^{-\alpha}}{\Gamma\left(1-\alpha\right)} Q_0 = -\lambda Q(t) + f\left(t\right), \qquad (6.78)$$

or in terms of the Caputo or MRL fractional derivative

$$\partial_t^\alpha \left[Q(t)\right] = -\lambda Q(t) + f\left(t\right), \qquad (6.79)$$

either of which is satisfactory. The fractional index α is determined by the IPL survival probability of the network, which determines how long an observer must wait between the occurrence of successive changes of global opinion. The zero-centered fluctuations of the random force $f\left(t\right)$ are determined by the statistics entailed by the finite size of the complex network. Consequently, the ensemble average of Eq.(6.78) and (6.79) yields Eq.(6.76) and (6.77), respectively.

6.6.4 The Subordinated Person

The solution to the noise-free (averaged) FLEs given above is the MLF, and consequently the average opinion of an individual is predicted to change much more slowly than the exponential rate of the isolated person. The time-dependent average opinion of a randomly chosen individual is presented in Figure 6.9, where the average is taken over 10^4 independent realizations of the dynamics. The DMM network dynamics have been shown [96] to undergo a phase transition as the control parameter is varied below, at and above its critical value. The results of the DMM calculation in the subcritical, critical and supercritical regimes are shown by the three panels in Figure 6.9 [113]. These properties are entailed by the fundamental assumption that the individuals within the DMM network imperfectly imitate one another.

The general solution to the FLE given by Eq.(6.79) is

$$Q(t) = Q(0)E_\alpha\left(-\lambda t^\alpha\right) + \int_0^t \lambda\left(t-t'\right)^{\alpha-1} E_{\alpha,\alpha}\left(-\lambda\left[t-t'\right]^\alpha\right) f\left(t'\right) dt' \quad (6.80)$$

where the homogeneous solution is given by a MLF and the kernel of the integral is the two-indexed MLF. Because the fluctuations are symmetric the average of the solution over an ensemble of realizations of the single individual's trajectories is the homogeneous term alone. Consequently, the dynamics of the average individual is the same as that of the individual in isolation, which when $\alpha = 1$ is the exponential shown by the dotted line in the three panels in Figure 6.9.

The parameters of the MLF are fit to the numerical calculations of a two-dimensional lattice network of ten thousand individuals undergoing nearest neighbor interactions [113]. Note that the average influence of the 9,999 other members of the highly nonlinear complex dynamic network

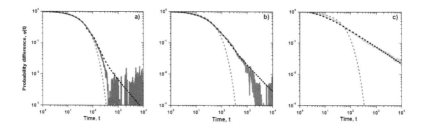

Figure 6.9 The dashed curve in the three figures is the exponential for the average opinion of an isolated individual. The dotted curve is the fit of the MLF to the erratic network dynamics calculated using the DMM for a network of 10^4 elements. Left: subcritical; Middle: critical; Right: supercritical. (From [113] with permission.)

on the individual of interest is here predicted by the MLF solution of a linear fractional differential equation without linearizing the dynamics. The response of the individual to the group mimics the group's behavior most closely when the control parameter is equal to or greater than the critical value as depicted in Figure 6.9. In other words, an isolated person with a Poisson personality, indicated by the dotted line in the figure, becomes an interactive person with a Mittag-Leffler personality asymptotically in time, as indicated by the numerical and analytic solutions to the FLE. In this way an individual is transformed by those with whom s/he interacts.

The results of this section provides a partial answer to a question in sociophysics identified by Kulakolwki and Nawojczyk [47] concerning how empirical regularities such as prejudice or tolerance can be derived from global social properties such as entropy or temperature. We can interpret their use of the nomenclature "temperature" as the control parameter in the DMM network. Here we have demonstrated how the state of the network, as described by the global dynamics, can influence the decision making behavior of the individual.

In the middle of the last century, when all the young men had returned home from World War II, obtained an education on the GI Bill, and entered the civilian workforce, the phenomenon of loss of individuality was identified in *The Organization Man* [115]. This was one of the most influential popular books ever written about American corporate life and its effect on the behavior of the individual. Whyte attempts to resolve two conflicting points of view. On the one hand, is the American tradition of the rugged individual overcoming all obstacles using individual creativity to realize his/her dreams. On the other hand, there is safety in numbers and it is believed by many that organizations make better decisions than do individuals. Consequently, working within an organization is preferable to working alone. An analysis of these two perspectives led Whyte to argue for the creativity of the individual to produce better results than those made by the anonymous risk-adverse managers of the corporation. The arguments are compelling, but whether or not one is convinced, one way or the other, depends on their predispositions.

In his argument Whyte adopted the social assumption that individuals have a "need to belong", which is in conflict with their independent nature, and often over rides it. Actualizing the need to belong results in the individual's personality becoming subservient to the expectations of the organization. We see a related effect resulting from the dynamics in the DMM social network. However the influence of the organization (network) on the individual in the DMM is not the result of some vague urge to belong, but is entailed by two cascading effects. At bottom is the interaction hypothesis that people imperfectly imitate one another. As a consequence of this interaction the DMM network undergoes a phase transition and it is this cooperative behavior that produces the change in the individual's behavior. Consequently, the organization can have a subtle, yet profound, influence on the decision making behavior of an individual. Furthermore, this influence is the result of the way people interact with one another and is not a consequence of any intended behavior inspired by the organization. The latter effect may in fact occur, but conformity in itself does not require it.

6.6.5   The Brain is Critical

Topology and criticality are two of the central concepts arising in the application of dynamic networks to the understanding of the measurable properties of the brain. The spatial organization or topology is related to the IPL PDF of such newly observed phenomena as neuronal avalanches [9] and criticality [33]. The second phenomenon, that of criticality, was observed in a number of networks consisting of a large number of structurally similar interacting elements with properties determined by local interactions. At a critical value of the control parameter the interactions suddenly change from local to long range and what had been the superposition of independent dynamic elements becomes dominated by long-range interactions and coordinated activity. The dynamical source of these latter properties in explaining criticality in the context of the brain is a topic of active investigation [80].

The approach of assessing a network's complexity solely by means of its topology has been widely adopted. We [111] adopted a somewhat different point of view and emphasized the emergence of temporal complexity through the intermittency of events in time, as well as through topological complexity entailed by the dynamics. An event is interpreted as a transition of a global variable from one critical state to another. In this way we identify two distinct forms of complexity; one associated with the connectivity of the elements of the network and the other associated with the variability of the time interval between events. Both forms of complexity have IPLs; one in terms of the number of connections, the other in terms of the time intervals between events.

Although independent of one another, the two measures of complexity are consequences of criticality.

A procedure widely adopted in neuroscience is to define functional connections between different brain regions [33, 90]. Numerous studies have shown the scale-free character of networks created by correlated brain activity as measured through electroencephalography [61], magnetoencephalography [90] or magnetic resonance imagining [29]. Fraiman *et al.* [33] used the Ising model to explain the origin of the scale-free neuronal network, and found the remarkable result that the brain dynamics operate at the corresponding critical state. Our own research [111] yielded the additional discovery that the emergence of consensus using a DMM network produces long-range connections as well as a scale-free topology.

The analysis of brain dynamics by Bonifazi *et al.* [15] established that, in a manner similar to other biological networks, neural networks evolve by gradual change, incrementally increasing their complexity, and rather than growing along the lines of preferential attachment, neurons tend to evolve in a parallel and collective fashion. The function of the neuronal network is eventually determined by the coordinated activity of many elements, with each element contributing only to local, short-range interactions. However, despite this restriction, correlation is observed between sites that are not adjacent to each other, a surprising property suggesting the existence of a previously incomprehensible long-distance communication [24]. The DMM dynamical approach affords the explanation that the local, but cooperative, interactions embed the elements in a phase-transition condition that is compatible with long-range interdependence.

What does this have to do with the fractional calculus?

The answer to this question relies on the fact that a proper description of critical behavior requires non-local interactions in space and time. One descriptor of this non-locality is provided by renormalization group relations and another by fractional derivatives. In the next chapter we show that the solution to the fractional partial differential equation, with α the order of the time derivative and β the order of the space derivative, has the following properties: (1) if $\alpha = 1$ and $0 < \beta < 2$ the asymptotic solution is IPL in space; (2) if $0 < \alpha < 1$ and $\beta = 2$ the asymptotic solution is IPL in time and (3) if $0 < \alpha < 1$ and $0 < \beta < 2$ the asymptotic solution is IPL in both space and time, with independent indices. Note that to make a complete connection with the network interpretation the space degree of freedom is given by the connectivity index [97].

The self-organization of the DMM implemented on the two-dimensional network nearest neighbor interactions, generates a scale-free topology with $\nu \approx 1.1$, as well as the long-range links essential for the collective mind of the network of self-organized elements of Couzin [24]. The exciting discovery of dynamic-hub neurons with $1.1 \leq \nu \leq 1.3$ [15] is a challenge for the dynamical derivation of the scale-free condition that is mainly confined to $\nu > 2$ [2, 27,

31]. It is remarkable that the DMM approach generates an IPL index in the range of the experimental results of Bonifazi *et al.* [15]. Aquino *et al.* [3] show that this kind of complex dynamical network shares the brain's sensitivity to $1/f$ noise. In short, the scale-free PDF of links is a consequence of dynamic self-organization rather than being the cause of it.

6.7 After Thoughts

Perhaps the most surprising result presented in this chapter was the fact that the detailed dynamics of a highly complex network with imitative interactions [96, 112] could be reduced to a FLE description. We have seen that one method of constructing fractional differential equations is through the process of subordination, whereby a deterministic rate equation in operational time is transformed to a fractional differential equation in chronological time. For the rate of change of a process this transforms a Poisson process into a ML process, that is, from an isolated individual without memory, to a member of a network with an IPL memory. This transition was particularly interesting in the application to complex networks, where the interaction of the individual with the network at criticality collapses the long-range influence that emerges at the phase transition into a fractional derivative. At the present time this collapse is a phenomenological result, meaning that we do not yet have an intuition as to how the two-state master equation for the DMM network can produce fractional dynamics. What seems clear is that the fractional operator is a consequence of criticality, but how to prove that in general remains an open question.

The construction of the FLE is phenomenological, as was Langevin's original argument. The test of the FLE is the quality of the predictions of the PDF given by the solutions when compared with the numerical simulation of the DMM network. The fit of the solutions to the data generated by numerical simulation is excellent [113]. This result suggests that the fractional calculus might be used in the future to replace the equations of motion for highly nonlinear dynamic networks, with an equivalent set of fractional linear dynamic stochastic equations. One of the benefits of the latter representation would be the ability to use control theory directly to model how one might intervene to guide the operation of the network, say to facilitate the work of a physician studying physiological networks.

One complex physical process whose simplest model benefits from the fractional calculus is homogeneous turbulence. The statistics of a homogeneous turbulent velocity field measured at a point has a Lévy PDF, which is to say the velocity fluctuations have a very long tail in the time interval between fluctuations of a given size. The velocity field observed in both the open ocean and in the laboratory are seen to display such non-Gaussian statistics. The velocity fluctuations have macroscopic time scales that are distinct from Brownian motion where the fluctuations are thermally generated and therefore have microscopic time scales. Consequently, the two phenomena have no physical mechanisms in common and they are rarely found in the same experiment, due to this wide separation in time scales. However the fractional operator does couple the dynamics over a wide variety of scales in space and time as we discuss more fully in the next chapter.

There is a trade-off between the fractional differential equation driven by white noise and a simple rate equation driven by a more complex noise field. Consider the Laplace transform of a fractional Langevin equation where $f(t)$ is white noise

$$\partial_t^\alpha [Q(t)] = f(t),\tag{6.81}$$

given by

$$s^\alpha \widehat{Q}(s) - s^{\alpha-1} Q(0) = \widehat{f}(s).\tag{6.82}$$

The equation in Laplace variables can be simplified to

$$s\widehat{Q}(s) - Q(0) = \frac{\widehat{f}(s)}{s^{\alpha-1}} \equiv \widehat{g}(s).\tag{6.83}$$

Inverse Laplace transforming this last equation yields

$$\frac{dQ(t)}{dt} = g(t).\tag{6.84}$$

Thus, the memory induced by the fractional response to the white noise driver in Eq.(6.81) is mathematically equivalent to a linear response to fluctuations with an IPL memory. However, the systems described by Eq.(6.81) and (6.84) are quite different. In the former case the internal dynamics of the system has a memory and selectively responds to fluctuations in the white noise with an ever receding strength in time. On the other hand, the latter case is one in which the system responds instantaneously to fluctuations that have a slowly fading memory of fluctuations that have occurred in the past.

Magin (in a private communication) suggested the situation where the inhomogeneous term is not noise, but is proportional to the dynamic variable, that is, $f(t) = \lambda Q(t)$. In this latter case the rate equation given by Eq.(6.84) becomes

$$\frac{dQ(t)}{dt} = \lambda(t)Q(t),\tag{6.85}$$

in which we have a relaxation rate that is IPL in time:

$$\lambda(t) \equiv \frac{\lambda}{t^{1-\alpha}}.\tag{6.86}$$

The relaxation rate clearly decreases with increasing time for $0 < \alpha < 1$ and the solution to the rate equation is a stretched exponential

$$Q(t) = Q(0)e^{-\lambda t^\alpha},\tag{6.87}$$

the early time approximation to the MLF.

A related argument was recently given by Thiel and Sokolov [94] to interpret the time-dependent diffusivity introduced by Batchelor into the study of homogeneous isotropic turbulence. The "scaled" Brownian motion (sBm) they constructed had a Langevin equation with white noise $\eta(t)$ multiplied by a monomial in time: $g(t) \propto t^\gamma \eta(t)$ in Eq.(6.84), for an appropriately chosen power-law index, as observed a decade earlier by Lim and Muniandy [50]. This is another of the many kinds of Brownian motions that investigators have found useful in studying a variety of partly deterministic and partly random phenomena that can be described using the fractional calculus.

References

[1] Agrawal, O.P. 2007. *J. Phys.* **40**, 6287.

[2] Albert, R. and A.-L. Barabási. 2002. *Rev. Mod. Phys.* **74**, 47.

[3] Aquino, G., M. Bologna, P. Grigolini and B.J. West. 2010. *Phys. Rev. Lett.* **105**, 040601.

[4] Baggaley A.W. and C.F. Barenghi 2011, *Phys. Rev. A* **84**, 020504.

[5] Bak, P. and C. Tang and K. Wiesenfeld. 1987. *Phys. Rev. Lett.* **59**, 381.

[6] Baronchelli, A. and F. Radicchi. 2013. *Chaos, Solitons & Fractals* **56**, 101.

[7] Barrat, A., M. Barthelemy and A. Vespignani. 2008. *Dynamical Processes on Complex Networks*, Cambridge University Press, Cambridge, UK.

[8] Bartumeus, F., F. Peters, S. Pueyo, C. Marrase and J. Catalan. 2003. *PNAS* **100**, 12771.

[9] Beggs, J. and I. Plenz. 2003. *J. Neurosci.* **23**, 11167-783 11177.

[10] Beggs, J.M. and N. Timme. 2012, *Frontiers Fractal Physiology* **3**, 163.

[11] Bénichou, O., C. Loverdo, M. Moreau and R. Voituriez. 2011. *Rev. Mod. Phys.* **83**, 81.

[12] Bennett, K.M., K.M. Schmainda, R.T. Bennett, D.B. Rowe, H. Lu and J.S. Hyde. 2003. *Magn. Reson. Med.* **50**, 727.

[13] Bennett, K.M., J.S. Hyde and K.M. Schmainda. 2006. *Magn. Reson. Med.* **56**, 235.

[14] Blackledge, J., E. Coyle and D. Kearney. 2011. *The Third Inter. Conf. on Resource Intensive Appl. and Services*, pp. 22–27, Venice, Italy.

[15] Bonifazi, P., M. Goldin, M.A. Picardo, I. Jorquera, A. Cattani, G. Bianconi, A. Represa, Y. Ben-Ari and R. Cossart. 2009. *Science* **4**, 5958, 1419.

[16] Boettcher, F., Ch. Renner, H.-P. Waldl and J. Peinke. 2003. *Boundry-Layer Meteor.* **108**, 163.

[17] Brockmann, D., L. Hufnagel and T. Geisel. 2006. *Nature* **439**, 462.

[18] Brown, C.T., L.S. Liebovitch and R. Glendon. 2007. *Human Ecology* **35**, 129.

[19] Callen, E. and D. Shapero. 1974. *Physics Today* **27**, 23.

[20] Caputo, M. 1967. *Geophys. J. R. Soc.* **13**, 529.

[21] Carreras, B.A., V.E. Lynch, I. Dobson and D.E. Newman. 2002. *Chaos* **12**, 985.

[22] Cavagna, A., A. Cimarelli, I. Giardina, G. Parisi, R. Santagati, F. Stefanini and M. Viale. 2010. *Proc. Natl. Acad. Sci. USA* **107**, 11865.

[23] Chialvo, D.R. 2010. *Nature Phys.* **6**, 744.

[24] Couzin, I.D. 2007. *Nature* **445**, 715; Couzin, I.D. 2009, *TRENDS in Cognitive Sciences* **13**, 36.

[25] del-Castillo-Negrete, D., B.A. Carreras and V.E. Lynch. 2005. *Phys. Rev. Lett.* **94**, 065003-1.

[26] Doob, J.L. 1953. *Stochastic Processes*, Wiley, NY.

[27] Dorogovtsev, S.N. and J.F.F. Mendes. 2002. *Adv. Phys.* **51**, 1079.

[28] Edwards, A.M., R.A. Phillips, N.W. Watkins, M.P. Freeman, E.J. Murphy, V. Afanasyev, S.V. Buldyrev, M.G.E. da Luz, E.P. Raposo, H.E. Stanley and G.M. Viswanathan. 2007. *Nature* **449**, 1044.

[29] Eguiluz, V.M., D.R. Chialvo, G.A. Cecchi, M. Baliki and A.V. Apkarian. 2005. *Phys. Rev. Lett.* **94**, 018102.

[30] Feller, W. 1968. *An Introduction to Probabilty Theory and Its Applications*, Wiley, NY.

[31] Ferrer, R. and R.V. Sole. 2003. In *Statistical Mechanics of Complex Networks*, Lecture Notes in Physics. R. Pastor-Satorras, M. Rubi and A. Diaz-Guilera (eds.). Springer, Berlin, pp. 114.

[32] Fisher, M.E. 1967. *Rep. Prog. Phys.* **30**, 615.

[33] Fraiman, D., P. Balenzuela, J. Foss and D.R. Chialvo. 2009. *Phys. Rev. E*, **79**, 061922.

[34] Gladwell, M. 2000. *The Tipping Point*, Little, Brown & Comp., Boston.

[35] Goldberger, A.L. and B.J. West. 1987. *Yale J. Biol. Med.* **60**, 421.

[36] Goldberger, A.L., D.R. Rigney and B.J. West. 1990. "Chaos and fractals in human physiology", *Sci. Am.* **262**, 42.

[37] González, M.C., C.A. Hidalgo and A. Barabási. 2008. *Nature* **453**, 779–782.

[38] Hayano, J., K. Kiyono, Z.R. Struzik, Y. Yamamoto, E. Watanabe, P.K. Stein, L.L. Watkins, J.A. Blumenthal and R.M. Carney. 2011. *frontiers in Physiology* **2**, Article65, 1.

[39] Holm, D.D. 2005. *Los Alamos Science* **29**, 172.

[40] Holm, S. and R. Sinkus. 2010. *J. Acoust. Soc. Am.* **127**, 542–548.

[41] Humphries, N.E., H. Weimerskirch, N. Queiroza, E.J. Southalla and D.W. Sims. 2012. *PNAS* **109**, 7169.

[42] Jarad, F., T. Abdeljawad and D. Baleanu. 2012. *Abs. & Appl. Analysis* **2012**, ID 890396.

[43] Jug, G. 1986. *Chem. Phys. Lett.* **131**, 94.

[44] Katz, Y., C.C. Ioannou, K. Tunstrom, C. Huepe and I.D. Couzin. 2011. *PNAS* **108**(46), 18720.

[45] Kimmich, R. 2002. *Chem. Phys.* **284**, 253.

[46] Kolmogorov, A.N. 1941. *Coptes Rendus (Dokl.) Akad. Sci. URSS* **26**, 115.

[47] Kulakowski, K. and M. Nawojczyk. 2008. "Sociophysics —an astriding science", arXiv:0805.3886v1.

[48] Kulish, V.V. and J.L. Lage. 2002. *J. Fluids Eng.* **124**, 803.

[49] Langlands, T.A.M., B.I. Henry and S.L. Wearne. 2009. *J. Math. Biol.* **59**, 761.

[50] Lim, S.C. and S.V. Muniandy. 2002. *Phys. Rev. E* **66**, 021114.

[51] Lomholt, M.A., T. Koren, R. Metzler and J. Klafter. 2008. *PNAS* **105**, 11055.

[52] Mackay, C. 1852. *Memoirs of Extraordinary Popular Delusions and the Madness of Crowds, 2nd* Ed., London, Office of the National Illustrated Library.

[53] Magin, R.L., O. Abdullah, D. Baleanu and X.J. Zhou. 2008. *J. Mag. Res.* **190**, 255.

[54] Magin, R., C. Ingo, L. Colon-Perez, W. Triplett and T.H. Mareci. 2013. *Microporous and Mesoporous Materials* **178**, 39.

[55] Mainardi, F. 1995. In Nonlinear Waves in Solids. J.L. Wegner and F.R. Norwood (eds.). pp. 93–97, ASME/AMR, Fairfield.

[56] Mainardi, F. 1996. *Appl. Math. Lett.* **9**, 23.

[57] Mainardi, F. 2010. *Fractional Calculus and Waves in Linear Viscoelasticity*, Imperial Coll. Press, UK.

[58] Meerschaert, M.M. and R.J. McGough. 2014. *J. Vib. & Acous.* **136**, 051004-1.

[59] Metzler, R. and J. Klafter. 2000. *Phys. Rept.* **339**, 1.

[60] McDougall, M.P. and S.M. Wright. 2005. *Magn. Reson. Med.* **54**, 386.

[61] Micheloyannis, S., E. Pachou, C.J. Stam, M. Vourkas, S. Erimaki and V. Tsirka. 2006. *Neurosci. Lett.* **402**, 273; Stam, C.J. and E.A. de Bruin 2004, *Hum. Brain Mapp.* **22**, 97.

[62] Montroll, E.W. and G. Weiss. 1965. *J. Math. Phys.* **6**, 167.

[63] Monin, A.S. and A.M. Yaglom. 1971. *Statistical Fluid Mechanics, Vol. 1 & 2*. J.L. Lumley (ed.). English Translation, MIT Press, Cambridge, MA.

[64] Montroll, E.W. and B.J. West. 1979. In *Fluctuation Phenomena*. E.W. Montroll and J.L. Lebowitz (eds.). *Studies in Statistical Mechanics, Vol. VII*, North-Holland, Amsterdam; Second edition 1987.

[65] Peng, C.K., J. Mistus, J.M. Hausdorff, S. Havlin, H.E. Stanley and A.L. Goldberger. 1993. *Phys. Rev. Let.* **70**, 1343.

[66] Pierantozzi, T. and L. Vazquez. 2005. *J. Math. Phys.* **46**, 113512.

[67] Podlubny, I. 1999. *Fractional Differential Equations, Mathematics in Science and Engineering Vol.* **198**, Academic Press, San Diego.

[68] Pramukkul, P., A. Svenkeson, P. Grigolini, M. Bologna and B.J. West. 2013. *Adv. Math. Phys.* **1**, 498789.

[69] Radicchi, F. and A. Baronchelli. 2012. *Phys. Rev. E* **85**, 061121.
[70] Radicchi, F., A. Baronchelli and L.A.N. Amaral. 2012. *Plus one* **7**, e29910.
[71] Ramos-Fernández, G., J.L. Mateos, O. Miramontes, G. Cocho, H. Larralde and B.Ayala-Orozco. 2004. *Beh. Ecol. and Sociobio.* **55**, 223.
[72] Rayleigh, Lord. 1964. In *Scientific Papers by Lord Rayleigh Vol.* **V** *1902–1910*, 540–546, Dover, NY.
[73] Rhee, I., M. Shin, S. Hong, K. Lee, S.J. Kim and S. Chong. 2011. *IEEE/ACM Trans. on Networking* **19**, 630.
[74] Rhodes, T. and M.T. Turvey 2007, *Physica A* **385**, 255.
[75] Riascos A.P. and J.L. Mateos. 2012. *Phys. Rev. E* **86**, 056110-1.
[76] Riascos, A.P. and J.L. Mateos. 2014. *Phys. Rev. E* **90**, 032809-1,.
[77] Richardson, L.F. 1926. *Proc. Roy. Soc. London A* **110**, 709.
[78] Sánchez, R., B.A. Carreras, D.E. Newman, V.E. Lynch and B.Ph. van Milligen. 2006. *Phys. Rev. E* **74**, 016305.
[79] Schmidt-Nielsen, K. 1984. *Scaling: Why is Animal Size So Important?*, Cambridge University Press, Cambridge.
[80] Schuster, H.G., D. Plenz anb E. Niebur (eds.). 2014. *Criticality in Neural Systems*, John Wiley, NY.
[81] Seshadri, V. and B.J. West. 1982. *Proc. Nat. Acad. Sci. USA* **79**, 4501.
[82] Shlesinger, M.F. and J. Klafter. 1986. In *On Growth and Forms*, pp. 279–283. H.E. Stanley and N. Ostrowski (eds.). Matinus Nijhof, Amsterdam.
[83] Shlesinger, M.F., B.J. West and J. Klafter. 1987. *Phys. Rev. Lett.***58**, 1100.
[84] Sims, D.W., E.J. Southall, N.E. Humphries, G.C. Hays, C.J.A. Bradshaw, J.W. Pitchford, A.James, M.Z. Ahmed, A.S. Brierley, M.A. Hindell, D. Morritt, M.K. Musyl, D. Righton, E.L.C. Shepard, V.J. Wearmouth, R.P. Wilson, M.J. Witt and J.D. Metcalfe. 2008. *Nature* **451**, 1098.
[85] Sokolov, I.M., J. Klafter and A. Blumen. 2002. *Physics Today*, Nov., 48.
[86] Sokolov, I.M. and J. Klafter 2005, *Chaos* **15**, 026103–026109.
[87] Solomon, T.H., E.R. Weeks and H.L. Swinney. 1993. *Phys. Rev. Lett.* **71**, 3975.
[88] Song, J., Q. Yu, F. Liu, and I. Turner. 2014. *Numerical Algorithms* **66**, 911.
[89] Sousa, E. 2013. *Int. J. Bifur. & Chaos* **23**, 1350166-1.
[90] Stam, C.J. 2004. *Neurosci. Lett.* **355**, 25.
[91] Struzik, Z.R. 2014. *8th Conf. Eur. Study Group Cardio. Oscill.* (ESCGO) 221.
[92] Svenkeson, A., M.T. Beig, M. Turalska, B.J. West and P. Grigolini. 2013. *Phys. A* **392**, 5663.
[93] Taylor, G.I. 1921.*Proc. London Math. Soc.* **20**, 196.
[94] Thiel, F. and I.M. Sokolov. 2014. *Phys. Rev. E* **89**, 12115.

[95] Torrey, H.C. 1956. *Phys. Rev.***104**, 563.

[96] Turalska, M., M. Lukovic, B.J. West and P. Grigolini. 2009. *Phys. Rev. E* **80**, 021110-1.

[97] Turalska, M., E. Geneston, B.J. West, P. Allegrini and P. Grigolini. 2012. *Front. Physio.* **3**. doi:10.3389/fphys.2012.00052.

[98] Vazquez, L. 2011. *Adv. Diff. Eq.* **2011**, Article ID 169421.

[99] Viswanathan, G.M., V. Afanasyev, S.V. Buldyrev, E.J. Murphy, P.A. Prince and H.E. Stanley. 1996. *Nature (London)* **381**, 413.

[100] Viswanathan, G.M., S.V. Buldyrev, S. Havlin, M.G.E. Da Luz, E.P. Raposo and H.E. Stanley. 1999. *Nature (London)* **401**, 911.

[101] Viswanathan, G.M. 2010. *Nature* **465**, 1018.

[102] Viswanathan, G.M., M.G.E. da Luz, E.P. Raposo and H.E. Stanley. 2011. *The Physics of Foraging*, Cambridge University Press, UK.

[103] Weiss, G.H. 1994. *Aspects and Applications of the Random Walk*, North-Holland, Amsterdam.

[104] West, B.J. and V. Seshadri. 1982. *Physica* **113A**, 203.

[105] West, B.J. and M.F. Shlesinger. 1984. *J. Stat. Phys.* **36**, 779–786.

[106] West, B.J. 1990. *J. Opt. Soc. Am. A* **7**, 1074–1100.

[107] West, B.J. and W. Deering. 1994. *Phys. Rept.* **246**, 1.

[108] West, B.J. 2006. *Where Medicine Went Wrong*, Studies of Nonlinear Phenomena in Life Science Vol. 11, World Scientific, Singapore.

[109] West, B.J. and P. Grigolini. 2010. *Physica A* **389**, 5706.

[110] West, D. and B.J. West. 2013. *Phys. of Life* **10**, 210.

[111] West, B.J., M. Turalska and P. Grigolini. 2014. In H.G. Schuster, D. Plenz and E. Niebur (eds.). *Criticality in Neural Systems*, John Wiley, NY.

[112] West, B.J., M. Turalska and P. Grigolini. 2014. *Network of Echoes; Imitation, Innovation, and Invisible Leaders*, Springer, NY.

[113] West B.J., M. Turalska and P. Grigolini 2015, New J. Phys. 17, 085009.

[114] White A.C., C.F. Barenghi, N.P. Proukakis, A.J. Youd and D.H. Wacks 2010, *Phys. Rev. Lett.* **104**, 075301.

[115] Whyte, W.H. 1956. *The Organization Man*, Simon & Schuster, NY.

[116] Widom, A. and H.J. Chen. 1998. *J. Phys. A* **28**, 1243.

[117] Wismer, M. 2006. *J. Acous. Soc. Am.* **120**, 3493.

[118] Wu, R.S. 1986. *J. Wave Mater. Int.* **1**, 79.

[119] Xia, H., N. Franciois, H. Punzmann and M. Shats. 2014. *Phys. Rev. Lett.* **112**, 104501.

[120] Yates, C.A., R. Erban, C. Escudero, I.D. Couzin, J. Buhl, I.G. Kevrekidis, P.K. Maini and D.J.T. Sumpter. 2009. *PNAS* **106**, 5464.

[121] Yu, Q., F. Liu, I. Turner and K. Burrage. 2013. *Cen. Eur. J. Phys.* **11**, 646.

CHAPTER 7

Strange Statistics

Historically the two ways of describing the changing behavior of complex phenomena are through the Langevin equation for the dynamic variables, or equivalently through the phase space equation for the PDF. The dynamic equations are constructed in the physical sciences from general principles, such as the variation of energy in conservative systems. Recall from Chapter 2 that the complete system is often partitioned into the network of interest and the rest of the universe (the environment). In its simplest form the environment contributes to the network dynamics a random force and a linear dissipation of the network's energy, that constitutes the Langevin equation description of the environment's dynamic complexity. Also in Chapter 2 the argument for how to replace the system's trajectory with an ensemble of trajectories characterized by the PDF in phase space was presented. The dynamic equations were replaced with the phase space equation of evolution for the PDF and the environmental complexity entailed a statistical description of the system's response.

In intervening chapters these arguments were generalized to truly complex phenomena using the fractional calculus. The utility of the fractional Langevin equation (FLE), through its explanation of certain space-time applications, was subsequently established. To complete the picture let us turn our attention to the fractional phase space equations of motion for the PDF. The fractional calculus has proven its value in being able to describe exotic scaling in a variety of non-traditional statistical phenomena [45]. A partial list of such anomalous diffusive phenomena, from the natural to the social sciences, include the advection of passive scalars in the turbulent atmosphere [57]; transport in amorphous solids [47, 59]; magnetic resonance imaging [35]; microsphere motion in living cells [9]; search behavior [7]; and the intermittent fluctuations in the profit of stocks [53]. All manner of complex statistical phenomena have had their phase space equations for the PDF replaced by a more exotic fractional phase space equation (FPSE), or when the spatial derivative is second order, by a fractional Fokker-Planck equation (FFPE). A review of the implications of incorporating fractional dynamics into the theory of stochastic processes has been given by Metzler and Klafter [45] and from the perspective of chaotic trajectories by Zaslavsky [87].

The influence of the environment on an individual's dynamics appear to be model independent, as we discovered in Chapter 6. This independence is adopted in this chapter by generalizing the arguments resulting in Hamilton's equations for mechanical systems, to include the interaction with a complex environment. This presentation is facilitated by reviewing an extension of the central limit theorem to show that the MLF form of the survival probability, that appeared in the numerical results of the DMM network calculations, is not model-specific, but is a consequence of the compatibility of the stochastic central limit theorem with the fractional calculus.

Ordinary diffusion gives rise to a mean-square displacement that grows linearly in time $\langle Q^2(t) \rangle \propto t$, whereas anomalous diffusion grows nonlinearly in time $\langle Q^2(t) \rangle \propto t^{2H}$ with $H \neq 1/2$. We shall ignore a great deal of the subtlety associated with the Hurst exponent H, being different from its Brownian motion value, and focus on the implications of the fact that such transport can be described by the fractional calculus using the FFPE. This leads to a discussion of the strange statistics associated with complex phenomena, which begins here with a brief review of anomalous transport.

Rather than presenting an exhaustive discussion of the derivation of the phase space equations for anomalous transport, I thought it would be more useful for readers with a science background broader than physics to see how the PDFs acquire those properties that make them so interesting in modeling complexity. We begin by exploring the implications of the integral chain condition for Markovian PDFs. It is actually straight forward to show that the chain condition entails the Lévy PDF as its most general stationary solution [47] and that under a well defined set of conditions, the Lévy PDF is the solution to a fractional phase space equation. The application areas discussed in the previous chapter established the utility of the Lévy PDF in physiology, in particular the modeling of the variability of the time interval between heart beats in both healthy and diseased individuals. This physiological phenomenon was one of the many that lead to the development of a new subdiscipline within medicine, *Fractal Physiology* [6, 72, 78], in which the fractional calculus is extremely useful [74].

A second gateway to the application of the fractional calculus in phase space dynamics is one that relies on the extension of random walk concepts to continuous time. The incorporation of the structure and dynamics of the environment into the evolution of the network's PDF was made in 1965 by Montroll and Weiss [46] and has proven to be foundational in interpreting

FPSEs. Much of the CTRW formalism is relegated to Appendix 7.9.1 in order to remain focused on the FPSEs and their solutions. However we do present a third way of deriving the FFPE, using an operator method. It is useful to include a discussion of an eigenfunction series expansion of the solution to FFPE, in terms of MLFs, with arguments consisting of eigenvalues and the MLF shown to replace the exponential in the eigen series expansion. The details for a fractional harmonically bound particle are presented in some detail as an example of how the technique has been used.

Yet a fourth method for deriving the FPSEs is based on averaging over ensembles of fractal trajectories; the fractional kinetic equation (FKE) approach championed by Zaslavsky [87]. This method is reviewed because the solutions to the FPSE he obtained, using renormalization group methods, have interpretations facilitated by reference to fractal trajectories. This explicit reference to the fractal trajectories provides additional insight into the network's complexity through its scaling behavior. In fact, the exact formal solution to the FKE is obtained using only its scaling properties.

Certain applications of the fractional calculus have required that extensions be made to the FPSE. One such extension was constructed in the context of physiology regarding the modeling of heart rate variability and cardiac death. A mechanism was introduced to suppress the extrema in HRV resulting in a truncated Lévy PDF. The FPSE is modified to incorporate this truncation into the fractional operator itself and the resulting truncated Lévy PDF solution subsequently 'explains' why some individuals in a cohort succumb to cardiac death and others do not [80]. As a matter of nomenclature the exponentially truncated PDF is also referred to as a tempered PDF.

7.1 Fractional Hamiltonian Formalism

The paradigm of science has been physics since the time of Newton and perhaps more specifically the archetype has been that of mechanics. The clockwork universe rests on Newton's laws of motion and universal gravitation, which in a modern setting relies on symmetry and variational principles. The complexity of many physical and virtually all social and life science phenomena suggest that the mechanical model of the world ought to be abandoned, or at least significantly modified. The modification we briefly review in this section is a generalization of Hamilton's equations of motion, which incorporate into the structure of the dynamic equations the complexity of the environment with which the system of interest interacts. Thus, unlike the Langevin approach, which explicitly accounts for the influence of the environment through the addition of a random force and a related dissipation, the fractional Hamiltonian formalism implicitly includes environmental effects through the very definition of space and time.

The first attempt to systematically extend the Hamiltonian and Lagrangian formalisms to the fractional calculus was made by Riewe [56]. He determined that a classical Lagrangian could be generalized to include friction by adding a fractional derivative of order one-half to the equations of motion. However he was apparently unaware of the work done earlier in hydrodynamics to generalize the concept of viscosity and thereby achieve the same end. Consequently, his use of a variational technique to generate the equations of motion resulted in dynamic equations with fractional dissipation, which were subsequently shown to violate causality.

Riewe's idea was right but the implementation was shown to be flawed by Stanislavsky [65]. In this section we follow Stanislavsky and adopt the subordination concept to develop the fractional Hamiltonian formalism. We remark that he [65] motivated his development of the formalism with the observation that space and time are not the continuous featureless processes first assumed by Newton. We do not develop that argument here because that would tie us too closely to physical modeling and we wish to suggest ways to generate dynamic equations for non-physical systems as well.

7.1.1 Stochastic Central Limit Theorem

In the previous chapter we used the results of numerical calculation for a DMM network to determine the influence of the environment on a single individual. There we introduced the generalized hyperbolic PDF, that empirically fits the survival probability used in Eq.(6.65), to relate the operational time of the individual to the experimental time of the network as a whole. Stanislavsky [65] used a different relation, but one that has the same behavior asymptotically in the experimental time t, that is, a function whose asymptotic form is IPL.

The approach we [55] adopt parallels that of Stanislavsky [65] in that it is stochastic rather than dynamic, but this is where the similarity ends. Our [55] method requires the development of a new form of the central limit theorem (CLT) and the proof of the new CLT is developed elsewhere [55], using the fractional calculus. The Normal PDF is entailed by the traditional CLT and we show that the MLF is entailed by the stochastic central limit theorem (SCLT) [55]. We use the observation made by Gorenflo and Mainardi [15] and others [42] that the MLF PDF is universal in the same sense as the limit distributions of the CLTs of Laplace and Lévy are universal. We denote the MLF survival probability

$$\Psi_{ML}(t) = E_\alpha\left(-(\lambda t)^\alpha\right), \tag{7.1}$$

and the corresponding waiting-time PDF as

$$\psi_{ML}(t) = -\frac{d\Psi_{ML}(t)}{dt}. \tag{7.2}$$

Pramukkul *et al.* [55] prove that the MLF is universal, as a consequence of the SCLT by introducing fluctuating trajectories interpreted as a proper representation of complex processes with memory. The universality is a consequence of a subordination process that requires the fractional derivative to be interpreted as an average over infinitely many random trajectories. The SCLT is based on keeping fixed the probability P_S of detecting an event, that is, the probability that an event is visible in an experiment is constant in a given process. The theorem focuses on the interval between two consecutive visible events and adopts a rescaling procedure to compensate for the incomplete measurement-induced survival probability. We [55] show that all waiting-time PDFs generating a non-integrable survival probability, as a consequence of the SCLT, yield $\psi_{ML}(t)$. The proof leads to the conclusion that $\Psi_{ML}(t)$ is universal.

We can adapt the subordination argument from the previous chapter to write for the survival probability between two visible events

$$\Psi_V(t) = \sum_{n=0}^{\infty} \int_0^t G_n(t, t') (1 - P_S)^n \, dt', \tag{7.3}$$

where again the function relating the experimental and operational times is given by Eq.(6.65). We note that the condition $P_S \to 0$ has the effect of turning the discrete index n into a virtually continuous time τ; the discrete power $(1 - P_S)^n$ into the exponential $exp(-P_S\tau)$ and the survival probability between visible events $\Psi_V(t)$ into $\Psi_{ML}(t)$. Thus, the MLF survival probability with $\lambda = (P_S)^{1/\alpha} \lambda_0$ is the counterpart in real time of the ordinary exponential function $exp(-P_S\tau)$:

$$E_\alpha(-P_S(\lambda_0 t)^\alpha) = \int_0^{\infty} F^{(S)}(t, \tau) \, exp(-P_S\tau) \, d\tau, \tag{7.4}$$

and $F^{(S)}(t, \tau)$ is the PDF of experimental times t corresponding to the continuous operational times τ. The form of the PDF can be written as the inverse Laplace transform over the Laplace variable complementing the experimental time

$$F^{(S)}(t, \tau) = \mathcal{LT}^{-1}\left\{ e^{-\tau s^\alpha} s^{\alpha-1}; t \right\}. \tag{7.5}$$

A straight forward way to obtain Eq.(7.4) is to insert Eq.(6.65) into (7.3) and take the Laplace transform for $P_S \to 0$ to obtain

$$\widehat{\Psi}_V(s) = \frac{1}{s + \widehat{\Phi}_V(s) P_S}, \tag{7.6}$$

where $\widehat{\Phi}_V(s)$ is the Laplace transform of the Montroll-Weiss memory kernel for visible events defined by

$$\widehat{\Phi}_V(s) = \frac{s\widehat{\psi}_V(s)}{1 - \widehat{\psi}_V(s)}. \tag{7.7}$$

One can, without too much difficulty, show, using the rescaled Laplace variable, with the detection probability $s = s' P_S^{1/\alpha}$, that

$$\lim_{P_S \to 0} \widehat{\Psi}_V (s') = \frac{1}{s' + s'^{1-\alpha} \lambda_0^\alpha} \tag{7.8}$$

whose inverse Laplace transform is the MLF on the left hand side of Eq.(7.4). This demonstrates why the MLF is ubiquitous in data sets. It is the distribution of visible, that is, measurable events.

We are free to interpret Eq.(7.6) as a double Laplace transform $\widehat{F}^{(S)}(s, u)$ with $u = P_S$. By inverse Laplace transforming $\widehat{F}^{(S)}(s, u = P_S)$ with respect to s we obtain Eq.(7.4) where the time t can be interpreted in terms of the displacement of a random walker that keeps jumping in the same direction. It is, in fact, a one-sided Lévy process. We [55] were therefore led to interpret the SCLT to be a consequence of the generalized CLT of Lévy. The condition $n \to \infty$ generates the Lévy stable PDF and the sum over infinitely many Lévy processes weighted by the exponential function $exp(-P_S\tau)$ generates the MLF survival probability which is consistent with the analysis made by Stanislavsky [65].

Somewhat earlier Stanislavsky [64] derived a fractional Langevin equation in experimental time from an ordinary Langevin equation in operational time. His argument employed subordination and was superficially different from the presentation based on a fractional heat bath developed in Section 5.2.4. However, the linear dissipation and random fluctuations in operational time for a particle in contact with a thermal bath is physically the same for the two arguments, so we do not present the subordination discussion here and refer the reader back to Stanislavsky's lucid discussion.

7.1.2   Fractional Hamilton Equations

Assume that Hamilton's equations for a unit mass particle can be expressed in operational time in their normal form

$$\frac{dq}{d\tau} = \frac{\partial \mathcal{H}}{\partial p}, \quad \frac{dp}{d\tau} = -\frac{\partial \mathcal{H}}{\partial q}, \tag{7.9}$$

or equivalently in terms of normal modes

$$\frac{da}{d\tau} = -i\frac{\partial \mathcal{H}}{\partial a^*}, \quad \frac{da^*}{d\tau} = i\frac{\partial \mathcal{H}}{\partial a^*} \tag{7.10}$$

Paralleling the discussion of Stanislavsky [65], the dynamical system for normal modes are assumed to satisfy the relation

$$a_\alpha(t) = \int_0^\infty F^{(S)}(t, \tau) a(\tau) d\tau, \tag{7.11}$$

and its complex conjugate. The Laplace transform of the last equation with respect to experimental time has the algebraic form determined by the functional form depicted in Eq.(7.5):

$$
\begin{aligned}
\widehat{a}_\alpha(s) &= \int_0^\infty e^{-st} a_\alpha(t)dt \\
&= \mathcal{LT}\left\{a(\tau)\mathcal{LT}^{-1}\left\{e^{-\tau u^\alpha} u^{\alpha-1};t\right\};s\right\} = s^{\alpha-1}\widehat{a}(s^\alpha), \quad (7.12)
\end{aligned}
$$

where the fractional index is in the range $0 < \alpha < 1$ and the argument of the functions on the right is a consequence of the Laplace transform with respect to the operational time τ. The nested Laplace transform operations can be interpreted to first perform the integration on the experimental time t to obtain a delta function in the Laplace variables. This leaves an integration on the operational time τ to yield the Laplace transform of the normal mode amplitude.

We now define a new Hamiltonian $\mathcal{H}_\alpha$ in the experimental phase space to replace $\mathcal{H}$, whose equations of motion are in terms of the fractional normal mode a_α and its complex conjugate. To accomplish this we transform Hamilton's equations in operational time to those in experimental time

$$
\frac{\partial\mathcal{H}_\alpha}{\partial a_\alpha^*} = \int_0^\infty F^{(S)}(t,\tau)\frac{\partial\mathcal{H}}{\partial a^*}d\tau. \quad (7.13)
$$

Introducing the inverse Laplace transform for the PDF, inserting the Hamilton equation given by Eq.(7.10) and interchanging the order of integration between t and τ in Eq.(7.13) results in

$$
\frac{\partial\mathcal{H}_\alpha}{\partial a_\alpha^*} = i\mathcal{LT}^{-1}\left\{\int_0^\infty s^{\alpha-1}e^{-\tau s^\alpha}\frac{da(\tau)}{d\tau}d\tau;t\right\}, \quad (7.14)
$$

which after an integration by parts and inserting Eq.(7.12) yields

$$
\frac{\partial\mathcal{H}_\alpha}{\partial a_\alpha^*} = i\mathcal{LT}^{-1}\left\{s^{2\alpha-1}\widehat{a}(s^\alpha);t\right\}.
$$

Again using Eq.(7.12) this last equation has the equivalent representation

$$
\begin{aligned}
\frac{\partial\mathcal{H}_\alpha}{\partial a_\alpha^*} &= i\mathcal{LT}^{-1}\left\{s^\alpha\widehat{a}_\alpha(s);t\right\} \\
&= i\partial_t^\alpha[a_\alpha], \quad (7.15)
\end{aligned}
$$

or expressed in canonical form

$$
\partial_t^\alpha[a_\alpha] = -i\frac{\partial\mathcal{H}_\alpha}{\partial a_\alpha^*}, \quad (7.16)
$$

which, taken together with its complex conjugate, provide Hamilton's equations for the fractional normal modes in real time.

Thus, the fractional Hamiltonian for the linear fractional harmonic oscillator in real time phase space is given by

$$\mathcal{H}_\alpha = \omega_0 a_\alpha a_\alpha^*, \qquad (7.17)$$

and the dynamics for the fractional normal mode are given by Eq.(5.81). An equivalent analysis was given by Stanislavsky [65] in terms of the fractional displacement and momentum rather than normal modes. The results are, of course, the same, including the fact that when $\alpha = 1$ the equations reduce to the classical Hamilton's equations. He [65] carries out the analysis for the fractional linear harmonic oscillator, both free and driven, and obtains the results already presented in Chapter 5. However, he also considers the case of two coupled fractional oscillators, free and driven, and investigates resonance effects, which we do not discuss further here. But we mention them to alert the reader to the existence of these first steps into a whole new area of statistical mechanics.

7.1.3 Fractional Variational Calculations

The fractional Hamiltonian analysis is only the beginning of the potential modifications of the classical description of the dynamics of mechanical systems generalized to incorporate the random interaction of the system of interest with the environment. The assumption is that each subsystem moves to the beat of its own drum, which is to say it operates according to its own internal clock. The operational time classical Hamiltonian equations of motion are replaced with the fractional Hamiltonian equation, with the transition to experimental time in the averaging process. Similar analyses have been done for non-Hamiltonian systems such as the Lotka-Volterra system in ecology [66].

So what does this have to do with the non-mechanical system in the social and life sciences. The fact that such general principles as Noether's theorem for the problems of the calculus of variation with fractional derivatives have been obtained [1] strongly suggests the existence of symmetry and/or conservation laws for general systems from which fractional equations of motion can be generated.

7.2 Anomalous Transport

The phenomenon of diffusion explained by Einstein was in fact first chronicled by the Dutch physician, Jan Ingen-Housz in 1785 [23], who observed that finely powdered charcoal floating on an alcohol surface executed a highly erratic random motion. However, as we mentioned, it is the Scottish botanist Robert Brown [8], who in 1827 noted the erratic motion of pollen grains suspended in fluids, and whose name is now attached to the erratic motion. As we [47] previously observed, it would have been awkward for the English speaking world to refer to this as Ingen-Houzian motion. Brownian motion or classical diffusion is characterized by Gaussian statistics and a variance in displacement of the Brownian particle that increases linearly in time. However, we now turn our attention to modeling complex phenomena that exhibit behavior that is more complex than that of simple diffusion.

We begin the discussion of anomalous transport with the work of P. Lévy, who in the 1920s and 1930s was concerned with the question of when a sum of identically distributed random variables has the same probability distribution as any one of the terms in the sum. This is, of course, a question of scaling and the paradigm of fractals, that is, when does a part have the same properties as the whole. Lévy completely solved this problem and the resultant PDFs are now called Lévy stable laws. In general, Lévy laws deal with PDFs that have infinite moments and, thus, do not possess a definite scale. Lévy PDFs have become a well established field of mathematics. On the other hand, the appearance of infinite moments has limited its application to physical phenomena, until fairly recently.

7.2.1 Lévy PDF

There are a variety of ways to derive FPSEs and each method of derivation emphasizes a different aspect of the complexity of the process being modeled. Here we sketch a scaling argument to obtain the FPSE from the chain condition for the PDFs given by Eq.(2.24). The logarithm of the product form of the solution for the characteristic function, the Fourier transform of the PDF, Eq.(2.26), is

$$\log \widetilde{P}(k, t - t_0) = \log \widetilde{P}(k, t - t') + \log \widetilde{P}(k, t' - t_0), \qquad (7.18)$$

from which it is clear from additivity that the characteristic function factors into a function of the Fourier variable k, say $g(k)$, and a function of time. In order for the intermediate time t' to vanish from the solution, as it does on the left hand side of Eq.(7.18), the function of time must be linear. Thus, the form of the solution to Eq.(2.26) obtained from Eq.(7.18) is

$$\widetilde{P}(k, t) = \exp\left[g(k)t\right]. \qquad (7.19)$$

Since the PDF is normalizable at all times, the real part of $g(k)$ must be negative definite and in order for the characteristic function to retain the product form at all spatial scales, it must be infinitely divisible as well. If we scale the Fourier variable k by a constant factor λ, then in order for the PDF to be infinitely divisible $g(k)$ must be a homogeneous function such that

$$g(\lambda k) = \lambda^\beta g(k). \qquad (7.20)$$

The homogeneity requirement implies that

$$g(k) = -b(\beta) \, |k|^\beta \, , \tag{7.21}$$

where $b(\beta)$ is a complex function dependent on the parameter β, with a positive definite real part.

Thus, we have for the characteristic function

$$\widetilde{P}(k,t) = \exp\left[-b(\beta)\,|k|^\beta\,t\right]. \tag{7.22}$$

The symmetric solution to the chain condition is obtained by setting the constant in the exponential to be independent of β, $b(\beta) = b$. The most general solution to the chain condition [17, 47] is obtained using

$$b(\beta) = b\left[1 + iC\omega\,(k,\beta)\,\frac{k}{|k|}\right], \tag{7.23}$$

where C is a real parameter, $\omega\,(k,\beta)$ is a real function, and the imaginary part of the coefficient determines the skewness of the distribution.

The form of the characteristic function Eq.(7.22) is therefore given by

$$\widetilde{P}(k,t) = \exp\left[-b\left[1 + iC\omega\,(k,\beta)\,\frac{k}{|k|}\right]|k|^\beta\,t\right], \tag{7.24}$$

so that the inverse Fourier transform sets the conditions: $0 < \beta \le 2$, in order that the PDF be positive definite; $b > 0$ so that the PDF is normalizable, and $-1 \le C \le 1$ indicates the degree of skewness of the PDF. Finally the function $\omega\,(k,\beta)$ is defined by

$$\omega\,(k,\beta) = \begin{cases} \tan\,(\beta\pi/2) & \text{if } \beta \ne 1 \\ \frac{2}{\pi}\ln|k| & \text{if } \beta = 1 \end{cases}, \tag{7.25}$$

whose derivation can be found in Gnedenko and Kolmogorov [17]. The equation of evolution for the PDF is obtained by taking the time derivative of the characteristic function Eq.(7.24) just as we did in the case of isotropic homogeneous turbulence

$$\frac{\partial \widetilde{P}(k,t)}{\partial t} = -b(\beta)\,|k|^\beta\,\widetilde{P}(k,t). \tag{7.26}$$

The inverse Fourier transform of this equation yields

$$\frac{\partial P(q,t)}{\partial t} = -\mathcal{F}\mathcal{T}^{-1}\left\{b(\beta)\,|k|^\beta\,\widetilde{P}(k,t);q\right\}. \tag{7.27}$$

In the symmetric case we set $C = 0$ in Eq.(7.23) and using the convolution property of the product of Fourier amplitudes in Eq.(7.27) yields

$$\frac{\partial P(q,t)}{\partial t} = \frac{b}{\pi}\Gamma\,(\beta+1)\sin\,(\beta\pi/2)\int_{-\infty}^{\infty}\frac{P\,(q',t)\,dq'}{|q-q'|^{\beta+1}}. \tag{7.28}$$

Note that the integral term in Eq.(7.28) is the one-dimensional Riesz-Feller fractional derivative $\partial^{\beta}_{|q|}$, so that the equation can be formally written as

$$\frac{\partial P(q,t)}{\partial t} = D_{\beta}\partial^{\beta}_{|q|}\left[P(q,t)\right], \tag{7.29}$$

which was first applied in the context of anomalous diffusion by Seshadri and West [62] and whose solution is the symmetric Lévy PDF.

The process described by Eq.(7.29) was introduced in the discussion of foraging in the previous chapter. It was pointed out that this phenomenon differs from classical diffusion in that step sizes of arbitrary length could be made in a random walk description. In the discussion a distinction was drawn between a Lévy random walk and a Lévy random flight. In a Lévy flight the steps in space can be arbitrarily long and they occur in uniform intervals of time. On the other hand, in a Lévy walk the step size is coupled to the time interval necessary to take a step of that length. The concept was introduced in the discussion of homogeneous turbulence [60] and is reviewed in some detail in Section 7.2.4.

It is also useful to note that others have adopted the fractional-order equations Eq.(7.29) to model diffusion in living systems [77]. Copot *et al.* [11] apply it in a biological framework to model diffusion in the respiratory system for gas exchange, as well as, for drug diffusion during anesthesia. They model respiratory diffusion with loss of energy and heat production with a fractional-order impedance. Pharmacokinetics studies drug disposition in the body, due to absorption, distribution and elimination and has been generalized to the fractional calculus [12, 54]. Copot *et al.* suggest that the fractional-orders models avoid the high degree of inter-patient variability and nonlinearity resulting in the use of linear, rather than nonlinear, control techniques.

7.2.2 CTRW

As noted in previous chapters simple diffusion can be modeled using RRWs. The details of a random walk on a lattice to obtain the probability of the walker being at a lattice site $\mathbf{q}$ after n steps are given in a number of places [47, 70]. The basic problem of calculating the PDF by replacing discrete steps with continuous time was solved by Montroll and Weiss [46], where they included random intervals between successive steps in the walking process to account for local structure in the environment, such as traps. In Appendix 7.9.1 we follow the discussion of West and Grigolini [75] to explain the Montroll-Weiss (MW) approach in which the time interval for successive steps $\tau_n = t_n - t_{n-1}$ becomes a random variable. The MW generalization is referred to as the CTRW model, since the walker is allowed to step at a continuum of times and the time interval between successive steps is considered to be a second random variable, the vector value of the step taken being the first random variable.

The CTRW has been used to model a number of complex phenomena, from the microscopic structure of individual lattice sites for heterogeneous media, to the stickiness of stability island in chaotic dynamic networks [83].

The selections of the waiting-time PDF and the step-length PDF can be made to accommodate a variety of complex phenomena. A number of investigators have studied the properties of the solution given by Eq.(7.134) in Appendix 7.9.1 under a variety of conditions [44, 47, 70], as well as, its generalizations [60, 69, 87]. The physical interpretation of the formal solution for this array of models can be expedited by introducing the Montroll-Weiss memory kernel $K(t)$ defined in Appendix 7.9.1 as the inverse Laplace transform of

$$\widehat{K}(s) \equiv \frac{s\widehat{\psi}(s)}{1 - \widehat{\psi}(s)}, \tag{7.30}$$

into Eq.(7.134). We can then rewrite the resulting equation for the Fourier-Laplace transform of the PDF $P^*(\mathbf{k},s)$ to obtain the expression

$$sP^*(\mathbf{k},s) - 1 = -\widehat{K}(s)\left[1 - \widetilde{p}(\mathbf{k})\right]P^*(\mathbf{k},s),$$

whose inverse Fourier-Laplace transform yields the integro-differential equation

$$\frac{\partial P(\mathbf{q},t)}{\partial t} = \int_0^t dt' K(t - t') \left\{ -P(\mathbf{q},t') + \int p(\mathbf{q} - \mathbf{q}')P(\mathbf{q}',t')d\mathbf{q}' \right\}. \tag{7.31}$$

This is the Montroll-Kenkre-Shlesinger master equation [25] and is clearly non-local in both space and time. In time the non-locality is determined by the memory kernel, which in turn is determined by the waiting-time PDF, that is, by the inverse Laplace transform of Eq.(7.30). The spatial non-locality is determined by the jump-length PDF. The generality of this expression suggests that it can be used to model a great variety of complex phenomena in the supporting medium and therein lies its utility.

The CTRW equation of motion for the PDF can be determined directly in the asymptotic limits of long times and large distances by considering the approximations to $\widehat{\psi}(s)$ as $s \to 0$ and $\widetilde{p}(\mathbf{k})$ as $k \to 0$. Consider the asymptotic form of the IPL waiting-time PDF $\psi(t) \propto 1/t^{\alpha+1}$

$$\widehat{\psi}(s) = 1 - (\tau s)^{\alpha} \quad ; \quad s \to 0, \tag{7.32}$$

and the asymptotic form of the IPL step-length PDF $p(|\mathbf{q}|) \propto 1/|\mathbf{q}|^{\beta+1}$

$$\widetilde{p}(\mathbf{k}) = 1 - (\sigma |\mathbf{k}|)^{\beta} \quad ; \quad |\mathbf{k}| \to 0, \tag{7.33}$$

which when inserted into Eq.(7.134) yields

$$\left[s^{\alpha} + \frac{\sigma^{\beta}}{\tau^{\alpha}} |\mathbf{k}|^{\beta} \right] P^*(\mathbf{k},s) = s^{\alpha-1}. \tag{7.34}$$

The details of these expansions are worked out in a number of places, including in West and Grigolini [75]. The inverse Fourier-Laplace transform of Eq.(7.34) yields the FPSE in terms of the RL-differential operator in time and the Riesz-Feller fractional derivative in space

$$D_t^\alpha \left[P(\mathbf{q}, t) \right] - \frac{t^{-\alpha}}{\Gamma(1-\alpha)} P_0(\mathbf{q}) = K_{\alpha,\beta} \partial_{|\mathbf{q}|}^\beta \left[P(\mathbf{q}, t) \right], \qquad (7.35)$$

where the initial condition is denoted by $P_0(\mathbf{q}) = P(\mathbf{q}, 0)$. The generalized diffusion coefficient

$$K_{\alpha,\beta} \equiv \frac{k_B T \sigma^{\beta-2}}{m \eta_\alpha}, \qquad (7.36)$$

has incorporated into its definition a generalized friction constant per unit time $\eta_\alpha = \eta \tau^\alpha$, such that the product with a particle mass $m \eta_\alpha$ has the dimension $[T]^{\alpha-2}$, as discussed by Metzler and Klafter [45].

One can also write the FPSE in terms of the Caputo fractional derivative in time

$$\partial_t^\alpha \left[P(\mathbf{q}, t) \right] = K_{\alpha,\beta} \partial_{|\mathbf{q}|}^\beta \left[P(\mathbf{q}, t) \right], \qquad (7.37)$$

to be solved subject to the initial condition $P_0(\mathbf{q})$. Note that these two descriptions are equivalent and are used interchangeably for algebraic convenience.

7.2.3 Morphogen Gradients

The *Just So Stories* of Rudyard Kipling explain such things as how the camel got his hump, the leopard his spots and the elephant his trunk. These stories are enough to entertain and satisfy a child's curiosity, but adults need less fanciful tales, such as Alan Turing's article: *The chemical basis of morphogenesis* [68]. This unassuming paper launched what is now the science of morphogens, which does answer the teleologic questions of how a cell knows to develop into a toe or tongue; how a leopard gets its spots and a peacock its glorious plumage. The spatial patterns emerge as the response of embryonic cells to special signaling molecules called morphogens. The generation and degradation of morphogens produce spatial gradients of morphogen concentration to which cells respond as specific "codes" for their subsequent development through the expression of the relevant genes. These gradients are quite important and aberrations can lead to undesirable consequences in tissue development, including mutations [4].

The equation for the dynamics of the morphogen concentration, obtained from the CTRW model, using an IPL in both the waiting time PDF and the step length PDF, are fractional in space and time. However in order to systematically incorporate morphogen degradation requires an extension of the theory developed up to this point. Yuste *et al.* [82] caution on the dangers of constructing heuristic equations and emphasize that modeling degradation be done at the mesoscopic level of description and outline a proper extension of the CTRW model. A detailed argument leads to the fractional reaction-subdiffusion equation:

$$\frac{\partial \rho\left(q,t\right)}{\partial t} = K_\gamma \frac{\partial^2}{\partial q^2}\left\{e^{-k(q)t}D_t^{1-\gamma}\left[e^{k(q)t}\rho\left(q,t\right)\right]\right\} - k(q)\rho\left(q,t\right) \qquad (7.38)$$

where the effective diffusion coefficient is K_γ, the position-dependent reaction rate is $k(q)$, and $D_t^{1-\gamma}\left[\cdot\right]$ is the RL fractional derivative in time. Additional analysis of the properties of the solution to Eq.(7.38), with nonlinear reaction rates are given by Fedotov and Falconer [13].

Yuste *et al.* [82] point out that the position dependence of the reaction rate may arise from several degradation pathways, which for practical purposes can be regarded as "death" or "evanescence" processes. In the case of a constant reaction rate $k(q) = k$ the long-time or stationary solution to the FPSE is the symmetric exponential profile

$$\rho_{ss}\left(q\right) = \frac{j_0}{2\lambda k}e^{-|q|/\lambda} \;\; ; \;\; \lambda = \sqrt{\frac{K_\gamma}{k^\gamma}} \qquad (7.39)$$

where j_0 is the morphogen production rate at the origin and λ is the decay length of the gradient. They go on to discuss the interaction between subdiffusion and space-dependent reaction rates, and the robustness of the profile to fluctuations in the environment or genetics.

They also point out that the discussion associated with this FPSE could be important to the problem of controlled-release and microfluidic drug delivery technologies in tissue engineering, as well as to point-source pollutants in groundwater when these pollutants can recombine and otherwise disappear. Recall the discussion in Section 6.2.2 in which the suggested extensions of phenomenological laws to fractional form all involved spatial gradients and therefore could be generalized in parallel with the strategy sketched in this section.

7.2.4 Lévy Walk and Turbulence

Here we apply the CTRW argument to the physical process of turbulence, following Shlesinger *et al.* [60]. The Fourier-Laplace transform of the PDF for the relative separation of two tracer particles in a turbulent fluid flow is given in Appendix 7.9.1 as

$$P^*(\mathbf{k}, s) = \frac{1 - \widehat{\psi}\left(s\right)}{s\left[1 - p^*\left(\mathbf{k}, s\right)\right]}, \qquad (7.40)$$

when the probability for making a transition of a given length in a given time interval does not factor. Consider a random walker who jumps with probability $p(\mathbf{R})$ between successively visited sites, however distant. When the mean square displacement $\langle R^2 \rangle$ per jump is finite then from the CLT the PDF for the position of the walker after many steps is Gaussian. When $\langle R^2 \rangle$ is infinite this random process possesses no characteristic length scale and the set of sites visited is fractal. Shlesinger *et al.* [60] named this random process a Lévy flight and the governing PDF is a Lévy stable law.

Very often one considers that the walker can wait, or pause, for a random duration at each site before making an instantaneous jump to another site [46]. As we just discussed these are the CTRWs, because the emphasis is on the time rather than on the number of steps. Shlesinger *et al.* [60] extended these concepts and introduced $p(\mathbf{R}, t)$ to be the PDF of making a step $\mathbf{R}$ that takes a time t to complete, thereby generalizing the notion of a Lévy flight. They go on to consider a random walker that visits the same sites as a random Lévy flight, but instead of having instantaneous jumps that lead to an infinite mean square displacement, they choose the joint space-time PDF to be

$$p(\mathbf{R}, t) = \psi(t\,|\,\mathbf{R})\,p(\mathbf{R}), \tag{7.41}$$

where $p(\mathbf{R})$ is the probability that a jump of displacement $\mathbf{R}$ occurs and $\psi(t\,|\,\mathbf{R})$ is the conditional PDF that, given that a jump $\mathbf{R}$ occurs, it takes a time t to be completed. To model turbulence the time of the step is tied to the step length by the fluid velocity

$$\psi(t\,|\,\mathbf{R}) = \delta\left(t - \frac{|\mathbf{R}|}{|\mathbf{V}(\mathbf{R})|}\right), \tag{7.42}$$

which ties the length of the step taken to the duration time through a velocity that depends on the length of the step. When the second moment $\langle R^2 \rangle$ of $p(\mathbf{R})$ in Eq.(7.41) is infinite no largest walk scale exists and this process is identified as a Lévy walk to distinguish it from a Lévy flight. The distinction is introduced by Eq.(7.42) which weights the jumps according to the time spent in each jump.

The mean square displacement at time t for the Lévy walk process is given by

$$\langle R^2; t \rangle = \int |\mathbf{R}|^2\, P(\mathbf{R}; t)d\mathbf{R} = -\mathcal{L}^{-1}\left\{\nabla_{\mathbf{k}}^2 P^*(\mathbf{k}, s); t\right\}\big|_{\mathbf{k}=\mathbf{0}}, \tag{7.43}$$

and up until now we have left Eq.(7.40) quite general so that it can be applied to any problem for which the correlation of steps is known.

We now make contact with turbulence and consider a random-walk description of a fluid medium possessing a wide distribution of correlation lengths and associated velocities generated by the spatial and temporal structure of fully developed turbulence. We use the Lévy walk as a statistical representation for this process and choose Eq.(7.41) for large $|\mathbf{R}|$ the PDF for large steps to be

$$p(\mathbf{R}) \propto \frac{1}{|\mathbf{R}|^{\beta+1}}\;;\; 0 < \beta < 1. \tag{7.44}$$

so that the distribution of walk distances has no characteristic mean scale, that is, walks of all lengths occur. We [60] choose $\psi(t \,|\mathbf{R})$ as given by the velocity-dependent delta function and use Kolmogorov's scaling [30] to choose the proper dependence of the velocity and the length scale R.

It is probably not very useful to provide more detail into the matter of turbulence; those that are interested can consult to literature and those that are not have seen the utility of the approach for other contexts. However, for completeness we record that for the turbulent flow to be confined to an active region of fractal dimension D_f , the velocity field was shown to scale with the separation distance between two tracer particles as the power law [60]

$$V(R) \propto R^\gamma; \; \gamma = \frac{1}{3} + \frac{3 - D_f}{6}. \tag{7.45}$$

Thus, the mean square separation of two particles in an active region of the flow as $t \to \infty$ calculated from Eq.(7.43) is

$$\langle R^2; t \rangle = \begin{cases} t^{12/(4-\mu)} & , \; \beta \leq \frac{1-\mu}{3} \\ t^{2 + \frac{6(1-\beta)}{4-\mu}} & , \; \frac{1-\mu}{3} \leq \beta \leq \frac{10-\mu}{6} \\ t & , \; \beta \geq \frac{10-\mu}{6} \end{cases}, \tag{7.46}$$

where $\mu = 3 - D_f$. Note that the scaling exponents depend on the index β, as well as, the fractal dimension, a result that had not before then been encountered and one which relates the spatial dynamics to the various diffusive regimes. The two limits in the middle case smoothly match the first and third cases. The first case corresponds to an infinite mean time spent in a correlated transition so that no characteristic transition time exists. Diffusion is the most enhanced in this domain. The last case is analogous to molecular diffusion. Note further that if $\mu = 0$, that is, there is no correction for intermittency and $\beta \leq 1/3$ the Richardson t^3-Law is recovered.

In summary, we may say that in a homogeneous turbulent flow field the relative velocity between two particles is determined by their separation distance, with fluid parcels a larger distance apart moving faster relative to one another than those that are closer together. Consequently, in such a background fluid the motion in space and time are tightly coupled. By way of contrast, in a Lévy flight the distribution of step sizes is IPL and a step of size q is drawn from the distribution $1/\,|q|^{\beta+1}$ for each unit of time. The generalized CLT can be shown to yield Eq.(7.29) as the equation of evolution for the PDF in the continuous limit. The Fourier-Laplace transform of the FPSE, given by Eq.(7.29), reduces to the ordinary diffusion equation when $\beta = 2$ in which case the parameter D_2 becomes the ordinary diffusion coefficient D.

7.3   Fractional Fokker-Planck Equation

It is interesting to consider how one proceeds from the FLE to the FFPE and in so doing we follow the derivation in West *et al.* [73]. The connection between the dynamics and the phase space applications of the fractional calculus is illuminating. Consider the FLE given by

$$\partial_t^\alpha [Q(t)] = -\frac{U'(Q)}{m\eta_\alpha} - \gamma^\alpha Q + f_\alpha(t) \tag{7.47}$$

where $U(Q)$ is the deterministic potential, γ^α is the dissipation, $f_\alpha(t)$ is the random force, and the parameter η_α is introduced to suitably adjust the units. The dynamics in phase space are followed by the phase space density

$$\rho(q, t | q_0) = \delta(q - Q(q_0, t)), \tag{7.48}$$

$q_0 = Q(0)$ and an arbitrary analytic function can be expressed as

$$F(Q) = \int F(q)\rho(q, t | q_0) \, dq. \tag{7.49}$$

The difference in the function at two nearby points in time is

$$F(Q[t + \Delta t]) - F(Q[t]) = \int F(q) \left[\rho(q, t + \Delta t | q_0) - \rho(q, t | q_0) \right] dq, \tag{7.50}$$

from which we can construct the limits from two forms of the differences. The first form is given by the fractional time derivative

$$\partial_t^\alpha [F] = \lim_{\Delta t \to 0} \frac{F(Q[t + \Delta t]) - F(Q[t])}{\Delta t^\alpha}; \tag{7.51}$$

the second form is given by

$$\lim_{\Delta t \to 0} \frac{F(Q[t + \Delta t]) - F(Q[t])}{\Delta t^\alpha} = \lim_{\Delta t \to 0} \frac{F(Q + \Delta Q) - F(Q)}{\Delta t^\alpha}$$
$$= \lim_{\Delta t \to 0} \frac{dF}{dQ} \frac{\Delta Q}{\Delta t^\alpha} = \frac{dF}{dQ} \partial_t^\alpha [Q], \tag{7.52}$$

where we have Taylor expanded the difference on the upper line and retained to lowest-order term in ΔQ on the second line.

The two expressions for the limits of the function can also be expressed in terms of operations on the phase space density. From Eq.(7.51) we have

$$\partial_t^\alpha [F] = \int dq F(q) \lim_{\Delta t \to 0} \frac{\rho(q, t + \Delta t | q_0) - \rho(q, t | q_0)}{\Delta t^\alpha}$$
$$= \int dq F(q) \partial_t^\alpha [\rho], \tag{7.53}$$

and from Eq.(7.52) we obtain

$$\frac{dF}{dQ} \partial_t^\alpha [Q] = \int dq \frac{dF(q)}{dq} \partial_t^\alpha [q] \rho(q, t | q_0).$$

Integrating this last equation by parts and setting the integrands to zero at the limits we have

$$\frac{dF}{dQ}\partial_t^\alpha\left[Q\right] = -\int dq F(q)\frac{\partial}{\partial q}\left[\partial_t^\alpha\left[q\right]\rho\left(q,t\,|q_0\right)\right]. \tag{7.54}$$

Consequently, since the analytic function $F(q)$ is arbitrary we can equate the coefficients within the integrands of Eqs.(7.53) and (7.54) to obtain

$$\partial_t^\alpha\left[\rho\left(q,t\,|q_0\right)\right] = -\frac{\partial}{\partial q}\left[\partial_t^\alpha\left[q\right]\rho\left(q,t\,|q_0\right)\right],$$

and inserting the FLE from Eq.(7.47) into this expression yields

$$\partial_t^\alpha\left[\rho\left(q,t\,|q_0\right)\right] = -\frac{\partial}{\partial q}\left[\left\{-\frac{U'(q)}{m\eta_\alpha}-\gamma^\alpha q+f_\alpha(t)\right\}\rho\left(q,t\,|q_0\right)\right] \tag{7.55}$$

Averaging over an ensemble of realizations of the fluctuating force in Eq.(7.55) and using the second-order approximation of this term as presented in Lindenberg and West [32], we obtain the FFPE

$$\partial_t^\alpha\left[P\left(q,t\,|q_0\right)\right] = \frac{\partial}{\partial q}\left[\left\{\frac{U'(q)}{m\eta_\alpha}+\gamma^\alpha q\right\}P\left(q,t\,|q_0\right)\right]+K_\alpha\frac{\partial^2 P\left(q,t\,|q_0\right)}{\partial q^2} \tag{7.56}$$

and K_α corresponds to the effective diffusion coefficient. Note that this is essentially the FFPE obtained by Metzler and Klafter [44] using an entirely different method.

The FFPE can be expressed in terms of the Fokker-Planck operator, after suppressing the dependence on the initial state

$$\partial_t^\alpha\left[P(q,t)\right] = \mathcal{L}_{FP}\left[P(q,t)\right], \tag{7.57}$$

with the FP operator $\mathcal{L}_{FP}$ for the non-dissipative case defined by

$$\mathcal{L}_{FP} \equiv \frac{\partial}{\partial q}\frac{U'(q)}{m\eta_\alpha}+K_\alpha\frac{\partial^2}{\partial q^2}. \tag{7.58}$$

Equation(7.57) is a linear fractional equation in time for the PDF and from the arguments given in Chapter 4 has the formal initial value solution

$$P(q,t) = E_\alpha\left(\mathcal{L}_{FP}t^\alpha\right)P_0(q), \tag{7.59}$$

and the MLF is expressed in terms of the FP operator.

7.3.1 Fractional Eigenvalue Expansions

The stationary or steady-state solution to the FFPE can be obtained from

$$\mathcal{L}_{FP}\left[P_{ss}(q)\right] = 0, \tag{7.60}$$

which is obtained from the FFPE in the $t \to \infty$ limit. Using the definition of the FP operator and the condition that the probability current is constant, the solution to Eq.(7.60) is the steady-state PDF

$$P_{ss}(q) = \exp\left[-\frac{U(q)}{m\eta_\alpha K_\alpha}\right]. \tag{7.61}$$

Note that the steady-state PDF is the equilibrium Boltzmann PDF with the requirement that the generalized Einstein-Stokes-Smoluchowski relation is recovered [44]

$$K_\alpha = \frac{k_B T}{m\eta_\alpha}. \tag{7.62}$$

A formal representation of the general solution to the FFPE is given by Eq.(7.59). A functional representation would be in terms of a series expansion over a complete set of eigenfunctions. Consequently, we write the general solution in the product form

$$P(q,t) = \sum_{n=0}^{\infty} \phi_n(q) T_n(t), \tag{7.63}$$

and inserting this series into Eq.(7.57) results in the two eigenequations

$$\mathcal{L}_{FP}\phi_n(q) = -\lambda_n \phi_n(q), \tag{7.64}$$
$$\partial_t^\alpha [T_n(t)] = -\lambda_n T_n(t), \tag{7.65}$$

where the eigenvalues are ordered $\lambda_0 < \lambda_1 < \lambda_2 < ...$ Here again the solution to the linear fractional rate equation is given by

$$T_n(t) = E_\alpha(-\lambda_n t^\alpha), \tag{7.66}$$

which can also be obtained from Eq.(7.59) by inserting Eq.(7.63) into the right hand side and using the orthonormality of the eigenfunctions $\phi_n(q)$. In this way Eq.(7.63) can be written as

$$P(q,t) = \sum_{n=0}^{\infty} \phi_n(q) E_\alpha(-\lambda_n t^\alpha), \tag{7.67}$$

and we see that the relaxation to the equilibrium state is IPL due to the asymptotic form of the MLF. Consequently, for a given set of eigenfunctions and eigenvalues, that is a given structure in the space variable, the relaxation to equilibrium is much slower than in the exponential case $\alpha = 1$.

7.3.2 Fractional Harmonic Oscillator

Metzler and Klafter [44] have reviewed the solution to the FFPE for a number of standard potentials including a harmonically bound particle when $\alpha \neq 1$.

It is, of course, possible to construct a discrete spectrum for the FP operator, however $\mathcal{L}_{FP}$ is in general not Hermitian so that, as pointed out by Zhang *et al.* [89], it may have various types of spectra governing the dynamics of the system of interest. In order to construct a self-adjoint FP operator we introduce the steady-state solution to the FPPE given by Eq.(7.61) and define the auxiliary function $\Phi(q,t)$ by the expression

$$P(q,t) = P_{ss}(q)\Phi(q,t). \tag{7.68}$$

The auxiliary function is substituted into the FFPE to obtain, after some algebra,

$$
\begin{aligned}
\partial_t^\alpha \left[\Phi(q,t)\right] &= -\frac{U'(q)}{m\eta_\alpha}\frac{\partial\Phi(q,t)}{\partial q} + K_\alpha \frac{\partial^2\Phi(q,t)}{\partial q^2} \\
&= -\mathcal{L}\Phi(q,t).
\end{aligned}
\tag{7.69}
$$

The function $\Phi(q,t)$ can be expanded in a complete set of eigenfunctions for the operator $\mathcal{L}$:

$$\mathcal{L}\phi_n(q) = \lambda_n\phi_n(q), \tag{7.70}$$

as follows

$$\Phi(q,t) = \sum_{n=0}^{\infty} a_n\phi_n(q)E_\alpha\left(-\lambda_n t^\alpha\right), \tag{7.71}$$

which when inserted into Eq.(7.69) yields a tautology. The coefficients in the eigenfunction expansion are given in terms of the initial PDF with an inner product defined with respect to the steady-state PDF as a weight function

$$a_n = \int P_0(q)\,P_{ss}(q)^{-1}\,\phi_n(q)dq, \tag{7.72}$$

indicating that the operator $\mathcal{L}$ is self-adjoint on the space spanned by the basis functions $\phi_n(q)$.

We can now rewrite Eq.(7.70) as the time-independent Schrödinger equation in terms of the variable $y = q/\sqrt{K_\alpha}$

$$\frac{\partial^2 \Psi_n(y)}{\partial y^2} + [\lambda_n - V(y)]\,\Psi_n(y) = 0, \tag{7.73}$$

where the wavefunction is given in terms of the eigenfunction

$$\Psi_n(y) \equiv \sqrt{K_\alpha P_{ss}(y)}\phi_n(y), \tag{7.74}$$

and the "potential" is

$$V(y) = \left[\frac{1}{2m\eta_\alpha K_\alpha}\frac{\partial U(y)}{\partial y}\right]^2 - \frac{1}{2m\eta_\alpha K_\alpha}\frac{\partial^2 U(y)}{\partial y^2}. \tag{7.75}$$

This means that we could have initiated the investigation with the fractional Schrödinger equation and introduced the effective potential .

For a harmonically bound particle the potential can be written

$$V(y) = \left(\frac{\omega^2 y}{2\eta_\alpha}\right)^2 - \frac{\omega^2}{2\eta_\alpha}, \tag{7.76}$$

so that in the new variable $z = y\sqrt{\omega^2/2\eta_\alpha}$ Eq.(7.73) reduces to

$$\Psi_n''(z) + \left[\varepsilon_n + 1 - z^2\right]\Psi_n(z) = 0, \tag{7.77}$$

where we have introduced the scaled eigenvalue

$$\lambda_n = \frac{\omega^2}{2\eta_\alpha}\varepsilon_n. \tag{7.78}$$

The solution to the "Schrödinger" equation Eq.(7.77) is obtained for the eigenvalues $\varepsilon_n = 2n$ and the normalized eigenfunction are given in terms of Hermite polynomials [3]

$$\Psi_n(z) = \frac{1}{\sqrt{2^n n!\sqrt{\pi}}} e^{-z^2/2} H_n(z). \tag{7.79}$$

Some algebra enables us to express the full PDF as a series over Hermite polynomials [43]

$$P(q,t\,|q_0,0) = \sqrt{\frac{m\omega^2}{2\pi k_B T}} \sum_{n=0}^{\infty} \frac{H_n(z_0)\,H_n(z)}{2^n\Gamma(n+1)} e^{-z^2} E_\alpha\left(-\lambda_n t^\alpha\right), \tag{7.80}$$

with the eigenvalues $\lambda_n = n\omega^2/\eta_\alpha$,

$$z = y\sqrt{\frac{\omega^2}{2\eta_\alpha}} = q\sqrt{\frac{\omega^2}{2\eta_\alpha K_\alpha}} = q\sqrt{\frac{m\omega^2}{2k_B T}},$$

and z_0 has the same definition as z only with q replaced by its initial value q_0. Eq.(7.80) is a hauntingly familiar result from quantum mechanics and suggests the possibility of applying the fractional calculus in the quantum domain.

The solutions for all moments of the displacement $\langle q^n; t\rangle$ can be obtained by multiplying the FFPE equation by q^n and integrating over q to obtain

$$\partial_t^\alpha\left[\langle q^n; t\rangle\right] = \int_{-\infty}^{\infty} q^n \mathcal{L}_{FP}\left[P(q,t\,|q_0)\right] dq. \tag{7.81}$$

Consequently, introducing the definition of the FP operator and integrating by parts, yields for $n = 1$

$$\partial_t^\alpha\left[\langle q; t\rangle\right] = -\lambda_1\langle q; t\rangle, \tag{7.82}$$

and $n = 2$

$$\partial_t^\alpha \left[\langle q^2; t \rangle \right] = -2\lambda_1 \langle q^2; t \rangle + K_\alpha. \tag{7.83}$$

The solution to the first-moment fractional rate equation is given by

$$\langle q; t \rangle = q_0 E_\alpha \left(-\lambda_1 t^\alpha \right) \qquad \lambda_1 = \frac{m\omega^2}{k_B T}. \tag{7.84}$$

Notice that the average displacement is consistent with the solution obtained in Chapter 5 for the fractional harmonic oscillator. An optimist might call this the *Fractional Correspondence Principle* in which the quantum average of a fractional quantum variable is equal to the classical average of a fractional oscillator system.

The solution to the second-moment fractional rate equation is

$$\langle q^2; t \rangle = \frac{1}{\lambda_1} + \left[q_0^2 - \frac{1}{\lambda_1} \right] E_\alpha \left(-2\lambda_1 t^\alpha \right), \tag{7.85}$$

and the properties of these solutions are discussed in detail by Metzler and Klafter [44]. Of course the relaxation of the solutions to thermal equilibrium are IPL as determined by the MLF and eventually

$$\lim_{t \to \infty} \langle q^2; t \rangle = \frac{1}{\lambda_1} = \frac{k_B T}{m\omega^2}.$$

The non-exponential relaxation of the fractional harmonically bound particle to equilibrium is different from the exponential relaxation to thermal equilibrium obtained for an Ornstein-Uhlenbeck process [44].

We emphasize that the solution to the FFPE has a relaxation to equilibrium even though there is no physical dissipation included in the dynamic description. We encountered this effect in the discussion of the fractional linear harmonic oscillator in Chapter 5. However, we subsequently interpreted the relaxation in the MLF solution to be the result of phase interference between members of the ensemble of trajectories contributing to the solution. This relaxation mechanism has not been universally accepted and cries out for further investigation.

7.4 Fractional Sturm-Liouville Theory

The discussion of the underlying Schrödinger equation of the last section is often initiated using one of the most remarkable mathematical schema of the nineteenth century; the Sturm-Liouville theory of ordinary differential equations. This theory has been the foundation for the development of spectral methods and the theory of self-adjoint operators. In physics Sturm-Liouville theory is motivated by the separation of the three-dimensional Helmholtz equation

$$\nabla^2 \psi + k^2 \psi = 0 \tag{7.86}$$

into the form of the *Liouville equation*

$$\mathcal{L}u(x) + \lambda w(x) u(x) = 0. \tag{7.87}$$

Here $w(x)$ is a positive definite weighting function, λ is a constant, and the Liouville operator $\mathcal{L}$ is defined by

$$\mathcal{L}u(x) \equiv \frac{d}{dx}\left[p(x)\frac{du(x)}{dx}\right] + q(x)u(x). \tag{7.88}$$

The three auxiliary functions p, q and w are characteristic of the coordinates used in the separation of variables and have a number of properties that I will not introduce here, but which can be found in a number of excellent texts [49] along with a discussion on the limitations of the theory. I merely wish to point out that the Sturm-Liouville problem is to determine the dependence of the eigenvalues λ on the homogeneous boundary conditions imposed on the eigenfunction solutions $u_\lambda(x)$. The polynomial eigenfunctions obtained in this way, for various choices of the auxiliary functions, and boundary conditions, are associated with the names Bessel, Chebyshev, Hermite, Jacobi, Laguerre and Legendre.

The purpose in this brief jogging of your memory is to bring to your attention the fact that Zayernouri and Karniadakis [88], among others, have generalized the Sturm-Liouville problem to the fractional calculus. They generalize the operator in Eq.(7.88) to fractional form, and extend the boundary conditions to include fractional derivatives and show that the eigenvalues of the problem are real, with orthogonal eigenfunctions corresponding to the real eigenvalues. In short they develop a spectral theory for what they call the regular and singular fractional Sturm-Liouville problems. Moreover, they demonstrate the utility of the spectral method by constructing explicitly proper basis for numerical approximations of fractional functions. They demonstrate that the eigenfunctions are of non-polynomial form, called Jacobi *Poly-fractonomials*. In addition, they show these eigenfunctions are orthogonal and form a complete basis in the Hilbert space.

There is ample opportunity for a young investigator to start from the fractional form of the Helmholtz equation and construct the corresponding fractional Sturm-Liouville theory. It is doubtful that the resulting theory is unique.

7.5   Fractional Kinetics Equation

Fractional kinetics (FK) is a term introduced by Zaslavsky [87] to describe the singular fractal representation of trajectories using renormalization group methods to describe the kinetics of chaotic dynamical systems. As he [87] points out, this was an outgrowth of the notion of exploiting the fractional calculus to represent the kinetics of systems described by fractional integro-differential equations to signals [38], kinetics of advected particles [81], and percolation [24] to name a few. FK has been interpreted as a way to describe the dynamics of systems sharing the characteristics of stochastic and regular behavior as done, for example, in the construction of the FFPE [44].

Herein we consider the dynamic equations for a PDF generalized in two distinct ways: 1) the set of points over which the PDF is evaluated is assumed to be fractal and 2) the dynamic equations are nonlinear. Zaslavsky [87] demonstrates how to relate the limit of a PDF $P(t)$ defined on a fractal set of points to the fractional derivative

$$\partial_t^\alpha [P(t)] \equiv \lim_{\Delta t \to 0} \frac{P(t + \Delta t) - P(t)}{\Delta t^\alpha} \qquad (7.89)$$

using renormalization group arguments. The parameter α is the critical exponent that characterizes the fractal structure of the trajectory in time.

In this section we introduce the kinetic equation interpretation for the evolution of the PDF of a dynamic process $Q(t)$ along a trajectory having a value in the interval $(q, q + dq)$ at time t. Zaslavsky [87] provides an excellent description of the relation between chaotic trajectories generated by nonlinear equations of motion and the fractional kinetics of the PDF. He considers the infinitesimal changes of $P(q, t)$ in time along chaotic trajectories whose local averages yield a fractional kinetic equation (FKE), which is to say an equation of motion for the PDF in phase space in terms of fractional derivatives. There are a number of alternative derivations of the FKE including the use of subordination [16] and the fractional generalized Langevin equation [33]. The form of the FKE is given by Eq.(7.37) and the parameters α and β are scaling exponents that characterize the fractal structure of the trajectories in Zaslavsky's approach. Of course it is also possible that the diffusion coefficient, as well as the fractional orders of the derivatives, are dependent on t and/or q, but we do not consider these cases here and instead refer the reader back to the literature [29, 87].

7.5.1   Renormalization Group Scaling

Zaslavsky [87] applied a renormalization group (RG) transformation R to the system dynamics such that the scaling properties of the incremental changes are

$$R: \quad \Delta q \to \lambda_q \Delta q \ , \quad \Delta t \to \lambda_T \Delta t, \qquad (7.90)$$

which apply, after some averaging, in a restricted space-time domain and (λ_q, λ_T) are scaling parameters. He goes on to say that a basic feature of renormalization group kinetics (RGK) is that the FKE given by Eq.(7.37) is invariant under the operation of a RG transformation

$$R\{\partial_t^\alpha [P(q, t)]\} = R\left\{K_{\alpha,\beta} \partial_{|q|}^\beta [P(q, t)]\right\}, \qquad (7.91)$$

implying that the FKE satisfies the RGK scaling behavior

$$\lambda_T^\alpha \partial_t^\alpha [P(q, t)] = \lambda_q^\beta K_{\alpha,\beta} \partial_{|q|}^\beta [P(q, t)]. \qquad (7.92)$$

This renormalization procedure may be applied an arbitrary number of times and consequently the resulting fractional differential equation remains valid only if the ratio of the scaling parameters satisfies

$$\left(\frac{\lambda_q^\beta}{\lambda_T^\alpha}\right)^n = 1, \tag{7.93}$$

for arbitrary integer n. The FKE is linear, so that the sum of the individual fixed-point solutions is also a solution. In this way the fixed point equation Eq.(7.93) has an infinite number of solutions

$$\frac{\lambda_q^\beta}{\lambda_T^\alpha} = e^{in2\pi} \quad ; \quad n = 0, \pm 1, \pm 2, \ldots \tag{7.94}$$

where the fixed-point solutions are denoted by integer n. Consequently, indexing the time parameter with the fixed-point index, the ratio of the time to space parameters becomes

$$\frac{\alpha_n}{\beta} = \frac{\ln \lambda_q}{\ln \lambda_T} + i\frac{2\pi n}{\beta \ln \lambda_T} \quad ; \quad \alpha_0 = \alpha. \tag{7.95}$$

Indexing the time parameter by n is arbitrary and is only intended to distinguish among the various fixed-point solutions to the FKE. This result of a complex ratio of parameters is one source of the complex fractal dimension and the log-periodicity introduced earlier.

In Section 7.5.3 we show that the solution to Eq.(7.37) has the general scaling form

$$P(q,t) = \frac{1}{t^\delta} F\left(\frac{q}{t^\delta}\right) \quad ; \quad \delta = \alpha/\beta, \tag{7.96}$$

using the zero-order solution to Eq.(7.95). Consequently the first moment of the dynamic variable takes the form

$$\begin{aligned}
\langle q; t \rangle &= \int q P(q,t) dq = \int \frac{q}{t^\delta} F\left(\frac{q}{t^\delta}\right) dq \\
&= D_{\alpha\beta}^{(0)} t^\delta,
\end{aligned} \tag{7.97}$$

where the coefficient can be calculated to be [87]

$$D_{\alpha\beta}^{(0)} = \int y F(y) dy = K_{\alpha,\beta} \Gamma(1+\beta). \tag{7.98}$$

However, if we include the higher-order fixed-point solutions from Eq.(7.95) in the average we obtain

$$\langle q; t \rangle = t^\delta \sum_{n=0}^{\infty} A_n \cos\left(2\pi n \frac{\ln t}{\ln \lambda_T}\right), \tag{7.99}$$

resulting in periodic variations in the average value, with variations in $\ln t$ having period $\ln \lambda_T$. Note that the log-periodicity of the solution results from the imaginary part of the complex fractal dimension, which were observed in a number of phenomena and discussed in Section 3.3.2. Zaslavsky [86] points out that such log-perodic oscillations had been observed in numerous phenomena including in anomalous transport [84, 85], in phase transitions [50], in dynamical systems [34, 19] and in geophysics [63] to name a few.

The expansion coefficients A_n in Eq.(7.99) can be explicitly calculated when the kinetic equation of motion has a discrete renormalization invariance, see, for example, [19, 48, 71]. The lowest-order coefficient is given by $A_0 = D_{\alpha\beta}^{(0)}$. On the other hand, these coefficients can also be used as heuristic expansion coefficients, much like what is done with empirical Fourier series fitting the expansion coefficients to time series data.

The PDF solution to the FKE in both space and time has been presented in this section with the view of justifying the statistical properties found in the applications discussed in earlier chapters along with some new ones.

7.5.2 Mittag-Leffler Statistics

The most direct explicit definition of the fractional operators is in terms of their Fourier and Laplace transforms. In Appendix 7.9.1 the solution to the FKE in Fourier-Laplace space is written

$$P^*(k, s) = \frac{s^{\alpha-1}}{s^\alpha + K_{\alpha,\beta} |k|^\beta}. \tag{7.100}$$

The inverse Laplace transform of the Fourier-Laplace solution to the FKE yields the MLF for the characteristic function, as we observed in Section 4.3:

$$\widetilde{P}(k, t) = E_\alpha \left(-K_{\alpha,\beta} |k|^\beta t^\alpha \right). \tag{7.101}$$

Consequently, the relatively benign statistics of Poisson occurs for $\alpha = 1$, where the MLF reduces to an exponential, and therefore by an inverse Fourier transform to a Lévy process in space. This latter PDF yields a power-law dependence on time for the θ-moment of the variate

$$\langle q^\theta; t \rangle \propto t^{\theta/\beta}, \tag{7.102}$$

reducing to a linear growth in the second moment when $\beta = \theta = 2$.

On the other hand, the process manifests intermittent IPL statistics for the index in the interval $0 < \alpha < 1$ due to the asymptotic behavior of the MLF expressed by Eq.(4.45). The complexity of the resulting statistics is captured in the IPL index, much like the allometry exponent captures the complexity of fractal structure of allometric phenomena [76].

Eq.(7.100) can be Fourier inverted for $\beta = 2$ to obtain

$$\widehat{P}(q, s) = \frac{s^{\frac{\alpha}{2}-1}}{2} \exp\left[-K_\alpha |q| s^{\frac{\alpha}{2}} \right] \quad ; \quad 0 < \alpha \le 1. \tag{7.103}$$

Mainardi [36] considered the inverse Laplace transform of Eq.(7.103) and obtained an analytic solution in terms of a Wright function that reduces to the Laplace PDF $\exp\left[-K_0\,|q|\right]/2$ when $\alpha = 0$ and when $\alpha = K_1 = 1$ to the Gaussian PDF $\exp\left[-q^2/4t\right]/\sqrt{4\pi t}$. It is evident by using a Tauberian theorem that asymptotically $(s \to 0)$ the PDF given by the inverse Laplace transform of Eq.(7.103) decays as an IPL in time $t^{-\alpha/2}$.

Thus, we see that the FFPE or equivalently the FPSE can generate the two most familiar forms of complexity that arise in the study of complex dynamic networks: *topological complexity* given by an IPL in an extensive variable such as the number of events, for example, where an event can be a new connection within a network [5]; *temporal complexity* given by an IPL in time, where the time denotes an interval between events, for example, the time between receiving and answering emails or letters [51].

It is interesting to note that with $\beta = 0$, the FKE reduces to a fractional rate equation, whose PDF solution is the MLF. In this case the probability that the number of events $N(t)$ that occur in a time t is determined by the convolution equation

$$P(N(t) = n) = \int_0^t \Psi\left(t - t'\right)\psi_n\left(t'\right)dt',\qquad(7.104)$$

which is the product of the probability that n events occur in the time interval $(0, t')$ and no events occur in the time interval (t', t). We [75] show that the Laplace transform of $P(N(t) = n)$ is given by

$$\widehat{P}_n(s) = \widehat{\Psi}(s)\left[\widehat{\psi}(s)\right]^n,$$

so that using the MLF as the waiting-time PDF yields

$$
\begin{aligned}
\widehat{P}_n(s) &= \frac{s^{\alpha-1}}{s^\alpha + \gamma^\alpha}\left[1 - \frac{s^\alpha}{s^\alpha + \gamma^\alpha}\right]^n \\
&= \frac{\gamma^{n\alpha}}{\left[s^\alpha + \gamma^\alpha\right]^{n+1}}s^{\alpha-1}.
\end{aligned}\qquad(7.105)
$$

The inverse Laplace transform of Eq.(7.105) is

$$P(N(t) = n) = \frac{(\gamma t)^{n\alpha}}{\Gamma(n+1)}E_\alpha\left(-(\gamma t)^\alpha\right),\qquad(7.106)$$

which is a clear generalization of the Poisson distribution of the number of events with parameter γt and was called the α − *fractional Poisson distribution* by its inventors Mainardi *et al.* [37]. Of course this probability becomes the ordinary Poisson distribution when $\alpha = 1$.

7.5.3   Scaling Solution

Uchaikin [69] directly inverse transformed Eq.(7.100) for arbitrary α and β and we refer the reader to the original literature for that level of mathematical detail. The desired insight for our purposes here is provided by utilizing the scaling properties of Eq.(7.101) and considering the PDF in the form of the inverse Fourier transform

$$P(q,t) = \mathcal{FT}^{-1}\left\{ E_\alpha \left(-K_{\alpha,\beta} |k|^\beta t^\alpha \right) ; q \right\}. \tag{7.107}$$

The series expansion for the MLF given by Eq.(4.43) when inserted into Eq.(7.107) allows us to write the scaling relation

$$P(\lambda_q q, \lambda_T t) = \sum_{n=0}^{\infty} \frac{(-K_{\alpha,\beta}\lambda_T^\alpha t^\alpha)^n}{\Gamma(n\alpha+1)} \cdot \frac{\Gamma(n\beta+1)}{|\lambda_q q|^{n\beta+1}}, \tag{7.108}$$

where the second factor in the summation is the result of applying the Tauberian theorem to the inverse Fourier transform of $|k|^{n\beta}$. A renormalization group scaling equation emerges when the parameters satisfy the equality $\lambda_q = \lambda_T^{\alpha/\beta}$, which is the lowest-order fixed point solution to Eq.(7.95), resulting in Eq.(7.108) reducing to

$$P(\lambda_T^{\alpha/\beta} q, \lambda_T t) = \frac{1}{\lambda_T^{\alpha/\beta}} P(q,t).$$

If we now select the time parameter to be $\lambda_T = 1/t$, so that the scaled variable $q/t^{\alpha/\beta}$ becomes the new dynamic variable we can write

$$P(q,t) = \frac{1}{t^{\alpha/\beta}} P\left(\frac{q}{t^{\alpha/\beta}}, 1 \right). \tag{7.109}$$

Finally, the PDF that solves the FKE in terms of the similarity variable q/t^{μ_q} satisfies the scaling equation

$$P(q,t) = \frac{1}{t^{\mu_q}} F_q\left(\frac{q}{t^{\mu_q}} \right) \quad \text{and} \quad \mu_q = \alpha/\beta. \tag{7.110}$$

The function $F_q(\cdot)$ in Eq.(7.110) is left unspecified, but it is analytic in the similarity variable q/t^{μ_q}. In a standard diffusion process $Q(t)$ is the displacement of a diffusing particle from its initial position at time t, and for vanishing small dissipation the scaling parameter is $\mu_q = 1/2$ and the functional form of $F_q(\cdot)$ is a Gauss PDF. However, for general complex phenomenon there is a broad class of distributions for which the functional form of $F_q(\cdot)$ is not Gaussian and the scaling index $\mu_q \neq 1/2$. For example, the alpha-stable Lévy process [47, 58, 62, 90] scales in this way and the Lévy index is in the range $[0,2]$, with the Gauss PDF resulting for the maximum value of the index. The scaling index in Eq.(7.110) is related to the Lévy index by $1/\mu_q$.

7.6 Physiology and Complexity Loss

The understanding of complexity has guided most of the formalism presented in this book. In medicine the purpose is somewhat different, because it is often the loss of complexity in disease that needs to be understood in order to reestablish the functionality associated with that complexity. One approach to understanding the loss of complexity is through the study of extreme value theory (EVT), that being the analysis of the statistics of unusual or rare events. EVT is the branch of statistics that deals with extreme deviations from the median of the PDF. Half the data is greater than the median and half is less than the median and consequently the median value is not the same as the mean or average value except when the PDF is symmetric, as in the case of the Normal PDF. For many complex phenomena the IPL PDF has replaced the Normal, in fact, this replacement is often used to define complexity. The existence of a long tail implies there are a great many more large magnitude events than in exponential processes. The importance of these rare events cannot be over stated, since extrema dominate such processes in general, for example, in determining mortality in heart beat irregularities; the frequency of stock market crashes; or the size of the next expected earthquake [80].

In a previous chapter the analysis of RR-interval for healthy individuals and those with heart disease were shown to have the same statistical distribution, that being a Lévy stable PDF [52]. This non-Gaussian behavior of the HRV intermittent time series has, a quarter century later, been shown to be more subtle than originally believed. The scale invariant fluctuations in the healthy human heart beat were examined under a variety of statistical assumptions by Kiyono *et al.* [26]. They proposed a data processing model for sudden cardiac death after atrial myocardial infarction (AMI) and found that a truncated Lévy PDF could not be ruled out as a proper descriptor of the HRV statistics [27]. The processing was done by aggregating the RR time series data using increasingly longer segments of the time series that eventually converged on a Gaussian distribution for sufficient coarse graining of the time series. Kiyono *et al.* [26] found it impossible to distinguish between a truncated Lévy PDF and the approximated PDF based on the analysis using the technique of Castaing *et al.* [10]. On the other hand, the cascading mechanism that is used in the interpretation of the intermittency of velocity fluctuation in turbulent fluid flow by Castaing *et al.* [10] and subsequently used in the interpretation of intermittency in HRV interval statistics could not be confirmed in the analysis of HRV time series by Kiyono and Belli [28].

7.6.1 Truncated Lévy Hypothesis

In Section 7.1.1 we determined that the solution to the FKE with $\alpha = 1$ is the Lévy PDF. It was noted that such PDFs have fat or IPL tails that decay in time more slowly than does the typical exponential. Such fat tails can generate diverging variances, which are not plausible for real physiologic data. Consequently, it is necessary to find a PDF that behaves as an IPL for intermediate amplitudes, but which manifests physiologic control to mitigate the occurrence of truly extreme events. For this reason we assume that the HRV statistics of healthy individuals are determined by a physiologic feedback mechanism in which the tail of a Lévy PDF is exponentially truncated. This is a useful assumption that needs to be rigorously tested against clinical data.

We hypothesize a physiologic feedback mechanism that produces an exponential suppression of very large fluctuations in the interbeat interval. This exponential decay of large fluctuations can be formally incorporated into the anomalous diffusion equation Eq.(7.29) in the following way

$$\frac{\partial P(q,t)}{\partial t} = D_\beta \left[\partial_{|q|} + \gamma \right]^\beta \left[P(q,t) \right], \tag{7.111}$$

thereby suppressing both positive and negative extrema. The solution to the modified FKE is again obtained using Fourier transforms. The Fourier transform of the shifted operator is determined by the binomial expansion in Appendix 7.9.2 resulting in the equation for the characteristic function

$$\frac{\partial \ln \widetilde{P}(k,t)}{\partial t} = -D_\beta \left[k^2 + \gamma^2 \right]^{\beta/2} \cos \psi \tag{7.112}$$

$$\psi \equiv \beta \left[\tan^{-1}(k/\gamma) - \frac{\pi}{2} \mathrm{sign} k \right]. \tag{7.113}$$

Integrating Eq.(7.112) yields the characteristic function after including a term to insure proper normalization of the probability density, that is, $\widetilde{P}(k=0,t) = 1$,

$$\ln \widetilde{P}(k,t) = -D_\beta t \left\{ \left[k^2 + \gamma^2 \right]^{\beta/2} \cos \psi - \gamma^\beta \right\}. \tag{7.114}$$

The inverse Fourier transform of Eq.(7.114) yields the truncated symmetric Lévy PDF

$$P(q,t;\gamma,\beta) = e^{-\gamma|q|} L_\beta(q, D_\beta t). \tag{7.115}$$

The truncated Lévy PDF was first studied numerically in the context of stock market fluctuations by Mantegna and Stanley [39, 40] and soon thereafter Koponen [31] provided a formal derivation of the characteristic function for a truncated Lévy flight. Matsushita *et al.* [41] explain that the resulting process is infinitely divisible by scaling q and γ with $(D_\beta t)^{1/\beta}$. Consequently, in terms of the scaled variables $q_s = q/(D_\beta t)^{1/\beta}$ and the scaled parameter $\gamma_s = \gamma (D_\beta t)^{1/\beta}$ the PDF becomes

$$P(q,t;\gamma,\beta) = \frac{e^{-\gamma_s|q_s|}}{t^{1/\beta}} L_\beta(q_s), \tag{7.116}$$

where $L_\beta(q_s)$ is the stable Lévy PDF with Lévy index β. The probability of being in the interval $(q, q+dq)$ at time t is the same as that of being in the scaled interval $(q_s, q_s + dq_x)$:

$$P(q,t;\gamma,\beta)dq = e^{-\gamma_s|q_s|} L_\beta(q_s) dq_s, \tag{7.117}$$

Consequently, the truncation suppresses the most extreme RR-intervals, as well as, the greatest velocity fluctuations in turbulent fluid flow.

7.6.2 Application to HRV

In Figure 7.1 the curves denote the HRV PDFs from a study [20] of a collection of 670 post-AMI (acute myocardial infarction) patients using 24 hour Holter monitor data sets, yielding heart beat interval variability from the time series. In this study a number of individuals suffered cardiac death, others died by non-cardiac causes and some survived. The PDFs determined by the time series for the three groups are indicated.

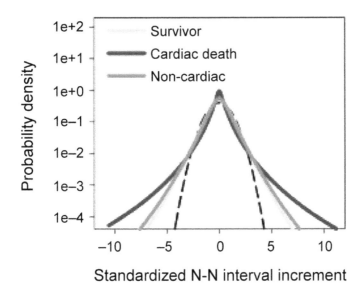

Figure 7.1 The HRV distributions from 24 hour RR interval time series for a group having suffered atrial myocardial infarcation. The patients are separated into those that suffer cardiac death, another with non-cariac death and third consisting of survivors. (Adapted from [20] with permission.)

The first thing to notice about the results in Figure 7.1 is that no group has Normal statistics, which in terms of the standardized variable would coincide with the dashed curve. Next the survivors and those succumbing to non-cardiac death have essentially the same variability PDF. The extrema for these processes would be called *black swans* by Taleb [67] and are unpredictable because they share the statistics of their smaller siblings. Here we would fit these two groups by the truncated Lévy PDF, whose tails far exceed those of the Gaussian PDF, but are less extreme than those experiencing cardiac death.

The extrema for those that suffer a cardiac death fall into the category of Sornette's *dragon kings* [63]. The variability PDFs of the cardiac death patients are very different from those that survive even though there is a great deal of overlap in the central regions of the distributions. This might indicate that the feedback mechanism producing the truncated Lévy PDF for those that survive is disrupted for those in the cardiac death group. The pathology of the HRV PDF being Lévy stable would then be the result of a suppression of a physiological control process [80] that suppress the unhealthy extrema. It remains for cardiologists to test the usefulness of this assumption.

7.7   Diffusion Entropy

A possible mechanism for identifying the difference between black swans and dragon kings may be obtained from an adaptation of a data processing technique created by Allegrini *et al.* [2].They assume the HRV time series consists of a mixture of random and pseudorandom events. In the random case the intervals are independent of one another and in the pseudorandom case the intervals are correlated. The experimental data are used to generate a diffusion process described by the diffusing variable q, using the data as input to a random walk process. It is expected that the diffusion process generated in this way has a PDF that satisfies the scaling property given by Eq.(7.96).

The diffusion entropy (DE) approach to analyzing the data is dependent on the Shannon entropy:

$$S(t) = -\int P(q,t)\ln P(q,t)dq. \tag{7.118}$$

Inserting Eq.(7.96) for the PDF into the entropy expression yields

$$S(t) \quad = \quad S_0 + \delta \ln t \ , \tag{7.119}$$

$$S_0 \quad \equiv \quad \int F(y)\ln F(y)dy. \tag{7.120}$$

In this way the scaling parameter δ can be obtained more efficiently than through a calculation of the second moment of the PDF. This is useful because when the PDF is IPL with index < 2, or is Lévy stable, the second moment diverges. For an experimental data set the second moment would always be finite due to unavoidable statistical limitations, even when the theoretical value diverges. Consequently, the misleading results generated by these statistical inaccuracies are circumvented by using Eq.(7.119).

The diffusion control process, with which to compare data, consists of a random walker who only steps at times determined by a waiting-time PDF that is IPL with index μ, and always in the same direction [2]. The resulting diffusion process has been shown to yield an asymmetric Lévy PDF fitting with Eq.(7.96). If the IPL index is in the interval $2 < \mu < 3$ the scaling parameter δ is [18]

$$\delta = \frac{1}{\mu - 1}. \tag{7.121}$$

Here we see that the scaling resulting from the IPL nature of the waiting-time PDF is $\delta > 0.5$; a value distinctly different from RRW that would yield $\delta = 0.5$ using the DE approach. Using a model with known computer generated random and pseudorandom intervals Allegrini *et al.* [2] establish that the DE measures the scaling of the pseudorandom intervals only.

A second measure is provided by the experimental autocorrelation function $C_{\text{exp}}(t)$ that is presumed to consist of a rapidly decaying piece with a weight $(1 - \varepsilon^2)$ for the random events and an IPL piece with a weight ε^2 for the pseudorandom events. Since the correlation between random events drops precipitously in one time step the weight parameter can be determined from data to be

$$\varepsilon^2 = C_{\text{exp}}(t = 1). \tag{7.122}$$

In this way experiment HRV data can be characterized by the parameter pair (δ, ε^2) and a distribution of patients can be compared to one another on the corresponding parameter plane of Figure 7.2.

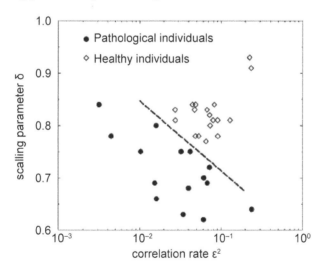

Figure 7.2 Values of the scaling parameters δ and ε^2 for the healthy and congestive heart failure individuals (from [2] with permission)

The surprising result is that from a population of known healthy and congestive heart failure individuals the two segregate themselves from one another in this parameter space. The dashed diagonal line partitions the two groups with the extreme $(\delta = 1, \varepsilon^2 = 1)$ denoting maximum correlation in data dominated by pseudorandom events. In the model space this would mean that health is a consequence of having pseudorandom events in the HRV data, which is to say, strong correlations between events are necessary to be healthy. This would be compatible with an interpretation in terms of black swans, since the correlations would not change the underlying statistics.

At the other extreme ($\delta = 0, \varepsilon^2 = 0$) there are only random events with the statistics being determined by a Gaussian basin of attraction. There is a complete loss of correlation, so that the heart beat time intervals are completely uncorrelated from one another. This loss of correlation may, in fact be what produces the dragon kings in Figure 7.1.

A healthy heart beat is a delicate balance between order and randomness, where the order is measured by the correlation. This could explain why the majority of the healthy individuals cluster in a narrow region of parameter values above the diagonal in Figure 7.2. Whereas the heart-failure subjects are significantly more broadly distributed below the diagonal, but they never achieve the balance necessary for health.

7.8   After Thoughts

This chapter, with the possible exception of Chapter 5, is probably the most dense mathematically. There were even the outlines of a couple of proofs and for this I apologize; by which I mean that I will explain why it was done. It is my impression that even something as mundane as changing the notation, for a vector in an equation of motion from an arrow over a variable to making the variable bold faced without an arrow, encountered resistance in the physical sciences. Why? I believe that, in addition to the all too human characteristic of opposing change of any kind, most scientists prefer to keep mathematics in the background even while using it to solve a scientific problem. This is no less strange than a writer not wanting to be distracted by individual words when writing the first draft of a story.   The mathematical consistency will come later for the scientist, just as the word selection for the polished prose will for the author. However, this chapter was less about solving problems than it was about how we think about a class of problems; those in which complexity is the barrier to understanding. In such cases the way the problem is framed often predetermines whether or not it can be solved. Moreover how a problem is formulated is often determined by the mathematics available to formalize the verbal description. Thus, the concern was on understanding what the fractional calculus offered by way of overcoming certain kinds of barriers.

Another surprise was the fact that the bedrock of classical mechanics, Hamilton's equations, could be generalized to the fractional calculus. This extension allows us to set aside the clockwork universe and accept a less preordained view of dynamics; one that could be adopted in the social and biological sciences without overly constraining them. To provide a context for this generalization we introduced an extension of the CLT, whose proof is based on a subordination argument and the visibility of events in a sequence of measurements. A consequence of the proof is the realization that the MLF is universal.

In this chapter it was recognized that complexity can only be addressed through the introduction of randomness into the dynamics. Consequently, even a technique as fundamental to physics as the variational calculus leading to Hamilton's equation was generalized to account for the interaction of system of interest with the environment. The extension of the Hamiltonian formalism rested on introducing a new interpretation of time, one local to the system of interest and the other available to the greater world of experiment. In this way the extended theory was shown to be compatible with the fractional calculus and to lead to fractional equations of motion. What was presented is barely an introduction to an area of mathematical physics that has exploded in the last decade. Hopefully it points the way to future areas of research, particularly in the nascent field of the fractional Sturm-Liouville theory.

We saw in earlier chapters that statistical fluctuations were necessary to capture the multi-modal nature of complexity, particularly in FLEs. In this chapter we have shown that the FLE has as its natural description in phase space fractal equations for the PDF. In various combinations of fractional derivatives in time, space or both the fractional calculus is able to explain all manner of anomalous transport using the CTRW, FKE and/or FFPE. It is interesting that the most general solution to the FKE in a potential-free region has a scaling solution for the PDF and the scaling parameter is complex. The imaginary part of the scaling parameter leads to a log-periodic modulation of the average value of the variate as experimentally observed in Chapter 2 and explained in a number of places [61, 63, 86].

The chapter closes with a suggestion of how using the fractional calculus one can generate an exponentially truncated Lévy PDF. The truncation suggests a mechanism to distinguish black swans and dragon kings in the analysis of HRV data for a post-AMI cohort.

7.9 Appendices Chapter 7

7.9.1 CTRW

In its original formulation the CTRW Montroll and Weiss [46] explicitly assume that the sequence of time intervals $\{\tau_n\}$ constitutes a set of independent, identically distributed, random variables. The probability that the random time variable lies in the interval $(t, t+dt)$ is given by $\psi(t) dt$ where $\psi(t)$ is the waiting-time PDF. The length of the sojourn is characteristic of the structure of the medium and is here considered to be a renewal process so that the PDF at step $n + 1$ can be expressed as a convolution over the times of the previous steps

$$\psi_{n+1}(t) = \int_0^t \psi(t - t') \psi_n(t')dt', \quad \psi(t) \equiv \psi_1(t).$$ (7.123)

Its name reflects the fact that the walker waits for a given time at a lattice site before stepping to the next site.

The waiting time PDF is used to define the probability that a step has not been taken in the time interval $(0, t)$. Thus, if the function $W(\mathbf{q}, t)$ is the probability that a walker is at the lattice site $\mathbf{q}$ at a time t immediately after a step has been taken, and $P(\mathbf{q}, t)$ is the probability of a walker being at $\mathbf{q}$ at time t, then in terms of the survival probability $\Psi(t)$:

$$\Psi(t) = \int_t^\infty \psi(t') \, dt',$$ (7.124)

we have

$$P(\mathbf{q}, t) = \delta_{\mathbf{q},0} \Psi_0(t) + \int_0^t \Psi(t - t') W(\mathbf{q}, t')dt'.$$ (7.125)

If a walker arrives at site $\mathbf{q}$ at time t and remains there for a time $(t - t')$, or the walker has not moved from the origin of the lattice, $P(\mathbf{q}, t)$ is the average over all arrival times with $0 < t' < t$. Note that the waiting time at the origin might be different, so that $\Psi_0(t)$ can be determined from the transition probabilities. It should be pointed out that the function $P(\mathbf{q}, t)$ is interpreted as a probability when $\mathbf{q}$ is a lattice site and as a PDF when it is is a continuum point is space.

The probability $W(\mathbf{q}, t)$ itself satisfies the recurrence relation

$$W(\mathbf{q}, t) = p_0(\mathbf{q}, t) + \sum_{\mathbf{q}'} \int_0^t p(\mathbf{q} - \mathbf{q}', t - t') W(\mathbf{q}', t')dt',$$ (7.126)

where $p(\mathbf{q}, t)dt$ is the probability that the time between two successive steps is in the interval $(t, t + dt)$, and the step length extends across $\mathbf{q}$ lattice points. The transition probability satisfies the normalization condition

$$\sum_{\mathbf{q}} p(\mathbf{q}, t) = \psi(t),$$ (7.127)

and the integral over time yields the overall normalization

$$\int_0^\infty \sum_{\mathbf{q}} p(\mathbf{q}, t) \, dt = \int_0^\infty \psi(t) \, dt = 1.$$ (7.128)

Montroll and Weiss made the explicit assumption that the length of a pause and the size of a step are mutually independent random variables, an assumption we implement through the product form of the waiting-time and the jump-length PDFs:

$$p(\mathbf{q}, t) = p(\mathbf{q})\psi(t), \tag{7.129}$$

in which case the random walk is said to be factorable.

We introduce the Fourier-Laplace transform, indicated by an asterisk:

$$p^*(\mathbf{k}, s) = \mathcal{FT}\{\mathcal{LT}\{p(\mathbf{q}, t); s\}; \mathbf{k}\} = \widetilde{p}(\mathbf{k})\widehat{\psi}(s), \tag{7.130}$$

where $\widetilde{p}(\mathbf{k})$ is used to denote both the discrete and continuous Fourier transform of the jump probability. Consequently, using the convolution form of Eq.(7.126), we obtain

$$W^*(\mathbf{k}, s) = \frac{p_0^*(\mathbf{k}, s)}{1 - p^*(\mathbf{k}, s)},$$

and for the double transform of the probability from Eq.(7.125) we have

$$P^*(\mathbf{k}, s) = \widehat{\Psi}_0(s) + \frac{\widehat{\Psi}(s)\, p_0^*(\mathbf{k}, s)}{1 - p^*(\mathbf{k}, s)}, \tag{7.131}$$

where we determined that

$$\widehat{\Psi}(s) = \frac{1 - \widehat{\psi}(s)}{s}, \tag{7.132}$$

from Eq.(7.124).

In the situation where the transition from the origin does not play a special role $p_0(\mathbf{q}, t) = p(\mathbf{q}, t)$, $\widehat{\Psi}_0(s) = \widehat{\Psi}(s)$ and Eq.(7.131) simplifies to

$$P^*(\mathbf{k}, s) = \frac{\widehat{\Psi}(s)}{1 - p^*(\mathbf{k}, s)}, \tag{7.133}$$

which can be further simplified when the space-time transition probability factors

$$P^*(\mathbf{k}, s) = \frac{1 - \widehat{\psi}(s)}{s\left[1 - \widetilde{p}(\mathbf{k})\widehat{\psi}(s)\right]}. \tag{7.134}$$

This is the Montroll-Weiss solution to the standard CTRW in Fourier-Laplace space.

7.9.2   Riesz Fractional Derivative

The modified anomalous diffusion equation Eq.(7.111) introduces the shifted Riesz fractional derivative, which in this Appendix we evaluate using a binomial expansion of the operator. Using the definition of the Riesz fractional derivative [21]:

$$\partial^{\beta}_{|q|}\left[g(q)\right] \equiv \Gamma\left(\beta - 1\right) \frac{\sin\left(\beta\pi/2\right)}{\pi} \int_{0}^{\infty} d\xi\ \xi^{1-\beta} \left\{g''\left(q+\xi\right) + g''\left(q-\xi\right)\right\},$$

(7.135)

where the prime denote derivatives with respect to q. In the text we wish to take the Fourier transform of the shifted operator so we substitute $g(q) = e^{ikq}$ into Eq.(7.135) to obtain

$$\partial^{\beta}_{|q|}\left[e^{ikq}\right] \equiv -k^2 e^{ikq} \Gamma\left(\beta - 1\right) \frac{\sin\left(\beta\pi/2\right)}{\pi} \int_{-\infty}^{\infty} d\xi\ |\xi|^{1-\beta} e^{ikq}. \qquad (7.136)$$

The integral is given by [14]

$$\int_{-\infty}^{\infty} d\xi\ |\xi|^{1-\beta} e^{ik\xi} = \frac{|k|^{\beta-2}}{2\Gamma\left(\beta - 1\right)\sin\left(\beta\pi/2\right)},$$

which when substituted into Eq.(7.136) yields

$$\partial^{\beta}_{|q|}\left[e^{ikq}\right] = -|k|^{\beta} e^{ikq}. \qquad (7.137)$$

Here we see that e^{ikq} is the eigenvector and $-|k|^{\beta}$ is the eigenvalue of the fractional operator.

We now apply the binomial theorem to the shifted Riesz operator interpreted as the real part of the operation

$$\left(\partial_{|q|} + \gamma\right)^{\beta}\left[e^{ikq}\right] = \sum_{n=0}^{\infty} \binom{\beta}{n} \gamma^n \partial^{\beta-n}_{|q|}\left[e^{ikq}\right]. \qquad (7.138)$$

Using the eigenvalue relation we write

$$\partial^{\beta-n}_{|q|}\left[e^{ikq}\right] = -|k|^{\beta} \operatorname{Re}\left\{(ik)^{-n}\right\} e^{ikq},$$

which enables us to write

$$\begin{aligned}
\left(\partial_{|q|} + \gamma\right)^{\beta}\left[e^{ikq}\right] &= -\sum_{n=0}^{\infty} \binom{\beta}{n} \gamma^n k^{\beta-n} \operatorname{Re}\left\{e^{-in\pi/2}\left(signk\right)^{\beta}\right\} e^{ikq} \\
&= -\operatorname{Re}\left\{\left(-i\ signk\right)^{\beta}\left(\gamma + ik\right)^{\beta}\right\} e^{ikq}. \qquad (7.139)
\end{aligned}$$

This last equation can be further simplified using

$$Ae^{i\phi} = \gamma + ik$$

to obtain

$$A = \left(\gamma^2 + k^2\right)^{1/2} \quad \text{and} \quad \phi = \tan^{-1}\left(\frac{k}{\gamma}\right)$$

which when substituted into Eq.(7.139) yields

$$\left(\partial_{|q|} + \gamma\right)^\beta \left[e^{ikq}\right] = -\left(\gamma^2 + k^2\right)^{\beta/2} \cos\beta \left[\tan^{-1}\left(\frac{k}{\gamma}\right) - \frac{\pi}{2}signk\right] e^{ikq}.$$

(7.140)

The solution given by Eq.(7.140) is easily checked by taking the limit as the shift parameter vanishes. In this case

$$\lim_{\gamma \to 0}\left(\partial_{|q|} + \gamma\right)^\beta \left[e^{ikq}\right] = -|k|^\beta e^{ikq} \lim_{\gamma \to 0} \cos\beta \left[\tan^{-1}\left(\frac{k}{\gamma}\right) - \frac{\pi}{2}signk\right]$$

and the final limiting term is unity so that

$$\lim_{\gamma \to 0}\left(\partial_{|q|} + \gamma\right)^\beta \left[e^{ikq}\right] = -|k|^\beta e^{ikq}$$

thereby reducing to the original eigenvalue relation Eq.(7.137).

References

[1] Agrawal, O.P., J.A. Machado and.J. Sabatier. 2004. *Nonlinear Dynam.* **38**, 1.
[2] Allegrini, P., P. Grigolini, P. Hamilton, L. Palatella and G. Raffaelli. 2002. *Phys. Rev. E* **65**, 041926.
[3] Arfkin, G. 1970. *Mathematical Methods for Physicists*, Academic Press, New York.
[4] Ashe, H.L. and J. Briscoe. 2006. *Development* **133**, 386.
[5] Albert, R. and A.-L. Barabási. 2002. *Rev. Mod. Phys.* **74**, 47.
[6] Bassingthwaighte, J.B., L.S. Liebovitch and B.J. West. 1994. *Fractal Physiology*, Oxford University Press, New York.
[7] Bénichou, O., C. Loverdo, M. Moreau and R. Voituriez. 2011. *Rev. Mod. Phys.* **83**, 81.
[8] Brown, R. 1829. *Phil. Mag.* **6**, 161. *Edinburgh J. of Sci.* **1**, 314.
[9] Caspi, A., R. Granek and M. Elbaum. 2000. *Phys. Rev. Lett.* **85**, 5655.
[10] Castaing, B., Y. Gagne and F.J. Hopfinger. 1990 *Physica D* **46**, 177.
[11] Copot, D., C.M. Ionescu and R. De Keyser. 2014. *19th World Congress, IFAC*, 9277.
[12] Dokoumetzidis, A. and P. Macheras. 2009. *J. Pharmacokinetics and Pharmacodynamics* **36**, 165.
[13] Fedotov, S. and S. Falconer. 2014. *Phys. Rev. E* **89**, 012107.

[14] Gel'fand, I.M. and G.E. Shilov. 1964. *Generalized Functions, Vol. 1*, translated by E. Saletan, Academic Press, New York.

[15] Gorenflo, R. and F. Mainardi. 2006. In *WEHeraeus-Seminar on Anomalous Transport: Experimental Results and Theoretical Challenges*, Physikzentrum Bad-Honnef.

[16] Gorenflo, R., F. Mainardi and A. Vivoli. 2007. *Chaos, Solitons and Fractals* **34**, 89.

[17] Gnedenko, B.V. and A.N. Kolmogorov. 1954. *Limit distributions for sums of independent random variables*, Addison-Wesley, Cambridge, MA.

[18] Grigolini, P., L. Palatella and G. Raffaelli. 2001. *Fractals* **9**, 439.

[19] Hanson, J.D., J.R. Cary and J.D. Meiss. 1985. *J. Stat. Phys.* **39**, 327.

[20] Hayano, J., K. Kiyono, Z.R. Struzik, Y. Yamamoto, E. Watanabe, P.K. Stein, L.L. Watkins, J.A. Blumenthal and R.M. Carney. 2011. *Frontiers in Physiology* **2**, Article 65, 1.

[21] Herrmann, R. 2013. *arXiv 1303:2939v3*.

[22] Hilfer, R. 2000. *Applications of Fractional Calculus in Physics*, World Scientific, Singapore.

[23] Ingenhousz, J. 1973. *Dictionary of Scientific Biology*. C.C. Gillispie (ed.). Scribners, New York, 11.

[24] Isichenko, M.B. 1992. *Rev. Mod. Phys.* **64**, 961.

[25] Kenkre, V.M., E.W. Montroll and M.F. Slesinger. 1973. *J. Stat. Phys.* **9**, 45.

[26] Kiyono, K., Z.R. Struzik, N. Aoyagi and Y. Yamamoto. 2006. *IEEE Trans. on Biomed. Eng.* **53**, 95.

[27] Kiyono, K., J. Hayano, E. Watanabe, Z. Struzik and Y. Yamamoto. 2008. *Heart Rhythm* **5**, 261.

[28] Kiyono, K. and N. Bekki. 2011. *Pacific Sci. Rev.* **12**, 185.

[29] Klafter, J., S.C. Lim and R. Metzler. 2012. *Fractional Dynamics*, World Scientific, NJ.

[30] Kolmogorov, A.N. 1941. *Coptes Rendus (Dokl.) Akad. Sci. URSS* **26**, 115.

[31] Koponen, I. 1995. *Phys. Rev. E* **52**, 1197.

[32] Lindenberg, K. and B.J. West. 1990. *The Nonequilibrium Statistical Mechanics of Open and Closed Systems*, VCH Publishers, NY.

[33] Lutz, E. 2012. In *Fractional Dynamics*. Klafter J., S.C. Lim and R. Metzler (eds.). World Scientific, NJ, pp. 265.

[34] MacKay, R., J. Meiss and I. Percival. 1984. *Physica D* **13**, 55.

[35] Magin, R.L., O. Abdullah, D. Baleanu and X.J. Zhou. 2008. *J. Mag. Res.* **190**, 255.

[36] Mainardi, F. and P. Pironi 1996, *Extracta Mathematicae* **11**, 140.

[37] Mainardi, F., G. Gorenflo and A. Vivoli. 2005. *Frac. Calc. and App. Anal.* **8**, 7.

[38] Mandelbrot, B.B. and J.W. Van Ness. 1968. *SIAM Review* **10**, 422.

[39] Mantegan, R.N. and H.E. Stanley. 1994. *Phys. Rev. Lett.* **73**, 2946.

[40] Mantegna, R.N. and H.E. Stanley. 2000. *An Introduction to Econophysics*, Cambridge University Press, Cambridge, UK.

[41] Matshshita, R., P. Rathie and S. Da Silva. 2003. *Physica A* **326**, 601.

[42] Meerschaert, M., E. Nane and P. Vellaisamy. 2011. *Electronic Journal of Probability* **16**, 1600.

[43] Metzler, R., E. Barkai and J. Klafter. 1999. *Phys. Rev. Lett.* **82**, 3563.

[44] Metzler, R. and J. Klafter. 2000. *Phys. Rept.* **339**, 1.

[45] Metzler, R. and J. Klafter. 2004. *J. Phys. A: Math. Gen.* **37**, R161.

[46] Montroll, E.W. and G. Weiss. 1965. *J. Math. Phys.* **6**, 167–181.

[47] Montroll, E.W. and B.J. West. 1979. In *Fluctuation Phenomena*. E.W. Montroll and J.L. Lebowitz (eds.). *Studies in Statistical Mechanics, Vol. VII*, North-Holland, Amsterdam; Second edition 1987.

[48] Montroll, E.W. and M.F. Shlesinger. 1984. In *Studies in Statistical Mechanics, Vol.* **11**, J. Lebowitz and E. Montroll (eds.). North-Holland, Amsterdam, pp. 1.

[49] Morse, P.M. and H. Feshbach. 1953. *Methods of Theoretical Physics*, *Chapter 6*, McGraw-Hill, NY.

[50] Niemeijer, T. and J. van Leeuwen. 1976. In *Phase Transitions and Critical Phenomena, Vol. 6*, pp. 425, Eds. C. Domb and M. Green, Academic Press, London.

[51] Oliveria, J.G. and A.-L. Barabási. 2005. *Nature* **437**, 1241.

[52] Peng, C.K., J. Mistus, J.M. Hausdorff, S. Havlin, H.E. Stanley and A.L. Goldberger. 1993. *Phys. Rev. Let.* **70**, 1343.

[53] Plerou, V., P. Gopikrishnan, L.A.N. Amaral, X. Gabaix and H.E. Stanley. 2000. *Phys. Rev. E* **62**, R3023.

[54] Popovic, K., M.T. Atnackovic, A.S. Pilipovic, M.R. Rapaic, S. Pilipovic and T.M. Atanackovic. 2010. *J. Pharmacokinetics and Pharmacodynamics* **37**, 119.

[55] Pramukkul, P., A. Svenkeson, P. Grigolini, M. Bologna and B.J. West. 2013. *Adv. Math. Phys.* **1**, 498789.

[56] Riewe, F. 1997. *Phys. Rev. E* **55**, 3581.

[57] Richardson, L.F. 1926. *Proc. Roy. Soc. London A* **110**, 709.

[58] Samorodnitsky, G. and M.S. Taqqu. 1994. *Stable Non-Gaussian Random Processes*, Chapman & Hall, NY.

[59] Scher, H. and E.W. Montroll. 1975. *Phys. Rev. B* **12**, 2455.

[60] Shlesinger, M.F., B.J. West and J. Klafter. 1987. *Phys. Rev. Lett.* **58**, 1100.

[61] Shlesinger, M.F. and B.J. West. 1991. *Phys. Rev. Lett.* **67**, 3200.

[62] Seshadri, V. and B.J. West. 1982. *Proc. Nat. Acad. Sci. USA* **79**, 4501.

[63] Sornette, D. 1998. *Phys. Rept.* **297**, 239.

[64] Stanislavsky, A.A. 2003. *Phys. Rev. E* **67**, 021111.

[65] Stanislavsky, A.A. 2006. *Eur. Phys. J. B* **49**, 93.

[66] Svenkeson, A., M.T. Beig, M. Turalska, B.J. West and P. Grigolini. 2013. *Phys. A* **392**, 5663.

[67] Taleb, N.N. 2007. *The Black Swan: the impact of the highly improbable*, Random House, NY.

[68] Turing, A.M. 1952. *Phil. Trans. Roy. Soc. London B: Biol. Sci.* **237**, 37.

[69] Uchaikin, V.V. 2000. *Int. J. Theor. Phys.* **39**, 3805.

[70] Weiss, G.H. 1994. *Aspects and Applications of the Random Walk*, North-Holland, Amsterdam.

[71] West, B.J., V. Bhargava and A.L. Goldberger. 1986. *J. Appl. Physiol.* **60**, 189.

[72] West, B.J. 1990. *Fractal Physiology and Chaos in Medicine*, World Scientific, Singapore.

[73] West, B.J., E.L. Geneston and P. Grigolini. 2008. *Phys. Rept.* **468**, 1.

[74] West, B.J. 2010. *Frontiers in Physiology* **1**, 1.

[75] West, B.J. and P. Grigolini. 2011. *Complex Webs; Anticipating the Improbable*, Cambridge University Press, UK.

[76] West, B.J. and D. West. 2012. *Frac. Calc. & App. Analysis* **15**, 1.

[77] West, B.J. 2010. *Frontiers in Physiology* **1**, 12. doi:10.3389/fphys.2010.00012.

[78] West, B.J. 2013. *Fractal Physiology and Chaos in Medicine*, 2nd Edition, World Scientific, Singapore.

[79] West, B.J., M. Turalska and P. Grigolini. 2014. *Network of Echoes; Imitation, Innovation, and Invisible Leaders*, Springer, NY.

[80] West, B.J. 2014. *Frontiers in Physiology* **5**, 1.

[81] Young, W., A. Pumir and Y. Pomeau. 1989. *Phys. Fluids A* **1**, 462.

[82] Yuste, S.B., E. Abad and K. Lindenberg. 2010. *Phys. Rev. E* **82**, 061123.

[83] Zaslavsky, G.M., R.Z. Sagdevv, D.A. Usikov and A.A. Chernikov. 1991. *Weak Chaos and Quasi-Regular Patterns*, Cambridge, Cambridge University Press.

[84] Zaslavsky, G.M., M. Edelman and B.A. Niyazov. 1997. *Chaos* **7**, 159.

[85] Zaslavsky, G.M. and B.A. Niyazov. 1997. *Phys. Rept.* **283**, 73.

[86] Zaslavsky, G.M. 2000. In *Applications of Fractional Calculus in Physics*. R. Hilfer (ed.). World Scientific, Singapore.

[87] Zaslavsky, G.M. 2002. *Phys. Rept.* **371**, 461.

[88] Zayernouri, M. and G.E. Karniadakis. 2013. *J. Comp. Phys.* **252**, 495.

[89] Zhang, D.A., G.W. Wei, D.J. Kouri and D.K. Hoffman. 1997. *J. Chem. Phys.* **106**, 5216.

[90] Zolotarev, V.M. 1986, *One-dimensional Stable Distributions*, Trans. Math. Mono. **65**, Am. Math. Soc., Providence, RI.

CHAPTER 8

What Have We Learned?

The physical sciences entered a period of dramatic change in the decades prior to the Second World War, in which physicists wrestled with and resolved many of the paradoxes presented by quantum mechanics. The second half of the twentieth century was no less dramatic, if somewhat less successful, in overcoming the challenges of complexity. Anomalous diffusion had been experimentally observed; the significance of nonlinear dynamics and the implications of chaos theory were being rigorously explored and rapidly diffusing through the science community, and the importance of fractal geometry was introduced, as a way to capture the complexity of a world that Euclid did not, and could not, have imagined. The loss of predictability in deterministic nonlinear dynamic phenomena was becoming widely accepted, and the notion that this would provide a useful way to describe biomedical phenomena was also being adopted, albeit at a slower pace.

One aspect of the data generated by complex phenomena that we mentioned, but did not dwell on, is the $1/f$ character of power spectra obtained from time series measurements with IPL correlations. The variety of mathematical models used to explain the generation of $1/f$-noise is as wide ranging as the disciplines in which the $1/f$−scaling has been observed. A recent strategy for modeling the dynamic variability captured by $1/f$−spectra appears to be independent of the specific generating mechanisms and involves fractal statistics, in one way or another. This approach actually began with fractionally integrated white noise [1]; it was extended to fractal Brownian motion [9]; it was further modified to fractal shot noise [6]; it was shown to be related to the statistical mechanisms of subordination [14, 16, 23]; and it was subsequently related to fractal renewal processes [7]. Grigolini *et al.* [5] argue that this historical trajectory has now targeted non-stationary and non-ergodic stochastic processes as the natural way to describe the complexity captured by $1/f$−noise [18]; the last target appears to be the ultimate destination of the arguments resulting in fractional differential equations of motion, whether for system dynamics or phase space equations of motion.

As we step from discipline to discipline, what we identify as complexity seems to change in significant ways; however, a common characteristic appears to be a balance between regularity and randomness. In sociology the formation of consensus in an electorate appears to be quite different from the agreement reached by aggregating the opinion of a collection of independent individuals; in ecology time is not an absolute, but due to the difference in metabolism, time passes more slowly for the elephant than it does for the shrew; the slowing down of activity at criticality seeming to contradict the flexibility observed in the critical dynamics of cognition. These and many other curious results were shown to be discipline-independent properties of the criticality manifest by complex phenomena.

Let us attempt to organize what we have learned into a manageably number of empirical truths if not scientific principles. However, before we draw those inferences let us review the arguments developed in support of the fractional calculus view of complexity. This recounting is worth doing because there is no one terse logical presentation that would convince any reasonable skeptic that fractional calculus will dominate the science of the future. It is more the cumulative weight of multiple sources that tip the scales in the direction of fractional dynamics.

8.1   Why the Fractional Calculus?

William of Ockham (1287–1347) was an English Franciscan friar and scholastic philosopher, who, to assist in answering weighty philosophical questions, formulated the following principle: *among competing hypotheses adopt the one with the fewest assumptions.* It may turn out that more complicated solutions are ultimately shown to be correct, but lacking the information necessary to establish the truth of these more complex solutions, the fewer assumptions one makes the better. Or phrased a little differently: *the most desirable solution is the one with the fewest assumptions consistent with all the available information.* This may well encapsulate the history of science, or at least the history concerned with physical models, built to understand the results of experiment.

We began this investigation with the first simplifying assumption made in physical science, that being, physical processes are adequately modeled as being linear, or approximately so. However even this simplest of assumptions was based on the notion of continuous isotropic time, with which to record the ordering of events, and homogeneous space of infinite extent, in which those events occur. This book has been one prolonged argument against these, and similar, long held beliefs, that were necessitated by the desire to exclude complexity from the scientific/mathematical models of the world. The modern direction of science is to understand and even to embrace complexity.

However, Ockham's principle should be applied even to the understanding of complexity.

The purpose of the first chapter was to suggest reasons for the preeminence of complexity in tomorrow's science. This goal was admittedly hampered by my reluctance to provide a limiting and necessarily idiosyncratic definition of complexity. Part of the presentation was a vision of science that is less the disciplined logical methodology, taught to us in school, and more a kind of bounded mental anarchy of a child having fun at recess. Just as a first responder runs into danger from which people are fleeing; the scientist is drawn toward the unknown and to the very uncertainty that most people spend their lives trying to avoid.

Chapter 2 begins with one of the dominant ideas in science: to a first approximation physical phenomena can be described by linear models. Implicit in this view is that the linear model is followed by a second approximation to a system's dynamics that is relatively weak and which forms the basis of perturbation theory. The linear view was further supported by the superposition principle, with its introduction of eigenvalues and eigenfunctions into the description of physical processes. This was the first treatment of a body as a whole and not as a point particle. Linear superposition was a giant step forward in the understanding of matter, using a decomposition of the general behavior of a system's dynamics into a set of discrete modes identified by the spectrum of eigenvalues. The spectral method is used every day throughout science to identify important behaviors concealed within fluctuating time series.

It took another century, with the development of Sturm-Liouville theory, to systematically extend the idea of linear superposition to continuous spatial phenomena. At the close of the nineteenth century it appeared to many scientists that the mechanical view of the physical world was complete, whether one was studying fluid mechanics, electromagnetism, or kinetic theory, they all could be described by the analytic functions that were solutions to partial differential equations. What, at the time, seemed to remain was for physical theory to explain a few anomalous experimental results and the perplexing predictions of black-body radiation.

But the successes of linear modeling were not confined to deterministic phenomena. Even when the number of particles became so large that it was impossible to solve the individual equations of motion, the giant intellects of the nineteenth and early twentieth centuries handled this complexity by extending the notion of a field to include PDFs. This first systematic treatment of complexity allowed individual trajectories to be bundled together into an ensemble characterized by a PDF and the underlying process to be described in terms of averages. The dynamics of such many-body systems could be described by Langevin equations for the dynamic variables or alternatively by Fokker-Planck equations for the PDF. The physical observables were given in terms of the central moments of the PDF. The many successes of this dual approach included the explanation of classical diffusion, along with the prediction of such fundamental quantities as Avogadro's number, as well as the Law of Frequency of Errors.

This attractive linear picture of the physical world with all its successes was called into question by the French mathematician H. Poincaré. He was able to prove that nonlinear interaction did not only result in small departures from the solutions to linear equations, as maintained by perturbation theory, but could qualitatively change a system's dynamics. The trajectories generated by such nonlinear dynamical systems could even be fractal, as determined nearly a century later, but was, in fact, first described by Poincaré. One of the fundamental limitations of linear theory was highlighted by the FPU problem; the failure of a nonlinear dynamical system to relax to a state in which each of the dynamic modes has the same amount of energy, therefore apparently violating the energy equi-partition theorem. The development of nonlinear dynamics occupied the attention of a significant number of scientists in the latter half of the twentieth century. These investigators, in large part, established that in every discipline the fundamental barrier to understanding was complexity, which, more often than not, was tied to the underlying nonlinear nature of the phenomenon being studied.

Chapter 3 begins with the recognition that the limitations of linearity conjoined with the focus of today's science on complexity, requires science to abandon the simplifying assumptions it has made historically. As society becomes more interconnected it also becomes more fragile, as entailed by the networking of communications, transportation and economic systems. In order to understand and control such fragility requires that science adopt a new way of viewing complexity. To dramatize this point, we use the literary device of Alice attempting to understand a world in which the rules were different from those of the world she left behind. The purpose is to emphasize that the mental models of the world determines how we understand new phenomena and getting things a little wrong was tolerable when the time scale for economic change was days or weeks, allowing time for mistakes to be corrected. However, these same errors can today be destabilizing with catastrophic implications, when that time scale is microseconds.

It is not that complexity was ignored in the past. It is more a matter of making complexity intellectually manageable. This was no where more evident than with the introduction of uncertainty into science, without leaving behind quantifiability. Making a statistical description compatible with the mechanical model of the universe was truly an amazing accomplishment, and enabled the random scatter of measurements to be explained, while preserving the notion of predictability. The Normal PDF was ubiquitous in the science of the nineteenth and early twentieth centuries, the only place from which it was absent was from experimental data. While Normalcy did appear in the statistical scatter of most measurements, it was absent from the variability characterizing most natural phenomena. In Tables 3.1 and 3.2

the distribution of variability associated with phenomena, within ten distinct disciplines, are shown to be IPL, with Normalcy being totally absent. This has been associated in large part with the loss of one of the fundamental properties of linearity, that being additivity.

The loss of linearity found in statistical fluctuations, also appears in those functions that describe measurable complex phenomena, but that loss in the latter case is the non-existence of tangent curves. The observation that the velocity field of turbulent fluids is non-differentiable is not unique, in fact such non-differentiability is characteristic of distributed complexity. This behavior was forced into the consciousness of the scientific community in the middle of the last century by Benoit Mandelbrot, with his insistence on the ubiquity and singular importance of fractals in all scientific disciples, from anthropology to zoology, and all the scientific disciplines in between. The recognition that a fractal process is non-differentiable, but still evolves over time, motivated the application of fractional equations of motion for such processes. This was further supported by establishing that a non-integrable microscopic Hamiltonian system could be represented by fractional equations of motion at the macroscopic level.

For two hundred years various disciplines have been identifying allometry relations that determine how the functionality of organs, organizations, and communities are related to their average size. In distinct cases the time series for such non-integrable systems gave rise to a number of bizarre parameters, including fractal dimensions that have real (generally non-integer) and imaginary parts. The real part of the dimension is related to the index of an IPL PDF and the imaginary part of the dimension is observed through a log-periodic modulation of some measurable property of the system. These effects are captured in scaling relations identified as solutions to renormalization group relations.

These early chapters establish the importance of non-differentiability for describing complexity and that fractional derivatives of non-differentiable functions converge. It was then appropriate to overlay the simpler mathematical properties of the fractional calculus with corresponding fractal phenomena. These properties were matched up with phenomena drawn from a variety of disciplines in Chapter 4. A topical exemplar is a discrete fractional representation of climate change, to determine if the average global temperature time series data is related to the number of sunspots time series data. The limits of the fractional discrete equations were shown to yield fractional derivatives and these were applied to fractal functions describing the observational data. This is only one of the controversies addressed and further indicates the need to adopt an overarching mathematics, such as the fractional calculus, to address the unresolved questions posed by complexity.

The fractional rate equation was used to model stress relaxation in viscoelastic materials, the in-between materials discussed in earlier chapters. The MLF solution to the fractional stress-strain relation was able to capture the two empirical limits for stress relaxation; those being the stretched exponential at early times and the IPL at late times. One of the most compelling applications of the simple fractional derivatives is to three-stage Brownian motion in the form of a fractional Langevin equation. This theory of Brownian motion has, in addition to the usual microscopic fluctuations and macroscopic dynamics, a mesoscopic dissipation. This fractional dissipation is based on a century old theory of the back-reaction of the fluid to the motion of a Brownian particle. The theory is timely because the technology has advanced to the point that experiments are now able to explicitly measure the deviation of predictions from the fractional Langevin equation to those of the traditional Langevin theory of Brownian motion. The predictions of the fractional Langevin equation are borne out by experiment.

In Chapter 5 the initial value solutions to simple linear systems of fractional rate equations are presented. The solution to a linear fractional rate equation is a MLF and the solution to a set of such equations is the MLF matrix, in direct parallel with the solution to a set of linear rate equations being an exponential matrix. The solution to the fractional harmonic oscillator equations are given in some detail and some of the properties of MLF discussed. The fractional Langevin equation, which was used as a phenomenological description of Brownian motion, is derived using a heat bath with a colored spectrum of oscillators. Here again experiments are available to test the theoretical predictions of the fractional dissipation of the autocorrelation function for a harmonically bound particle. The MLF relaxation of the correlation function is vindicated by an excellent fit to experimental data.

Perhaps the most significant idea of this chapter is that of generalizing control theory to incorporate the memory entailed by the fractional calculus into the controller. A complex phenomenon characterized by a fractal time series can be described by a fractal function. Such functions were shown to have divergent integer-order derivatives. Consequently, traditional control theory, involving as it does integer-order differential and integral operators, cannot be used to determine how feedback is accomplished. However control theory can be generalized to fractional-order operators. Therefore, it seems reasonable that one strategy for modeling the dynamics and control of such complex phenomena is through the application of the fractional calculus.

We have discussed how three kinds of complexity can be incorporated into the dynamics: fluctuations, nonlinearity and fractionality. Nonlinearity has been shown to be representable by an infinite-order linear description, so that in general it might be possible to represent classes of complexity by infinite-order fractional Langevin equations. An example of how to solve a quadratically nonlinear fractional rate equation by means of an operator technique was given in this chapter and the analytic solution matches the numerical calculation quite well.

Chapter 6 is devoted to fractional generalizations of the equations describing the dynamics of well studied collective phenomena. The chapter content differs from a substantial number of recent papers that offer fractional alternatives to dynamic equations, but which do not present, either a theoretical rational, or an attempt to explain anomalous experimental data. The most surprising result of this chapter has to do with how the dynamics of a complex network influences the behavior of an individual within a large network. The numerical solution for the state of an individual who is part of a ten thousand member DMM complex network, was shown to be well represented by a fractional Langevin equation. This is the first explicit demonstration that the highly nonlinear erratic dynamics of a complex network can be represented by fractional linear rate equations.

The collapse of the ten thousand dimensional system of equations onto a linear fractional equation is the result of two general features of the dynamics: (1) the network dynamics subordinate the dynamics of the individual by transforming the dynamics in operational time into dynamics in chronological time and (2) the dynamics of the network are critical, which is to say, the nearest neighbor interactions become long-range for a critical value of the control parameter. Subordination and criticality are intertwined, if not the two sides of one coin. Consequently, network science may have been enjoying the success of the past ten years due to the fact that the cooperative behavior observed in so many physical, social and ecological phenomena is the result of criticality and its natural expression through complex network dynamics. This suggests that the fractional equations of motion are a natural consequence of criticality of the underlying dynamics.

Another surprise was the fact that the bedrock of classical mechanics, Hamiltonian theory and the variational calculus, could be generalized to the fractional calculus. This extension sets aside the clockwork universe and indorses a less preordained view of dynamics; one that could be adopted in the social and biological sciences without overly constraining them. A context for this generalization was provided by an extension of the central limit theorem, whose proof is based on a subordination argument and the visibility of events in a sequence of measurements. A consequence of the proof is the realization that the MLF is universal.

Of course, fractional equations in physical systems can result from inhomogeneities within the media supporting the dynamics. We saw this earlier in the hydrodynamic-induced fractional dissipation of Brownian motion. Disturbances in fractal media may produce fractional wave propagation resulting from fractional gradients inducing a fractal flux of heat, chemical concentration, electrical current or mass density. The fractional dissipation of propagating waves can also arise due to IPL distributions of scatters, such as light scattering from the water droplets in fog, elastic waves scattering from grain boundaries in a material, or radar waves scattering from clouds in the atmosphere.

Certain properties of homogeneous isotropic turbulence in water or plasmas can be described by fractional equations. The Lévy statistics of velocity fluctuations in the wind field are described by fractional time derivatives, whereas the spatial energy cascades are described by fractional derivatives in space. On the other hand, plasma turbulence requires fractional derivatives in both space and time. In this regard the fractional generalization of the Bloch-Torrey equations for magnetization in living tissue, requires fractional derivatives in both space and time, as well. These generalized equations have been very useful in diagnosing neurodegenerative, malignant and ischemic diseases using fMRIs.

Finally, the fractional random walks introduced in Chapter 4 are used in Chapter 6 in the fractional search hypothesis, to incorporate either memory in time, large steps in space, or both, into a search algorithm. These ideas were used in describing the search of prey by predators, the cognitive searching for a particular memory, and the constraints on human mobility.

The formal arguments that pull together all the applications discussed in this book, and many many more that were not, are recorded in Chapter 7. It is useful to emphasize here that all the fractional equations describing the dynamics of phenomena presented so far are empirical. A phenomenological description of a complex phenomena may be sufficient for certain purposes, such as to guide the dynamics to a desired outcome; but in itself it is not the end of scientific investigation. Recall that thermodynamics is a phenomenological theory that captures the interrelation of the intensive and extensive properties of physical systems and its laws encapsulate centuries of experimental observation. Any more fundamental theory, such as statistical mechanics, must, in the end, yield results that are consistent with the predictions of thermodynamics. Phenomenological theory provides constraints that fundamental theory must satisfy.

One of the first cracks in the foundation to the statistical description of complex physical phenomena was anomalous diffusion. The nonlinear growth of the variance of a diffusive process with time was explained in terms of Lévy statistics, whose PDF was determined to be the solution to a FFPE. A version of this argument is used to show how the mean square separation of tracers in turbulent flow explains Richardson's dispersion law. This is traced to the CTRW of Montroll and Weiss, in which the structure of the medium supporting the process is modeled in terms of an IPL for both the waiting times PDF for each step and the step length PDF. In this general case the form of a phase space equation for the PDF is fractional in both space and time. One example of the application of the fractional behavior is to "how a leopard got his spots" given in terms of the fractional diffusion of morphogen concentration. Another example is the fractional Schrödinger equation for a harmonically bound particle suggesting the possible existence of a Fractional Correspondence Principle.

Fractional kinetic theory, based on the fractional trajectories of individual realizations of nonlinear dynamic process, provides an alternate justification for fractional diffusion in both space and time. The solution to the FKE uses renormalization group theory to establish the general scaling form of the PDF and to introduce Mittag-Leffler statistics, which includes the Lévy PDF as a special case.

As we have argued repeatedly in this monograph, the utility of the fractional calculus lies in the complexity of the phenomena that can be explained using it. The variability in heart rate has been used as an indicator of the health of the cardiovascular system and most recently the PDF of HRV has been shown to be given by a truncated Lévy PDF. The truncation observed in the data suggests a new physiological mechanism that suppresses potentially fatal, exceptionally long, time interval between heart beats.

8.2 Conformation Bias and Scientific Truth

For the purpose of encapsulating the high points of the discussion, the conventional wisdom of the physical sciences can be collected into a small number of traditional scientific truths. These truths, through a combination of repetition over time and our human weakness for simple explanations, have become so obvious that they are all but impossible to question. They are associated with the traditional descriptions scientists use to explain natural phenomena, and consistency between theory and experiment, all too often, is a manifestation of conformation bias. I briefly review these truths here to expose their weaknesses, as uncovered by scientists during the last quarter century in their pursuit of understanding complexity.

I adopt a format exploited elsewhere [20], that being to first present a traditional truth and in counterpoint introduce a corresponding non-traditional truth. This is done for a sequence of four traditional truths along with their non-traditional counterparts to clarify a new vision of science, one that is more amenable to our present understanding of complexity. The non-traditional truths presented have been uncovered over the last quarter century or so, beginning with the shift in thinking about nonlinearity and complexity in science in the early 1980s [15]. Since that time my own perspective has undergone a number of fundamental changes [17] culminating in the fractional calculus presentation made in this book.

8.2.1 Quantifiable

Nearly every physical scientist would say that traditionally the first scientific truth is that scientific theories are quantitative and that this is not a matter of choice, but of necessity. The Noble Laureate E. Rutherford had a view of science that is fairly typical of the physical scientist. He was dismissive of nearly everything that was not physics as evidenced by his comment [2]:

All science is either physics or stamp collecting. Qualitative is nothing but poor quantitative.

This was a criticism of the tradition in such disciplines as developmental biology, clinical psychology and medicine, where large amounts of data were gathered, but no quantitative predictions were made due to the absence of quantifiable theories. By the end of the nineteenth century there were two main groups in natural philosophy: those that followed the physics of Descartes and those that followed the dictates of Newton [12]:

Descartes, with his vortices, his hooked atoms, and the like, explained everything and calculated nothing. Newton, with the inverse square law of gravitation, calculated everything and explained nothing. History has endorsed Newton and relegated the Cartesian constructions to the domain of curious speculation (p. 763).

Thus, the first scientific truth is:

If it is not quantitative, it is not scientific.

That science should be quantitative is a visceral belief that began with physics and has, in large part, shaped the development of the disciplines of life science and sociology over the last century. These disciplines and others such as psychology and ecology have, by and large, accepted the exclusive need for quantitative measures. It would be counterproductive to say that the quantitative perspective is wrong, particularly since there is so much evidence to the contrary. There is, however, evidence mounting for adopting the more cautious assertion that the quantitative axiom is overly restrictive.

The non-traditional truth in counterpoint to the quantitative axiom is what you might expect, that being, that scientific theories need to be qualitative, as well as, quantitative. One of the leading proponents and the most successful implemented of this view in the last century was D'Arcy Thompson [13]. His interest in biological morphogenesis stimulated a new way of thinking about change; not the smooth, continuous quantitative change familiar in many physical phenomena, but the abrupt, discontinuous, qualitative change, familiar from the experience of "getting a joke," "having an insight" and the "bursting of a bubble" (physical or economic). The abrupt qualitative change in the dynamics of a system is a mathematical bifurcation.

One example of a mathematical discipline whose application in science emphasizes the qualitative over the quantitative is the bifurcation behavior of certain nonlinear dynamical equations. A bifurcation is a qualitative change in the solution to an equation of motion obtained through the variation of a control parameter. We encountered this behavior in Section 6.6 in our all too brief discussion of consensus formation in the DMM network model of social interactions. An extended discussion of the influence of a dynamic social group on the decision making behavior of an individual is given elsewhere [21], but it was shown that the influence of the group on the individual is given by the solution to an averaged fractional Langevin equation.

More broadly, in a nonlinear periodic system of unit period, changing the control parameter might generate a sequence of bifurcations in which the period of the solution doubles, doubles again, and so on producing more and more rapid oscillations with increases of the control parameter. In such systems the solution eventually becomes irregular in time and the system is said to have made a transition from regular rhythmic oscillations to chaotic motion. These kinds of transitions are observed in fluid flow going from laminar to turbulent behavior. The áperiodic behavior of the final state in such bifurcating systems has suggested a new paradigm for the unpredictable behavior of complex systems. This successive bifurcation path is only one of the many routes to chaos.

Another route to chaos is the onset of intermittency, which also arises through the change in a control parameter for a nonlinear system. In this transition to chaos a nonlinear system's motion is regular (periodic) when the control parameter is below a critical value. When the control parameter is above this critical value the system has long intervals of time in which its behavior is periodic, but this apparently regular behavior is intermittently interrupted by a finite-duration burst of irregular activity, during which the dynamics are qualitatively different. The time intervals between bursts are apparently random with an IPL PDF. As the control parameter is increased above the critical value, the bursts become more and more frequent, until finally only the bursts remain. A mathematical model with this type of intermittent transitions was used by Freeman [3] to describe the onset of low-dimensional chaos experimentally observed in the brain during epileptic seizures.

These observations along with many others suggest the form of the first non-traditional truth:

Qualitative descriptions of phenomena can be as important, if not more important, than quantitative descriptions in science.

8.2.2 Mathematical

The second scientific truth, according to tradition, has to do with how scientists use mathematics to model naturally occurring phenomena. Celestial mechanics accepted that the dynamics of physical systems are described by smooth, continuous, differentiable and unique functions, often referred to collectively as analytic functions. This belief permeated physics because the evolution of physical processes are modeled by sets of dynamical equations and the solutions to such equations were thought to be continuous and to have finite slopes almost everywhere. Consequently scientific understanding comes from prediction and description and not from the identification of teleological causation, which Galileo condensed into the three tenants:

Description is the pursuit of science, not causation. Science should follow mathematical, that is, deductive reasoning. First principle comes from experiment, not the intellect

Thus, the stage was set for Newton who embraced the natural philosophy of Galileo and in so doing inferred mathematical premises from experiments, rather than from physical hypotheses. This way of proceeding from the data to the model is no less viable today than it was over 300 years ago and the second traditional truth emerges as:

Physical observables are represented by analytic functions.

Of course, there are a number of phenomena that cannot be described by analytic functions. For example, ice melts, water freezes, rain evaporates, water vapor condenses into droplets and laminar fluid flow becomes turbulent, as experienced by even the infrequent airline passenger. In these critical phenomena, scaling is found to be crucial in the descriptions of the underlying processes, and analytic functions have not been very useful in developing their understanding. Implicit in the idea of an analytic function is the notion of a fundamental scale that determines the variability of the phenomenon being described by that function. When examined on a relatively large scale the function may vary in a complicated way. However, when viewed on a scale smaller than the fundamental scale, an analytic function is smooth and differentiable.

On the other hand, all complex phenomena contain scales that are tied together in time, in space or in both. The idea that either the structure or the behavior has a characteristic scale is lost, and most things have multiple scales contributing to them. This is the situation for a function to be scale-free and be a mathematical representation of data that can be traced on a graph as well. If such a function scales, the structure observed at one scale is repeated on all subsequent scales, the structure cascades downward and upward, never becoming smooth, always revealing more and more structure. Thus, we have structure, within structure, within structure. As we magnify such functions there is no limiting smallest scale size and therefore there is no scale at which the variations in the function subside. Of course this is only true mathematically; a physical or biological phenomenon would always have a smallest scale. However, if this cascade of interactions covers sufficiently many scales, it is useful to treat the natural function as if it shared all the properties of the mathematical function, including the lack of a fundamental scale. Such phenomena are called fractal, and they are described by fractal (non-analytic) functions. This was the argument used by Zaslavsky [22] to bundle into an ensemble the individual chaotic trajectories described by a PDF and replace the nonlinear dynamic equations with a linear fractional kinetic equation for the PDF.

Everything we see, smell, taste, and otherwise experience are in a continual process of change. If we experienced such changes linearly, meaning that our response is in direct proportion to the change, we would be continuously reacting to a changing environment, rather than acting in a way to modify that environment. Such behavior would leave us completely at the mercy of outside influences. But the changes in the world are not experienced linearly. We know from experiments conducted in the nineteenth century and repeated in the twentieth century, on the sensation of sound and touch that people do not respond to the absolute level of stimulation, but rather they respond to the percentage change in stimulation. These and other similar experiments became the basis of a branch of experimental psychology called psychophysics. These experiments supported a mathematical postulate made much earlier by D. Bernoulli in 1738 involving utility functions, which were intended to characterize an individual's social well being. See, for example, Roberts' volume in the *Encyclopedia of Mathematics* on measurement theory for a complete discussion [10].

When an erratic time series lacks a fundamental scale, it is called a statistical or random fractal. This concept has been used to gain insight into the fractal nature of a number of complex physiological systems: the chaotic heart, in which the cardiac control system determining the intervals between successive beats is found to have coupling across a broad range of dynamical scales; fractal respiration, in which breathing intervals are not equally spaced, but are basically intermittent in nature, is also seen to have a mixture of deep regular breaths with short irregular gasps; and for the first time scientists have seen that the fluctuations in human gait contain information about the motorcontrol system guiding our every step.

Thus, the second non-traditional truth is:

Many phenomena are singular (fractal) in character and cannot be represented by analytic functions.

8.2.3 Predictable

The third scientific truth has to do with prediction. The time evolution of physical systems are generally determined by systems of deterministic nonlinear dynamical equations, so the initial state of the system completely determines the solution to the equations of motion and correspondingly, the final state of that physical system. Consequently the final state is predicted from a given initial state using the dynamic equations. Until recently, this view prompted physical scientists to incautiously assert that the change of these observables in time, are absolutely predictable in the strict tradition of the mechanical clockwork universe. Consequently, the third traditional truth is:

The final state of a physical systems is predictable by means of its equation of motion and the given initial state.

This mechanistic view of the universe, with its absolute certainty about the future, has probably more than any other single factor, made the lay person uncomfortable with science. In large part this discomfort arises because such certainty about what is to come is inconsistent with most people's experience of the world. It is true that Newton's laws provide a valid description of the motion of material bodies under many reasonable conditions. But do his laws imply absolute predictability? The particular results with which we are concerned are the áperiodic (irregular) solutions to his equations of motion, that is, chaotic trajectories. Since chaotic processes are irregular, they have limited predictability and call into question whether phenomena are quantitative, analytic and predictable. Indeed a number of scientists have proposed that the statistical properties observed in complex phenomena do not arise from the large number of variables in the system, such as was argued in Chapter 2, but arise instead from the intrinsic uncertainty associated with chaotic dynamics, see Section 2.3.

The third non-traditional truth is related to the fact that most phenomena are described by nonlinear equations of motion and consequently their paths from the past into the future are chaotic. Chaos implies that the phenomena of interest are fundamentally unstable. Visualize an archer with his arm drawn back about to release an arrow towards a target. A slight change in the angle of the bow, a small increase in the tension of the bowstring, a modest deviation in the direction or speed of the wind, would divert the arrow from its course. Therefore a cluster of multiple arrows gives rise to a bell-shaped distribution of arrows embedded in the target, perhaps centered on the bull's eye. What if the smallest change in the angle of the arrow caused it to completely miss the target, or a puff of wind turned the arrow around in mid-flight?

Such behavior appears to be absurd, since it is completely outside our experience. This is however the nature of chaotic behavior; an unexpected dynamics that often runs counter to intuition. We are not often conscious of such behavior in the physical world, at least not on the time scale we live our lives. But don't forget that earthquakes, volcano eruptions and tornadoes, are natural phenomena that are large scale, unpredictable and devastating. At an earlier time it would have been argued that our inability to predict these things was a consequence of incomplete knowledge and given sufficient information we could anticipate the occurrence of such events with extreme accuracy. However, given the chaotic solutions to complex dynamical systems, we now know that predicting the exact time of occurrences of earthquakes, volcano eruptions and tornadoes is beyond the predictive capability of science. In the absence of solutions to the equations of motion, the evidence for the unpredictability of these unstable phenomena is provided by the IPL nature of the time intervals between catastrophic events.

No less dramatic phenomena come to us from the biological and social realms, which are filled with examples of instabilities. Such instabilities have been labeled tipping points [4], the sociologist's version of a critical point. At a tipping point two qualitatively distinct kinds of behavior are connected through a bifurcation that is accessed by the change in a control parameter in the phase space for the system. Just as the central limit theorem is implied by the FPE description of simple physical phenomena, so too the stochastic central limit theorem is implied by the FPSE description of relatively simple phenomena in complex environments as described in Chapter 7.

All this is summarized in the third non-traditional scientific truth:

Nonlinear deterministic equations of motion, whether discrete or continuous, do not necessarily have predictable final states due to the sensitivity of the solutions on initial conditions.

8.2.4   Scalable

It is probably apparent by now that these traditional scientific truths are not independent of one another, as they would be if they had a mathematical and not an experiential basis. However, each truth, like each of da Vinci's sketches of an elderly person, although derived from similar circumstances emphasizes very different experiences of those circumstances. Thus, scaling is lifted out of the background and into the foreground, as being central to our understanding of complexity. The general discussion on scaling given in Chapter 2 combined with the discussion in the intervening chapters suggests the fourth traditional scientific truth:

Physical systems can be characterized by fundamental scales, such as those of length and time.

Such scales provide the fundamental units in the physical sciences, without which measurements could not be made and quantification would not be possible. One way in which these scales manifest themselves is in the average values of distributions. This is consistent with the bell-shaped PDF of Gauss, where we have a well-defined average value of the variable of interest and a width of the PDF to indicate how well the average characterizes the result of experiment. Observe that a Gaussian PDF is completely characterized in terms of its width and average value, but what of PDFs whose average values diverge? IPL PDFs, for example, are known far and wide outside the physical sciences, see the tables in Chapter 3. The IPL is fundamentally different from the bell-shaped curve of Gauss. Its long tail extends far beyond the region of the bell and often indicates a lack of characteristic scale for the underlying process. This lack of scale can be manifest in the divergence of the average value of the dynamical variable.

In the same way that the bell-shaped curve can be related to Newton's laws and simple dynamical systems, so that the curve describes the distribution of errors, the IPL PDF can be related back to unstable dynamical systems. Very often the heavy-tailed PDF is a consequence of intermittent chaos in the underlying dynamics, so that large values of the variable occur more often than would have been expected based on the stable dynamics of simple systems. This does not happen to our friend the archer in the real world, because the distribution of his misses is bell-shaped. However, in the much more complex world of business, the forces determining a person's level of income interact in varied and sundry ways to produce an income level over which you and I have little or no control most of the time. In the case of income we might have the illusion of control, but if that were in fact true, the IPL PDF of income, discovered by Pareto over a century ago, would not be universal, which it appears to be.

Classical scaling principles are based on the notion that the underlying process is uniform, filling an interval in a smooth, continuous fashion as assumed in Chapter 2. The new principle is one that can generate richly detailed, heterogeneous, but self-similar structure at all scales. The non-traditional truth incorporates the observation that the structure of complex structures are determined by the scale of the measuring instrument, and such things as the length of a curve are a function of the unit of measure. For example, the length of a fractal curve depends on the size of the ruler used to measure it.

The fourth non-traditional truth is:

Natural phenomena do not necessarily have fundamental scales and may be described by scaling relations.

8.3   Final Thoughts

Complexity and consequently complex networks have the characteristics of both regularity and randomness, often producing order out of apparent chaos, so that quantitative measures of complexity may be constructed. These potential measures run the gamut from the replacement of Euclidean with fractal distances to non-stationary PDFs in function space, all of which find application in tomorrow's science, particularly in social/cognitive and informational networks. Phenomena with multiple scales in space and time require dynamic descriptions that extend beyond the differential and partial differential calculi into the fractional calculus, since past history and non-local interactions determine future evolution.

In a network of networks, every scale is linked to every other scale, either directly or indirectly, through phase transitions. A formalism of complex networks capable of predicting that influence of change, across multiple scales, is being sought. The mathematical challenge, as discussed in Chapter 5 is to recognize control systems as heterogeneous collections of physical, social and information networks, with intricate nonlinear interconnections and interactions. In this we may be guided by the ability of a fractional operator to absorb a great deal of dynamics into its fractional index.

At the beginning of this book I made a number of claims regarding complexity in science and the promise of the fractional calculus. The first was that most of the unsolved mysteries of modern science are the result of complexity. Second, complex phenomena remain ill-understood because the traditional methods of mathematical analysis are insufficient to overcome the barrier of complexity. Third, the sheer weight in the number and type of applications of the fractional calculus already made to complex phenomena is sufficient to convince a reasonable skeptic that tomorrow's science will be dominated by the fractional calculus. You are now in a position to gauge the degree of success I have had in realizing these goals.

Even if I have not convinced you that the fractional calculus is the best thing since sliced bread and that it will form the foundation of tomorrow's science, I hope to have convinced you of its utility. I believe the next time you encounter a time series whose average value becomes less well defined with an increasing number of measurements, or whose fluctuations come in bursts separated by long quiescent periods, or whose time average looks nothing like its ensemble average, you will find yourself wondering whether or not fractional kinetics can explain the confusion.

I close where I began, with an observation made by Leonardo [8]; this time concerning society's acceptance of new ideas:

> I know that many will call this a useless work, and they will be those of whom Demetrius said that he took no more account of the wind that produced the words in their mouths than the wind that came out of their hinder parts: men whose only desire is for material riches and luxury and who are entirely destitute of the desire of wisdom, the sustenance and the only true riches of the soul. For as the soul is more worthy than the body so much are the soul's riches more worthy than those of the body. And often when I see one of these men take this work in hand I wonder whether he will not put it to his nose like the ape, and ask me whether it is something to eat.

References

[1] Barnes, J.A. and D.W. Allan. 1966. *Proceedings of the IEEE* **54**, 176.

[2] Birks, J.B. 1962. *Rutherford in Manchester*, Ed. Heywood, London.

[3] Freeman, W.J. 1994. *Physica D* **75**, 151.

[4] Gladwell, M. 000. *The Tipping Point*, Little, Brown & Co., Boston, MA.

[5] Grigolini, P., G. Aquino, M. Bologna, M. Lukovic and B.J. West. 2009. *Physica A* **388**, 4129.

[6] Lowen, S.B. and M.C. Teich. 1990. *IEEE Trans. on Info. Theory* **36**, 1302.

[7] Lowen, S.B. and M.C. Teich. 1993. *Phys. Rev. E* **47**, 992.

[8] MacCurdy, E.(ed.). 1960. *The Notebooks of Leonardo da Vinci*, Konecky & Konecky, Old Saybrook, CT.

[9] Mandelbrot, B.B. and J.W. Van Ness. 1968. *SIAM Review* **10**, 422.

[10] Roberts, F.S. 1979. *Measurement Theory*, in *Encyclopedica of Mathematics and its Applications*, Ed. G. Rota, Addison-Wesley Publishing Co., Reading, MA.

[11] Schottky, W. 1918. *Annalen der Phsyik* **362**, 541.

[12] Thom, R. 1975. *Structural Stability and Morphogenesis*, Benjamin/Cummings, Reading, MA.

[13] Thompson, D.W. 1915. *On Growth and Form*; unabridged ed., Cambridge University Press, Cambridge, MA; Dover Pub., NY, 1992.

[14] Weissman, M.B. 1981. *Rev. Mod. Phys.* **53**, 497.

[15] West, B.J. 1985. *An Essay on the Importance of Being Nonlinear*, Lect. Notes in Biomath. **62**, Springer-Verlag, Berlin.

[16] West, B.J. and M.F. Shlesinger. 1989. *Internation J. of Mod. Phys. B* **395**-819; B.J. West and M.F. Shlesinger 1990, *Am. Sci.* **78**, 40.

[17] West, B.J. 2006. *Where Medicine Went Wrong*, Studies of Nonlinear Phenomena in Life Science Vol. 11, World Scientific, Singapore.

[18] West, B.J., E.L. Geneston and P. Grigolini. 2008. *Phys. Rept.* **468**, 1.

[19] West, B.J. and P. Grigolini. 2010. *Physica A* **389**, 5706.

[20] West, D. and B.J. West. 2013. *Phys. of Life* **10**, 210..

[21] West, B.J., M. Turalska and P. Grigolini. 2014. *Network of Echoes: Imitation, Innovation, and Invisible Leaders*, Springer, NY.

[22] Zaslavsky, G.M. 2002. *Phys. Rept.* **371**, 461.

[23] Ziel, A. van der. 1950. *Physica* **16**, 359.

Index

For Product Safety Concerns and Information please contact our EU
representative GPSR@taylorandfrancis.com
Taylor & Francis Verlag GmbH, Kaufingerstraße 24, 80331 München, Germany